AF540120

System Simulation and Agromet Advisory Services

NIPA® GENX ELECTRONIC RESOURCES & SOLUTIONS P. LTD.

New Delhi-110 034

About the Authors

BV Ramana Rao taught Agricultural Meteorology in the University of Agricultural Sciences, Bangalore during the years 1966 to 1979 and later joined the Indian Council of Agricultural Research and superannuated from service during April, 2000. He served as Head, Division of Climatology, Wind Power and Solar Energy Utilization at ICAR-Central Arid Zone Research Institute, Jodhpur (Rajasthan); He acted as founder Project Coordinator, All India Coordinated Research Project on Agrometeorology at ICAR Central Research Institute for Dryland Agriculture, Hyderabad (Telangana) during December 1984 to March 1994. His major contributions in the field of Agrometeorology include Agroclimatic Characterization, Delineation of Efficient Cropping Zones, Crop Weather Modeling, Classification and Management of Agricultural Droughts, Operational Agrometeorology etc. He published about 150 research papers in peer reviewed journals with 50 of his papers appearing in high impact International Journals. He authored more than 15 books and monographs and a number of scientific and technical bulletins. He served as Editor in Chief, Journal of Agrometeorology during the years 2009 to 2023.

He is the recipient of British Council Fellowship during the year 1986-87 for advanced training at University of Reading, United Kingdom. He was conferred as Honorary Fellow (2011) and Life Time Achievement Award (2025) by the Association of Agrometeorologists for his valuable contribution to the growth and development of Agricultural Meteorology in India.

Surender Singh has about thirty five years of experience with PhD in the field of Agrometeorology. Dr Singh worked in Agrometeorology encompassing teaching, research, extension and administration at various capacities viz., Principal Scientist, Professor & Head, Senior Scientist, Assistant Scientist, Advisor (Recruitments) and Associate Director (Students' Counseling and Placement) at CCS Haryana Agricultural University, Hisar, India. He is recipient of Two Gold Medals for Best Research Work during PhD Program and for Best PhD Dissertation (2001-02) awarded by the then President of India Dr APJ Abdul Kalam. He was offered UK Universities Vice Chancellors' Scholarship for Higher Studies twice (1991 & 1993) under Commonwealth Fellowship Program. He was Post Doctoral Fellow at International Centre for Theoretical Physics, Trieste, Italy; IAC – Instituto Agronômico, Campinas, Brazil and BOKU (Met) - Institute of Meteorology and Climatology, Vienna, Austria.

He is life member of Association of Agrometeorologists, Indian Meteorological Society, Indian Society of Remote Sensing, International Society for Agrometeorology, Society for Sustainable Agric& Resource Management, Environment Science Published for Everybody Round the Earth (ESPERE), European Drought Centre, South Asian Forum on Agriculture Meteorology (SAFOAM) and The Geographical Society of India. Organized quite a few trainings/symposia in Agrometeorology and associated domains at National level. Published more than 250 research papers, popular articles, manuals/teaching aids and research bulletins/book chapters etc. Performed responsibilities as Managing Editor - Journal of Agrometeorology; Research Collaborator, Monsoon Asia Agro-Environmental Research Consortium (MARCO), Japan; SPARC Research Collaborator, ETH, Zurich, Switzerland; Member, Expert Team - CAgM-WMO, Geneva, Switzerland (2018-20). Core Committee Member on Agriculture – Drought at WMO, Geneva, Switzerland (2021-24 & 2024-27). External Expert for evaluation of research projects, COST-Association, European Union, Brussels, Belgium. Many students (UG, PG and PhD) have been trained under him and placed across the country and well settled in various positions in the field of Agrometeorology. He is widely travelled researcher/expert across the globe on many academic pursuits. He has been conferred upon Fellow of Association of Agrometeorologists (FAAM) for the year 2024 in recognition of his exemplary achievements and enduring impact on agrometeorological science.

V Uma Maheswara Rao has forty years of experience with PhD in the broader field of Agrometeorology. Worked in Agrometeorology encompassing teaching, research, extension and administration at various capacities viz., Project Coordinator, Principal Scientist, Professor & Head, Agrometeorologist, and Assistant Scientist at CCS Haryana Agricultural University and Indian Council of Agriculture Research in India.

He is life member of Indian Meteorological Society, The Arid Zone Research Association of India, Indian Society of Remote Sensing, Association of Agrometeorologists, International Society for Agrometeorology, and The Indian Society of Dryland Agriculture. Organized several trainings/workshops in Agrometeorology at the International and National level. Published more than 345 research papers, popular articles, manuals/teaching aids and research bulletins/book chapters. Member of the CCI-ET-IIC, WMO, Geneva (2014-2017). Member of XI Plan High Powered Review Committee of Ministry of Earth Sciences, Government of India (2012). In 2014, the prestigious Chaudhary Devi Lal Award, ICAR's Best AICRP on Agrometeorology was received under his tenure as Project Coordinator. Many students (UG, PG and PhD) have been trained under his guidance are placed across the country and settled in various positions in the field of Agrometeorology. He was conferred as Fellow (2011) by the Association of Agrometeorologists. He visited numerous countries in official capacity representing Indian delegations.

System Simulation and Agromet Advisory Services

BV Ramana Rao
MSc Tech; FAAM
Founder Project Coordinator - AICRP on Agrometeorology
ICAR-CRIDA, Hyderabad, Telangana, India

Surender Singh
MSc; PhD; FAAM
Professor/Principal Scientist (Agrometeorology)
CCS HAU, Hisar, Haryana, India

V Uma Maheswara Rao
MSc; PGDIAS; PhD; FAAM
Project Coordinator-Rtd (AICRP on Agrometeorology)
ICAR-CRIDA, Hyderabad, Telangana, India

NIPA® GENX ELECTRONIC RESOURCES & SOLUTIONS P. LTD.
New Delhi-110 034

NIPA® GENX ELECTRONIC RESOURCES & SOLUTIONS P. LTD.

101,103, Vikas Surya Plaza, CU Block
L.S.C.Market, Pitam Pura, New Delhi-110 034
Ph : +91 11 4386 0225, 9717133558, 9540816132
E-mail: newindiapublishingagency@gmail.com
Website: www.niparesources.com

Print ISBN: 978-93-72196-19-1

ebook ISBN: 978-93-72197-83-9

Composed and Designed by NIPA®.

Dedicated to

Dr William G Duncan

(1909-1991)

Dr William G Duncon, *the Pioneer of Dynamic Crop Growth Modeling; in grateful recognition of his visionary work in integrating plant physiology, agronomy, and environmental science into dynamic simulation models that laid the foundation for modern crop modeling systems.*

His pioneering efforts in the 1960s not only advanced scientific understanding of crop-environment interactions but also inspired the development of tools that continue to guide agrometeorological advisories and decision support systems across the globe.

Through this dedication, we honor his legacy and lasting impact on agricultural systems science and its service to farmers and food security.

Dr LS Rathore

Former Director General of Meteorology
MoES-India Meteorological Department, Govt of India

Former Permanent Representative of India
Former Chair Intergovernmental Board on Climate Services
Former Vice-President Commission on Agriculture Meteorology
World Meteorological Organization, United Nations

Foreword

Agriculture, the backbone of the Indian economy and the lifeline for millions of its citizens, has always been inextricably linked to the vagaries of weather. For generations, the Indian farmer has faced an unending battle against the unpredictable forces of nature – the erratic monsoon, devastating droughts, sudden floods, and increasingly, extreme weather events that defy historical patterns. These uncertainties in crop production lead to significant economic losses, social distress, and pose a constant threat to national food security. While the resilience and perseverance of our farming communities are commendable, a sustainable future for agriculture demands more than just endurance; it requires proactive strategies, scientific foresight, and precise interventions.

It is in this critical context that the importance of Agrometeorological (Agromet) Advisories cannot be overstated. These advisories, rooted in accurate weather forecasting and built upon sound scientific hypotheses, serve as invaluable guides for farmers, enabling them to make informed decisions regarding every stage of the crop cycle – from land preparation and sowing to irrigation, pest management, and harvesting. The timely dissemination of such scientific intelligence can significantly mitigate risks, optimize resource utilization, and ultimately enhance agricultural productivity and profitability.

The paradigm shift required in modern agriculture is beautifully addressed by the integration of **System Simulation coupled with Decision Support Systems (DSS)**. Systems simulation provides a robust framework for understanding the complex interactions within agro-ecosystems, allowing us to model the behavior of crops, pests, and soil under varying environmental conditions. When this powerful analytical tool is integrated with Decision Support Systems, it transcends mere prediction, offering actionable insights and scenario-based guidance. Such a synergy can revolutionize farm management, transforming uncertainty into calculated risk, and traditional practices into scientifically optimized operations. This holistic approach empowers farmers with the knowledge to adapt to climate change, improve efficiency, and secure their livelihoods.

I applaud the profound wisdom and foresight of the Indian Council of Agricultural Research (ICAR) in recognizing this imperative. The formulation of '**System Simulation and Agromet Advisory Services**' as an elective course for BSc (Hons) Agriculture and allied streams students, based on the recommendations of the 6th Deans' Committee Report, is a monumental step. It reflects a commitment to nurturing a generation of agricultural professionals who are not only well-versed in traditional agronomy but are also adept at leveraging cutting-edge technologies and analytical methodologies for sustainable and climate-resilient agriculture. This course will undoubtedly equip future agricultural leaders with the essential tools to tackle the challenges of the 21st century.

This book provides a comprehensive and accessible exploration of these vital subjects. It elucidates the principles of systems simulation, delves into the intricacies of weather forecasting, and critically, details the methodologies for generating and effectively disseminating agro-advisories. The clarity, depth, and practical relevance of the content are commendable.

I am particularly delighted to see this treatise authored by a stellar team of highly respected scholars: **Professors BV Ramana Rao, Surender Singh, and V Uma Maheswara Rao**. Their names are synonymous with pioneering contributions and profound expertise in the discipline of agricultural meteorology. Their collective wisdom, extensive research, and practical experience shine through every chapter, making this book an authoritative and indispensable resource for students and practitioners alike. Their dedication to advancing agricultural science is evident in the meticulous detail and thoughtful organization of this text.

Having had the privilege to go through the pages of "**System Simulation and Agromet Advisory Services**," I am filled with immense satisfaction. This book is more than just a textbook; it is a beacon guiding the future of Indian agriculture towards a path of scientific precision, resilience, and prosperity. It is an essential read for anyone aspiring to make a tangible difference in the lives of farmers and in the broader agricultural landscape. I am confident that it will serve as a foundational text for years to come and significantly contribute to building a climate-smart agricultural sector in India.

Best wishes to the authors and the readers for a successful learning journey.

(LS Rathore)
President, Vigyan Bharti, Rajasthan
President, Society for Rural Improvement
Member, Earth System Science Council, GOI
Chairman, Centre for Environment & Development Studies, Jaipur

Preface

The agricultural landscape is continuously evolving, demanding a new generation of professionals equipped with interdisciplinary knowledge and advanced analytical tools. It is with this vision that we present **"System Simulation and Agromet Advisory Services,"** a comprehensive textbook designed to meet the requirements of an elective course for B.Sc. (Hons) students. This book is meticulously aligned with the recommendations of the Indian Council of Agricultural Research (ICAR) 6th Deans Committee Report, aiming to bridge the gap between theoretical knowledge and practical application in agricultural sciences.

In an era where climate variability poses significant challenges to food security and farm profitability, the ability to understand, predict, and adapt becomes paramount. This textbook delves into crucial domains that empower students to navigate these complexities. We begin with an in-depth exploration of **Systems Simulation**, a powerful methodology that enables the modeling and analysis of intricate agricultural processes. By understanding how various components of an agro-ecosystem interact, students will gain insights into optimizing resource use, predicting crop performance, and mitigating risks under different scenarios.

Complementing this, the book provides a thorough grounding in **Weather Forecasting**. Accurate and timely weather information is the cornerstone of effective agricultural planning. We cover the principles, techniques, and tools involved in generating reliable weather forecasts, equipping students with the knowledge to interpret meteorological data and understand its implications for farming operations.

The culmination of these insights lies in the **Generation and Dissemination of Agro-Advisories based on Weather Forecasts.** This section focuses on the practical application of simulated models and weather predictions to formulate actionable advice for farmers. From precise irrigation schedules and pest management strategies to optimal sowing and harvesting times, these agro-advisories are vital for enhancing farm productivity, reducing losses, and building climate resilience within the agricultural community.

The knowledge imparted through this course will undoubtedly serve as a critical asset for aspiring agricultural scientists, extension workers, and policymakers. It will enable them to contribute meaningfully to sustainable agricultural development, promote climate-smart farming practices, and empower farmers with the information necessary for informed decision-making. We are grateful to **Dr LS Rathore**, former Director General of Meteorology, IMD, MoES, New Delhi and Vice President, Commission for Agriculture Meteorology & Chair, Inter-governmental Board on Climate Services (IBCS), WMO, United Nations who has been a central figure in shaping and strengthening India's Agromet Advisory Services (AAS) for having provided a very inspiring foreword blended with great wisdom.

We believe this textbook will not only enrich the academic journey of BSc (Hons) and allied streams students but also inspire them to become leaders in leveraging scientific and technological advancements for the betterment of Indian agriculture. We extend our gratitude to ICAR for their visionary framework, which has shaped the essence of this important course.

Authors

Acknowledgement

The journey of compiling this book, System Simulation and Agromet Advisory Services, has been both intellectually enriching and professionally rewarding. We express our sincere gratitude to all those who have contributed, directly or indirectly, to the development of this work.

First and foremost, We are deeply thankful to the academic and research institutions, particularly the Indian Council of Agricultural Research (ICAR), the India Meteorological Department (IMD), and the various Agricultural Universities, whose initiatives and contributions in the field of agrometeorology and system simulation provided the foundation for this book.

We extend heartfelt appreciation to our colleagues, research collaborators, and students whose critical inputs, discussions, and field-level insights have greatly shaped the content and relevance of this volume.

Special thanks are due to the pioneers in system simulation and agromet advisory services whose pioneering research laid the groundwork for the concepts presented herein. We acknowledge the support of the World Meteorological Organization (WMO) and national agrometeorological networks for their continuous efforts in enhancing climate services for agriculture.

We are also grateful to the reviewers and mentors who provided constructive feedback, ensuring the clarity, accuracy, and practical utility of the chapters.

Authors are also grateful to the Association of Agrometeorologists (AAM) for their support and encouragement for the work. The open source text material/ images/references used in the book are also duefully acknowledged.

To our family and friends, their patience, understanding, and unwavering support have been the bedrock upon which this work stands.

Finally, We dedicate this work to the farming community—the real beneficiaries of agromet advisory services—whose resilience and reliance on timely weather information underscore the importance of integrating science with practice.

This book is a humble contribution toward strengthening agricultural resilience through advanced simulation tools and improved weather-based decision-making.

Authors

Contents

List of Figures

1

System Approach for Soil Plant Atmospheric Continuum

1.1 Introduction

A system approach to understand the soil-plant-water continuum fundamentally shifts our perspective from viewing these elements in isolation to recognizing them as an intricately woven, dynamic whole. It's about appreciating that every change in one part of this natural tapestry—be it the soil, the plant, or the water itself—reverberates through the others, creating a cascade of interconnected effects. This holistic lens allows us to dissect the continuous exchange of water and energy, unraveling the complex feedback loops and emergent behaviors that characterize this vital agricultural system. Instead of merely tracking individual inputs and outputs, a systems view compels us to comprehend how these components influence each other over time, leading to a far more nuanced understanding of their collective behavior.

At the heart of this continuum lie three primary, yet inseparable, components: the soil, the plant, and the water itself, all profoundly influenced by the atmosphere. The soil acts as both a foundational anchor and a dynamic reservoir. Its very composition—the proportion of sand, silt, and clay, its structure, and its organic matter content—dictates how readily water infiltrates, how much it can hold against the pull of gravity, and how quickly it can release water to plant roots. Processes like infiltration, percolation, and capillary rise describe water's journey within the soil matrix, while evaporation from the soil surface represents a significant loss pathway. Crucially, the soil also serves as the medium for nutrient availability, with water acting as the solvent that transports these vital elements to the roots.

The plant, in turn, is the primary biological conduit for water's journey from the soil back to the atmosphere. It actively absorbs water through its roots, moving it upward through the xylem to the leaves, where it is transpired through tiny pores called stomata. This transpiration process is not merely a loss; it's the driving force behind the plant's physiological engine, facilitating nutrient transport, cooling the plant, and supporting photosynthesis. A plant's characteristics, such as its root architecture, the total surface area of its leaves (Leaf Area Index or LAI), and its stomatal conductance, directly influence its water demand and its ability to cope with water stress. When water becomes scarce, the plant exhibits stress responses like wilting or stomatal closure, highlighting its dynamic interaction with the water supply.

Water itself is the central medium that binds these components, flowing dynamically through the system. It enters as precipitation or irrigation, infiltrates the soil, and is stored as soil moisture. From this reservoir, it is absorbed by plant roots, transported through the plant, and released as transpiration. Any excess water may become runoff, flowing over the surface, or deep percolation, draining beyond the root zone. The continuous movement and transformation of water are driven by energy, primarily solar radiation, which fuels both evaporation from the soil and transpiration from the plant.

The atmosphere acts as the overarching environmental driver. Solar radiation provides the energy for evaporation and transpiration, while atmospheric conditions—air temperature, humidity, wind speed, and vapor pressure deficit—collectively determine the evaporative demand that pulls water from both soil and plant surfaces. Furthermore, atmospheric CO_2 concentration directly influences photosynthetic rates, which in turn affect the plant's growth and its water requirements.

These components are not static; they engage in a constant, intricate dance of dynamic interactions and flows. For instance, precipitation or irrigation delivers water that infiltrates the soil, replenishing the soil moisture storage. As soil moisture levels change, so does the rate at which plants can absorb water. The water absorbed by roots fuels plant growth, which in turn increases the plant's leaf area, leading to a higher rate of transpiration and thus a greater demand for soil water. This creates a powerful feedback loop: more water leads to more growth, which leads to higher water demand, influencing how quickly the soil water is depleted. Conversely, if soil water becomes limited, plants reduce transpiration, impacting their growth and potentially conserving the remaining soil moisture, though at the cost of yield. Similarly, nutrient cycling in the soil is inextricably linked to water availability, as water mobilizes nutrients for plant uptake.

Representing this continuum effectively, especially for predictive purposes, invariably leads to the use of dynamic simulation models. These computational tools translate our understanding of the system into mathematical equations and algorithms. Within these models, the continuum is characterized by state variables—quantities that evolve over time, such as soil moisture content at various depths, daily plant biomass, or leaf area index. The processes of water movement and transformation are captured by fluxes or rates, which describe the speed at which water moves (e.g., infiltration rate, transpiration rate, root uptake rate). These rates are often complex functions of the current state variables, intrinsic parameters (like a specific soil's hydraulic conductivity or a crop's maximum stomatal conductance), and external driving variables provided by the atmosphere (e.g., daily rainfall, solar radiation, air temperature). Continuous simulation, often employing system dynamics principles, is a particularly powerful paradigm for modeling these ongoing flows and intricate feedback mechanisms, commonly found in sophisticated crop growth models like DSSAT or APSIM. Some models also incorporate spatial dimensions using GIS, allowing for a more accurate representation of heterogeneous field conditions.

The profound importance of this system approach for weather based agroadvisory cannot be overstated. By understanding the integrated dynamics of the soil-plant-water continuum, agroadvisors can move beyond generic advice to provide highly precise and context-specific recommendations. This includes optimizing irrigation scheduling, ensuring water is applied precisely when and where the plant needs it, thereby maximizing water use efficiency. It enables accurate prediction of drought stress, allowing farmers to take proactive measures. Furthermore, it supports the assessment of how various management practices—such as different tillage methods, the use of cover crops, or the adoption of drought-resistant crop varieties—will interact with the local soil and climate to affect crop performance and water balance. Ultimately, a systems approach empowers agroadvisory services to offer data-driven, sustainable solutions that enhance productivity, mitigate risks from climate variability, and promote long-term agricultural resilience.

1.2 System Approach: Concept and Principles

The System Approach" offers an indispensable lens through which to understand the intricate workings of the soil-plant-atmosphere (SPA) continuum, a quintessential example of a complex natural system. Rather than dissecting soil, plants, or the atmosphere into isolated subjects of study, this approach compels us to perceive them as a single, living, breathing entity where every component is in constant, dynamic interaction.

1.2.1 Concept of the Systems Approach in SPA

At its heart, the systems concept applied to the SPA continuum champions holism. It posits that the collective behavior of the soil, the plant, and the surrounding atmosphere is far more than the simple sum of their individual attributes. Consider, for instance, crop yield. It's not merely a function of how much water is available in the soil, nor solely dependent on the plant's genetic potential, nor exclusively dictated by atmospheric temperature. Instead, yield is an emergent property arising from the intricate interplay of soil moisture dynamics, nutrient availability, root growth, plant physiological responses like photosynthesis and transpiration, and the overarching atmospheric conditions that drive these processes (Fig.1.1). You cannot fully comprehend why a crop thrives or fails by studying just the soil's water holding capacity or the plant's stomatal conductance in isolation; you must understand how they mutually influence each other.

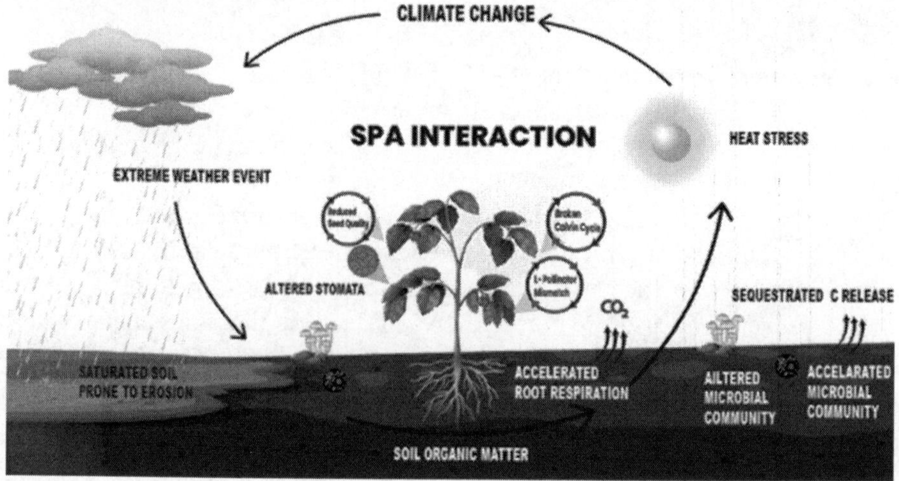

Fig. 1.1: Soil-plant-atmosphere interaction

This perspective emphasizes the profound relationships and interconnections between these elements. Water, the central medium, acts as a crucial link. Its movement from the atmosphere (as precipitation) into the soil, its absorption by plant roots, its transport through the plant, and its eventual release back into the atmosphere (as transpiration) defines a continuous, cyclical flow. Any change in one part of this cycle—say, a prolonged dry spell in the atmosphere—will trigger a cascade of effects: reduced soil moisture, decreased water uptake by plants, stomatal closure to conserve water, and ultimately, a potential reduction in plant growth and yield. The dynamic exchange of energy (e.g., solar radiation driving evaporation and photosynthesis) and mass (water, carbon dioxide, nutrients) between these components is what defines the SPA system.'

1.2.2 Definition of the System Approach in SPA

Applied to the soil-plant-atmosphere continuum, the systems approach is a holistic analytical framework focused on understanding the integrated functioning of soil properties, plant physiological processes, and atmospheric conditions as an interdependent unit. It involves delineating the system's operational boundaries (perhaps a specific field or watershed), identifying the key components (soil layers, root systems, plant canopy, atmospheric air layers), and meticulously analyzing the multifaceted relationships and dynamic interactions that govern the flow of water, energy, and nutrients within and across these components. The ultimate objective is to comprehend how the entire continuum behaves as a cohesive entity, facilitating more accurate predictions of crop performance, water use efficiency, and ecosystem resilience under various environmental and management scenarios.

1.2.3 Principles of the System Approach in SPA

Several fundamental principles illuminate how the systems approach unravels the complexities of the SPA continuum:

- **Holism and Synthesis:** Rather than analyzing soil physics, plant physiology, and atmospheric science as separate academic disciplines, this principle urges us to synthesize their knowledge into a unified understanding of the SPA continuum. Studying only the infiltration rate of soil or only the stomatal resistance of a leaf provides an incomplete picture. The holistic view considers how a specific soil texture influences water availability, which in turn affects a plant's ability to transpire, and how that transpiration subsequently influences the local atmospheric humidity. The synthesis reveals the emergent behavior of a healthy, productive agricultural system.
- **Interconnectedness and Interdependence:** The SPA continuum is a web of mutual reliance. For example, the rate of transpiration from plant leaves is not just a function of the plant's internal state; it is profoundly influenced by the atmospheric demand for water (determined by temperature, humidity, wind speed). Simultaneously, the plant's ability to transpire is limited by the soil's water availability, which itself is replenished by precipitation and depleted by both plant uptake and soil evaporation. Furthermore, the plant's root development directly impacts its access to soil water, and robust root systems can in turn improve soil structure, influencing future water infiltration. This illustrates the tight, reciprocal relationships between all components.

- **Emergence:** Concepts like crop yield potential,""drought resilience," or ecosystem productivity" are classic emergent properties of the SPA continuum. No single factor—neither a specific soil type, nor a particular crop variety, nor a given rainfall pattern—can solely define these. Instead, they arise from the complex, non-linear interactions of all soil, plant, and atmospheric elements over time. A fertile soil interacting with a genetically robust plant under optimal atmospheric conditions *emerges* as high yield, a property that cannot be predicted by summing the isolated characteristics of the soil, the plant, and the air.
- **Boundaries and Environment:** When applying the systems approach to a farmer's field, we might define the system's boundary as the soil profile down to the deepest root penetration, encompassing the plant canopy, and extending into the atmospheric boundary layer immediately above the field. However, this system is an open system deeply embedded within a larger environment. It receives inputs like solar radiation, precipitation, and carbon dioxide from the regional climate and atmosphere, and human interventions like irrigation, fertilizer application, and tillage. It provides outputs such as crop biomass, water vapor through evapotranspiration, and potentially water runoff or deep drainage to groundwater. Understanding these exchanges across the system's boundary is crucial for effective management.
- **Feedback Loops:** Feedback mechanisms are pervasive in the SPA continuum. A prime example is the plant water stress response. If atmospheric demand for water is high, and soil water is limited, the plant's stomata may close to conserve water. This negative feedback loop reduces water loss (a beneficial effect for survival) but simultaneously reduces carbon dioxide uptake, thereby slowing photosynthesis and growth (a negative effect on productivity). Conversely, ample soil water and high atmospheric demand can lead to increased transpiration, which can cool the plant and enhance nutrient flow, creating a positive feedback where efficient water use supports vigorous growth, leading to higher water uptake and transpiration rates.
- **Hierarchy and Subsystems:** The SPA continuum can be understood as a hierarchy of interconnected subsystems. The soil profile can be viewed as a subsystem composed of distinct horizons, each with unique hydraulic properties. The plant is a subsystem made of roots, stems, and leaves, each performing specialized functions. The atmosphere can be segmented into microclimates around the plant canopy and the broader regional climate. Each of these subsystems operates with its own internal

dynamics, yet they are all integrated within the larger SPA continuum, influencing and being influenced by other parts of the system.

- **Purpose and Function:** From a natural perspective, the inherent purpose of the SPA continuum is to sustain life, cycle water, and produce biomass. From an agricultural standpoint, the human-imposed purpose is often to maximize crop yield, optimize water use efficiency, or build resilience against climate variability. By defining this purpose, the systems approach guides the analysis towards identifying how best to manipulate or manage the interactions within the continuum to achieve desired outcomes, such as increasing food production while minimizing environmental impact.

By adopting this systems approach, we move beyond a fragmented view of soil, plant, and atmosphere to a sophisticated understanding of their integrated behavior. This holistic insight is absolutely vital for developing sustainable agricultural practices and providing effective, context-specific agroadvisory services.

1.3 Key Components of the System

The intricate dance of the soil-plant-atmosphere (SPA) continuum is orchestrated by four fundamental and inextricably linked key components, each playing a critical role in the system's overall function and behavior. These are the soil, the plant, water itself, and the atmosphere. Their individual properties and the dynamic processes they facilitate define the very essence of this vital natural system (Fig.1.2).

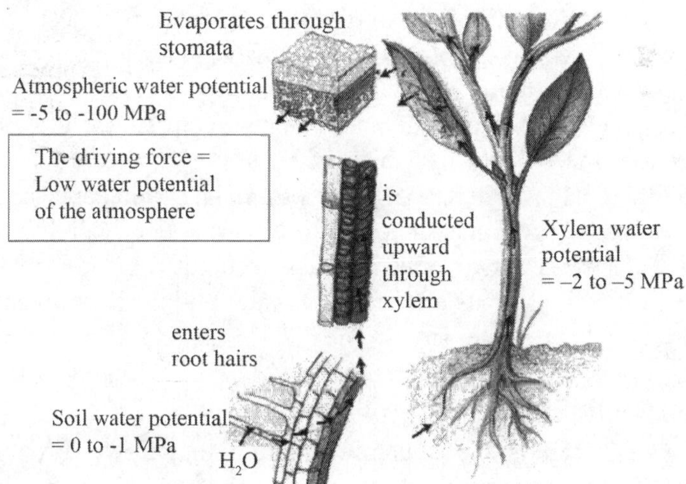

Fig. 1.2: Soil-plant-atmosphere continuum system

- First, the soil serves as the foundational anchor and a crucial reservoir within the continuum. Its properties are paramount to the system's functioning. The soil's texture, determined by the proportion of sand, silt, and clay particles, dictates its pore space, influencing how quickly water infiltrates and how much it can hold. Soil structure, the arrangement of these particles into aggregates, further affects aeration and water movement. The organic matter content is vital for water retention, nutrient cycling, and overall soil health. Other critical properties include bulk density, which reflects compaction and root penetration, and hydraulic conductivity, defining the ease with which water moves through the soil. The soil's water holding capacity, specifically concepts like field capacity (the maximum water it can hold against gravity) and wilting point (the moisture level at which plants can no longer extract water), directly govern water availability to plants. Within this component, key processes include the infiltration of water from the surface, its downward movement through percolation, the upward movement due to capillary rise, and direct evaporation from the soil surface. The soil also dictates the availability and transport of nutrients to plant roots, as water acts as the primary medium for their dissolution and movement.
- Second, the plant acts as the living intermediary, connecting the soil's water supply to the atmospheric demand. As the primary biological component, the plant is a sophisticated water transporter and consumer. Its root architecture, including depth, density, and distribution, determines its access to soil moisture and nutrients. The leaf area index (LAI), representing the total area of leaves per unit ground area, is a critical factor influencing light interception for photosynthesis and the rate of water loss through transpiration. Key physiological processes occurring within the plant include photosynthesis, where solar energy converts carbon dioxide and water into sugars for growth, and transpiration, the process by which water vapor is released from leaf stomata into the atmosphere. The plant's stomatal conductance regulates this water vapor exchange, adapting to environmental conditions and internal water status. Its overall biomass accumulationand growth rate are direct reflections of its efficiency in water and nutrient uptake, and its ability to manage water stress. The plant's responses to water deficit, such as wilting, reduced growth, or stomatal closure, highlight its active role in dynamically interacting with its water supply.
- Third, water itself is the dynamic medium that flows through and connects all other components. It is the lifeblood of the continuum, constantly moving and transforming. Its inputs into the system primarily

come from precipitation (rainfall, snowfall) and human-controlled irrigation. Within the system, water exists in various forms of storage, predominantly as soil moisture content within the root zone and, to a lesser extent, as plant water content. The flows of water are central to the system: movement through soil pores, uptake by roots, transport through the plant's vascular system (xylem), and its release back into the atmosphere. Water's outputs from the system include transpiration from plants, evaporation from the soil surface (collectively termed evapotranspiration), runoff over the land surface, and deep percolation or drainage below the root zone. The continuous cycling of water drives nutrient transport and energy exchange across the entire system.

- Finally, the atmosphere serves as the overarching environmental driver that profoundly influences the entire continuum. It acts as the ultimate sink for water vapor and the source of energy. Solar radiation is a critical atmospheric input, providing the energy that powers both evaporation from soil and transpiration from plants, as well as fueling photosynthesis in plants. Air temperature directly affects plant metabolic rates and the evaporative demand. Humidity (or vapor pressure deficit, VPD) dictates the pull" on water from surfaces, influencing transpiration rates. Wind speed also plays a role in removing humid air from leaf surfaces, thereby enhancing water loss. The atmosphere also provides the crucial input of carbon dioxide (CO_2), which plants utilize during photosynthesis, linking the carbon cycle directly to the water cycle within the system. Together, these atmospheric conditions dictate the evaporative demand, pushing the plant to regulate its water use.

These four components are not static, isolated entities; their continuous, interdependent interactions define the very essence of the soil-plant-atmosphere system. Understanding the unique role and properties of each, and critically, how they dynamically influence one another, is fundamental to any systems-based analysis or management of agricultural and ecological processes.

1.4 Interacting Processes and Flows

The soil-plant-atmosphere continuum is a symphony of ceaseless interacting processes and flows, where water, energy, and mass constantly move and transform, orchestrating the system's dynamic behavior. These interactions are not linear, but rather a complex web of reciprocal influences and feedback loops that govern everything from a plant's growth to regional water cycles.

- The journey often begins with precipitation, whether in the form of rainfall or the deliberate act of irrigation. As water lands on the soil surface, it initiates the process of infiltration, where it penetrates the soil profile. The rate and volume of this infiltration are profoundly influenced by the soil's properties, such as its texture and structure; a sandy soil will typically absorb water much faster than a compacted clay soil. Once infiltrated, this water contributes to the soil moisture storage, filling the pore spaces within the soil matrix. The amount of water held in storage is critical, as it dictates the immediate supply available to the plant.

- From this subterranean reservoir, the plant begins its active role through root uptake. Plant roots, extending through the soil profile, draw water from the soil pores, a process driven by the water potential gradient between the drier root cells and the wetter soil. This absorbed water is then transported upwards through the plant's vascular system, the xylem, reaching every leaf and stem. This movement is a continuous, internal flow, a testament to the plant's vital connection to its water source.
- The culmination of this upward journey is transpiration, the process where water vapor is released from the plant's leaves into the atmosphere, primarily through tiny pores called stomata. This is not just a passive release; transpiration is the driving force behind the plant's internal water transport and plays a crucial role in cooling the plant and facilitating nutrient uptake. The rate of transpiration is fundamentally linked to the atmospheric demand for water, which is a composite of solar radiation (providing the energy for evaporation), air temperature, humidity, and wind speed. High solar radiation and low humidity, for instance, create a strong evaporative pull, accelerating transpiration. This creates a significant feedback loop: as the atmosphere demands more water, the plant pulls more from the soil; if soil moisture is limited, the plant might respond by closing its stomata, thereby reducing transpiration but also limiting the uptake of carbon dioxide for photosynthesis.
- Concurrently with plant uptake and transpiration, water also undergoes evaporation directly from the soil surface. This process, often significant in bare or sparsely vegetated areas, also consumes water from the soil moisture storage, competing with plant roots for the available supply. Together, plant transpiration and soil evaporation constitute evapotranspiration, the major pathway for water returning from the land surface to the atmosphere.
- Beyond these evaporative losses, excess water in the system can lead to other flows. If the infiltration rate is overwhelmed by heavy rainfall or irrigation, water may flow over the land surface as runoff, potentially carrying away valuable topsoil and nutrients. Alternatively, if water

infiltrates beyond the root zone, it undergoes deep percolation, moving downwards to recharge groundwater reserves or eventually contribute to streamflow.

- The flow of energy is equally dynamic and intertwined with water. Solar radiation acts as the primary energy input, driving both photosynthesis within the plant and the phase change of water from liquid to vapor during evaporation and transpiration. The plant's ability to capture this solar energy for photosynthesis directly impacts its growth and, consequently, its demand for water. A larger, more vigorously growing plant (higher leaf area) will intercept more solar radiation, conduct more photosynthesis, and generally transpire more water, creating another interconnected feedback.
- Finally, the flow of mass, particularly carbon and nutrients, is intrinsically linked to water. Water acts as the solvent that mobilizes nutrients in the soil, making them accessible for uptake by plant roots. Similarly, the plant's uptake of carbon dioxide from the atmosphere during photosynthesis, which is then converted into plant biomass, is directly dependent on stomatal opening for water vapor exchange. If stomata close to conserve water during drought, carbon uptake is simultaneously inhibited, demonstrating the tight coupling between water, carbon, and nutrient flows throughout the SPA continuum.

In essence, these interacting processes form a highly responsive and self-regulating system. A change in atmospheric conditions (e.g., a hot, dry day) will trigger increased atmospheric demand for water, leading to higher transpiration rates. This will then deplete soil moisture more rapidly, potentially inducing plant water stress. The plant, in turn, may respond by closing stomata, reducing growth, or altering root development, all of which feedback to influence the rates of water flow and storage in the soil, demonstrating the continuous, dynamic interplay that defines the soil-plant-atmosphere continuum.

1.5 Concept of System Boundaries

A system boundary acts as a conceptual fence, delineating the elements, processes, and interactions that are directly included in the analysis from those that are considered external. Everything within the boundary is part of the system of interest," while everything outside is part of the environment" or supersystem." This distinction is vital because while the environment influences the system (through inputs), the system itself may or may not directly influence the environment within the scope of the study (outputs are typically observed at the boundary).

Defining a boundary is inherently an act of simplification. Real-world systems are endlessly interconnected. For instance, a single farm field is part of a watershed, which is part of a river basin, which is part of a continent, and so on. Without clear boundaries, the analytical task becomes overwhelming, as every interaction in the universe would theoretically need to be considered.

1.5.1 Purpose and Importance of System Boundaries:

- **Manage Complexity:** The primary purpose of defining boundaries is to reduce the overwhelming complexity of real-world phenomena into a manageable scope for analysis and modeling. It allows researchers and practitioners to focus on the most relevant interactions without getting lost in an infinite web of connections.
- **Define Scope of Analysis:** Boundaries clearly articulate what questions the system is designed to answer and what phenomena it is intended to explain or predict. If the objective is to understand water use efficiency in a specific crop, the boundary would encompass the plant, the soil it grows in, and the immediate atmospheric conditions. If the objective is to analyze regional water resource management, the boundary would expand to an entire watershed.
- **Identify Inputs and Outputs:** Once a boundary is drawn, it becomes much easier to identify the inputs" (elements, energy, or information crossing into the system from the environment) and outputs" (elements, energy, or information leaving the system to the environment). These exchanges across the boundary are critical for understanding how the system functions and interacts with its surroundings.
- **Resource Allocation:** In practical applications like developing simulation models or implementing agroadvisory programs, boundaries help in allocating resources. They dictate what data needs to be collected, what processes need to be modeled, and what interventions are feasible within the defined scope.

1.5.2 Characteristics of System Boundaries

- **Arbitrary but Logical:** System boundaries are rarely absolute or physically impenetrable. They are conceptual constructs chosen by the analyst, but they must be logically justified based on the study's objectives. A well-defined boundary reflects a clear understanding of the problem being addressed.
- **Dynamic and Context-Dependent:** The correct" boundary is not universal; it depends entirely on the specific question being asked, the

scale of analysis, and the available data and resources. The same real-world phenomenon might be encapsulated by different boundaries for different analytical purposes.

- **Permeable:** Most real-world systems, especially biological and environmental ones, are open systems," meaning their boundaries are permeable. They constantly exchange matter, energy, and information with their environment. The boundary defines where these exchanges occur and how they are accounted for.

1.5.3 Application to the Soil-Plant-Atmosphere (SPA) Continuum

Applying the concept of system boundaries to the SPA continuum is crucial for developing effective simulation models and providing targeted agroadvisory. The definition of the boundary will vary significantly depending on the scale and objective:

• Field-Level System (Micro-Scale for Crop Management)

Objective: To predict crop yield, optimize irrigation, or manage nutrients for a specific crop in a single field.

Boundary Definition: The system might be defined as the soil profile down to the typical rooting depth of the crop (e.g., 1-2 meters), the individual plant canopy above the ground, and the immediate atmospheric layer (e.g., up to 2-5 meters above the canopy) that directly interacts with the plants.

Inputs Across Boundary: Precipitation (rain, irrigation), solar radiation, air temperature, humidity, wind speed, atmospheric CO_2, fertilizer application, pesticide application.

Outputs Across Boundary: Crop yield (biomass harvested), evapotranspiration (water vapor to atmosphere), runoff (water leaving field surface), deep percolation (water draining below root zone), nutrient leaching (nutrients leaving root zone), greenhouse gas emissions.

• Watershed-Level System (Meso-Scale for Water Resource Management)

Objective: To understand regional water balance, assess impacts of agricultural practices on water quality, or manage water allocation among multiple farms.

Boundary Definition: The system encompasses all agricultural fields, natural vegetation, surface water bodies (rivers, lakes), and groundwater aquifers within a defined geographical watershed. The boundary would follow the topographical divides of the watershed.

Inputs Across Boundary: Regional precipitation, water inflow from upstream watersheds, human water diversions (e.g., from external rivers).

Outputs Across Boundary: Water outflow from the watershed (river discharge), groundwater recharge/discharge, evapo(transpi)ration from all land surfaces, nutrient and sediment loads carried by runoff to downstream bodies.

- **Farm-Enterprise Level System (Management Scale)**

Objective: To optimize overall farm profitability, labor allocation, or sustainability across diverse crops and livestock.

Boundary Definition: The entire farm property, including all cultivated fields, livestock operations, infrastructure, and associated resources.

Inputs Across Boundary: Financial capital, labor, seeds, fertilizers, machinery fuel, external water sources, market prices for inputs and outputs.

Outputs Across Boundary: Agricultural products sold, waste products, labor used, profits/losses, environmental impacts (e.g., carbon footprint).

1.5.4 Implications for Weather Based Agroadvisory

The careful definition of system boundaries is paramount for agromet advisory because it directly influences the **scope, relevance, and actionability** of the advice provided:

Tailored Advice: By defining the boundary at a specific field or farm, agroadvisors can offer highly tailored recommendations that consider local soil conditions, crop varieties, and prevailing microclimates, rather than generic advice.

Data Requirements: The boundary dictates what data needs to be collected (e.g., field-specific soil tests vs. regional climate data) and what scale of data integration is necessary.

Predictive Accuracy: Simulation models built with appropriate boundaries can provide more accurate predictions for the specific context, enabling better foresight for farmers.

Intervention Efficacy: Understanding the system within its boundaries helps identify the most effective leverage points for intervention. For example, if the boundary excludes off-farm water sources, the advisory would focus solely on rainwater harvesting or on-farm water conservation.

Avoiding Unintended Consequences: A well-defined boundary, combined with systems thinking, helps to anticipate how changes within the system (e.g., increasing fertilizer application) might affect elements outside the immediate focus but still within the defined environment (e.g., nutrient runoff impacting nearby water bodies).

In summary, system boundaries are not merely lines drawn on a diagram; they are crucial conceptual tools that enable the structured analysis of complex systems like the SPA continuum, making them tractable for scientific study, simulation modeling, and the delivery of effective agroadvisory services.

1.6 Representation using Simulation Models

Representing the complex and dynamic soil-plant-atmosphere (SPA) continuum in simulation models is a sophisticated endeavor that translates the natural system's intricate processes into a computational framework. These models are essentially simplified, mathematical approximations designed to mimic the real-world behavior of the continuum, allowing us to predict its responses under various conditions and explore what-if" scenarios. At the core of how the SPA continuum is represented in these models are four fundamental elements: state variables, fluxes/rates, parameters, and driving variables.

State variables are the quantities within the system that change over time and define its current condition. They are like snapshots of the system's state" at any given moment. For the soil-plant-atmosphere continuum, common state variables include:

Soil moisture content: Often represented for different depths within the soil profile, indicating how much water is stored at various levels.

Plant biomass: The total dry weight of the plant, which accumulates over time as the plant grows. This might be further broken down into above-ground biomass (stems, leaves) and root biomass.

Leaf area index (LAI): The ratio of total leaf area to ground area, which evolves as the plant grows and senesces, influencing light interception and transpiration.

Root depth and distribution: Representing how deep and widely the roots have explored the soil, which affects their access to water and nutrients.

Plant water content: Though often implicitly handled, some models might track the water status of different plant organs.

Fluxes, or rates, represent the movement or transformation of matter and energy within and out of the system. These are the dynamic processes that cause the state variables to change. They are typically calculated at discrete time steps (e.g., hourly, daily) based on the current state variables and external influences. Key fluxes in SPA models include:

Infiltration rate: How quickly water enters the soil from precipitation or irrigation.

Transpiration rate: The amount of water vapor released from the plant's leaves to the atmosphere.

Soil evaporation rate: The amount of water vapor lost directly from the soil surface.

Root water uptake rate: The rate at which plants absorb water from different soil depths.

Photosynthesis rate: The rate at which carbon dioxide is converted into plant sugars, driving biomass accumulation.

Nutrient uptake rate: The rate at which plants absorb essential nutrients from the soil.

Deep percolation rate: The rate at which water drains below the root zone.

These fluxes are almost always described by mathematical functions that incorporate the values of the state variables. For instance, the transpiration rate might be a function of the plant's LAI, the stomatal conductance, the soil moisture content (as water stress will reduce it), and atmospheric vapor pressure deficit.

Parameters are constant values within the model that characterize specific properties of the soil, plant, or environment being simulated. Unlike state variables, parameters do not change during a single simulation run, though they might be adjusted during model calibration to better fit observed data. Examples include:

Soil hydraulic properties: Values like saturated hydraulic conductivity, residual water content, and pore-size distribution index, which are specific to a particular soil type.

Crop coefficients: Values that describe a crop's specific water use efficiency or its maximum photosynthetic rate under ideal conditions.

Radiation use efficiency: A parameter indicating how efficiently a plant converts intercepted solar radiation into biomass.

Root growth coefficients: Describing the rate and pattern of root elongation.

Finally, **driving variables** are external inputs to the system that dictate its behavior but are not themselves simulated by the model. These are typically measured environmental data that compel the fluxes and changes in state variables. Common driving variables for SPA models include:

Daily or hourly rainfall: The amount of precipitation received.

Solar radiation: The amount of incoming solar energy.

Air temperature: Both maximum and minimum temperatures, influencing physiological rates and evaporative demand.

Relative humidity or vapor pressure deficit: Quantifying the dryness" of the air and its pull on water.

Wind speed: Affecting the boundary layer conductance for water vapor.

Carbon dioxide concentration: Increasingly important for simulating plant growth responses to changing atmospheric CO_2 levels.

These elements are integrated within specific simulation paradigms. Continuous simulation models, often based on principles of system dynamics, are widely used for the SPA continuum. They track the continuous accumulation and depletion of state variables over time, using differential or difference equations to describe the rates of flow. Crop growth models like **DSSAT** (Decision Support System for Agrotechnology Transfer), **APSIM** (Agricultural Production Systems sIMulator), and **WOFOST** (WOrld FOod STudies) are prime examples. These models simulate daily or sub-daily processes of water balance, nutrient cycling, and biomass accumulation. More advanced models might also incorporate spatially explicit components, using Geographic Information Systems (**GIS**) to represent variations in soil type, topography, or rainfall across a landscape, allowing for a distributed simulation of the SPA continuum.

By carefully defining these elements and their mathematical relationships, simulation models provide a powerful means to unravel the complexities of the soil-plant-atmosphere continuum, offering insights crucial for research, education, and practical agroadvisory.

To effectively represent the soil-plant-atmosphere (SPA) continuum in a computational environment, various simulation paradigms are employed. Each paradigm offers a distinct approach to modeling time, interactions, and system behavior, making them suitable for different types of questions and system complexities within the SPA context. Here are some of the key simulation paradigms:

1.6.1 Continuous Simulation (System Dynamics)

Core Idea: In continuous simulation, the state variables of the system are assumed to change smoothly and continuously over time. The system is typically described by a set of differential or difference equations that define the rates of change of these variables. This paradigm is particularly adept at modeling stocks (accumulations) and flows (rates of change) and capturing feedback loops, which are abundant in natural systems.

- **How it applies to SPA**

Modeling Flows and Accumulations: This paradigm is excellent for tracking the continuous flow of water through the soil profile (e.g., infiltration, percolation, evaporation) and its accumulation as soil moisture content. Similarly, it tracks the continuous increase in plant biomass, leaf area, and root growth as a result of physiological processes like photosynthesis and nutrient uptake.

Feedback Loops: It naturally incorporates the numerous feedback loops in the SPA continuum. For example, higher soil moisture leads to increased transpiration, which in turn depletes soil moisture; or increased plant biomass leads to higher photosynthetic rates, demanding more CO_2 and water.

Time Step: Calculations are performed over small, fixed time steps (e.g., daily, hourly), and the state variables are updated based on the calculated rates during that interval.

- **Examples in SPA/Agriculture**

Crop Growth Models (e.g., DSSAT, APSIM, WOFOST): These are prime examples. They simulate daily processes like water balance (inputs from rain/irrigation, losses from evapotranspiration, drainage), nutrient uptake, biomass accumulation, and phenological development, all through continuous equations that update the system's state over time. They help predict yields, assess irrigation needs, and evaluate fertilizer strategies.

Soil Water Balance Models: Simulating the dynamics of soil moisture content, runoff, and deep percolation based on rainfall, evapotranspiration, and soil hydraulic properties.

Nutrient Cycling Models: Tracking the continuous transformation and movement of nitrogen, phosphorus, and other nutrients through the soil, plant, and atmosphere.

1.6.2 Discrete Event Simulation (DES)

Core Idea: In contrast to continuous simulation, DES focuses on specific events" that occur at discrete, often irregular, points in time, causing instantaneous changes to the system's state. Between these events, nothing of importance is assumed to happen. It's ideal for systems where the flow of entities and the interactions with resources are key, often involving queues and delays.

- **How it applies to SPA**

While less common for simulating the continuous biophysical processes within the SPA continuum itself, DES is highly relevant for modeling the management and operational aspects related to it.

Resource Allocation: It can model agricultural operations that interact with the SPA continuum. For example, scheduling irrigation events (when and how much water is applied), harvesting operations (arrival of machinery, processing of crops), or pest scouting and spraying.

Supply Chain Logistics: Modeling the flow of produce from field to market, including storage, transportation, and processing, where events" like truck arrivals, loading, and unloading occur at specific times.

- **Examples in SPA/Agriculture:**

Farm Management Optimization: Simulating the sequence of field operations (e.g., plowing, planting, fertilizing, harvesting), considering machinery availability, labor, and weather windows. This helps optimize resource use and identify bottlenecks.

Irrigation System Scheduling: Modeling the demand for water by different fields and the capacity of the irrigation system, including pump operation, pipe flow, and sprinkler application events.

Post-Harvest Processing: Simulating the flow of harvested crops through sorting, cleaning, packaging, and storage facilities, identifying inefficiencies in the process.

1.6.3 Agent-Based Modeling (ABM)

Core Idea: ABM simulates the actions and interactions of autonomous agents" (individuals or groups) within an environment. Each agent follows a set of rules, and the overall system behavior, including emergent properties, arises from these individual interactions. ABM is particularly powerful for systems where individual variability, spatial location, and local interactions are important.

- **How it applies to SPA**

Individual Plant Behavior: While not common for single plants, ABM can simulate the interactions of multiple plants in a field, considering competition for resources (light, water, nutrients) or disease spread.

Pest and Disease Dynamics: Agents could represent individual pests, pathogens, or beneficial insects, whose movements, reproduction, and interactions with host plants (also agents) and the environment lead to population dynamics or outbreak patterns.

Farmer Decision-Making: Agents can represent farmers who make decisions about planting, irrigation, or fertilizer application based on their local observations, economic incentives, and social networks, influencing aggregated agricultural outcomes.

- **Examples in SPA/Agriculture**

Spread of Pests/Diseases: Simulating how a particular disease or insect pest spreads through a field or across a landscape, considering individual movement patterns, host plant susceptibility, and environmental factors.

Crop-Weed Competition: Modeling individual crop plants and weed plants as agents, competing for light, water, and nutrients, to understand how weed management strategies might affect crop yield.

Land-Use Change: Simulating how individual landholders' decisions about crop selection, irrigation adoption, or conservation practices aggregate to regional land-use patterns influenced by environmental and socio-economic factors.

1.6.4 Monte Carlo Simulation

Core Idea: Monte Carlo simulation is not a system-level paradigm in itself but rather a **technique used within other simulation paradigms** (continuous, discrete event, or agent-based) to incorporate and analyze uncertainty and variability. It involves running a model many times, each time using randomly sampled values for uncertain input parameters or variables from their defined probability distributions. The collection of outputs from these many runs provides a probability distribution of the system's response.

• How it applies to SPA

Uncertainty Quantification: The SPA continuum is inherently uncertain due to variable weather, soil heterogeneity, and biological variations. Monte Carlo simulation is used to quantify the uncertainty in model predictions (e.g., yield forecasts, water deficits).

Risk Assessment: It helps assess the risk associated with different management decisions under uncertain conditions.

Examples in SPA/Agriculture

Yield Probability Distribution: Running a crop growth model (continuous paradigm) multiple times with randomly sampled daily weather data (e.g., rainfall, temperature) from historical distributions to generate a probability distribution of potential yields, rather than a single point estimate.

Irrigation Schedule Robustness: Simulating an irrigation strategy under various uncertain future weather scenarios to assess its robustness in ensuring sufficient water availability and minimizing water stress.

Environmental Risk Assessment: Estimating the probability of nutrient leaching or pesticide runoff under variable soil and rainfall conditions by repeatedly sampling from distributions of these uncertain inputs.

By strategically choosing and combining these simulation paradigms, researchers and agroadvisors can build powerful models that reflect the nuanced realities of the soil-plant-atmosphere continuum, leading to more informed decisions and sustainable agricultural practices.

1.7 Importance in Agromet Advisories

The integration of a systems approach, particularly through the use of simulation models, is not merely an academic exercise; it is of paramount importance for modern agroadvisory services, transforming them from reactive responses to proactive, data-driven strategies. This approach fundamentally enhances the ability of agricultural advisors to provide precise, timely, and actionable guidance to farmers, leading to more resilient, productive, and sustainable agricultural systems.

Firstly, a systems approach vastly improves predictive capabilities. By capturing the dynamic interactions within the soil-plant-atmosphere (SPA) continuum, simulation models can forecast future outcomes under various conditions. For example, rather than simply stating current soil moisture levels, a systems-based agroadvisory can predict when a crop might experience water stress based on anticipated weather patterns, the specific soil's water holding capacity, and the crop's developmental stage and current water demand. This enables advisors to provide early warnings of potential droughts, nutrient deficiencies, or even pest outbreaks, allowing farmers to take preventive or mitigating actions before significant damage occurs. This foresight is invaluable in an increasingly volatile climate, empowering farmers to adapt their practices to optimize resource use and minimize losses.

Secondly, this approach facilitates optimized resource management. Traditional advisory might recommend a blanket application of water or fertilizer. However, a systems-informed advisory, leveraging simulation models, can provide highly precise recommendations tailored to the specific field, crop, and prevailing environmental conditions. For instance, by simulating the water balance in a particular soil type and the crop's daily water uptake requirements, advisors can suggest optimal irrigation scheduling – precisely when and how much water to apply – thereby maximizing water use efficiency, reducing water waste, and lowering input costs. Similarly, nutrient models can guide precision fertilizer application, ensuring nutrients are available when the plant needs them most, minimizing runoff, and enhancing profitability while reducing environmental impact. This precision agriculture is a direct benefit of understanding the interconnectedness of the SPA components.

Thirdly, the systems approach is crucial for risk assessment and mitigation. Agriculture is inherently susceptible to a multitude of risks, from extreme weather events to pest and disease infestations. Simulation models, when run with varying environmental inputs (e.g., Monte Carlo simulations with different rainfall scenarios), can assess the probability and severity of different risks. An agroadvisory service employing this approach can inform farmers about the likelihood of a late-season frost impacting yield, the potential spread of a specific crop disease under predicted humidity levels, or the economic viability of a particular crop choice given market volatility. This allows farmers to make more informed decisions about crop selection, insurance, and risk diversification strategies, fostering greater resilience in their operations.

Finally, integrating the systems approach empowers informed decision-making at all levels, from individual farmers to regional policymakers. For farmers, it means receiving advice that is not just prescriptive but also explanatory, helping them understand *why* certain recommendations are made based on the underlying biophysical processes. This deeper understanding builds trust and facilitates better adoption of climate-smart and sustainable practices. For policymakers, it provides a powerful tool to evaluate the potential impacts of different agricultural policies or interventions—such as promoting drought-resistant varieties or investing in new irrigation infrastructure—on regional food security, water resources, and environmental sustainability, before large-scale implementation.

In essence, by embracing the systems approach, agroadvisory services transcend basic guidance, offering a sophisticated, dynamic, and integrated framework that helps farmers navigate complex challenges, optimize their inputs, mitigate risks, and ultimately cultivate more productive, resilient, and environmentally responsible agricultural landscapes. It is a fundamental shift towards truly intelligent and adaptive agriculture.

1.8 Conclusion: The Path Forward for Agroadvisory

Having traversed the foundational concepts of system simulation and agroadvisory, and explored their profound integration, the subsequent chapters are designed to consolidate this knowledge and propel the reader towards the cutting edge of the field. The journey through the earlier parts of this book has meticulously laid out the principles, methodologies, and practical applications necessary to understand the dynamic interplay within the soil-plant-atmosphere continuum and its relevance to modern agriculture.

The concluding chapters will build upon this comprehensive understanding by delving into advanced topics and future directions. These sections are critical, as they will explore how the convergence of system simulation with emerging technologies and evolving challenges is reshaping the landscape of agroadvisory. The book will examine the transformative potential of Artificial Intelligence and Machine Learning in enhancing predictive capabilities and decision support. It will also highlight the integral role of Remote Sensing and GIS in providing real-time, spatially explicit data crucial for fine-tuning simulations and delivering hyper-local advice. Furthermore, the discussion will extend to the formidable implications of Climate Change, showcasing how simulation models are becoming indispensable tools for developing adaptive strategies and building resilience in agricultural systems facing unprecedented environmental shifts. Finally, these chapters will look ahead, identifying future trends and research opportunities, such as digital twins and the Internet of Things, that promise to further revolutionize the precision and reach of agroadvisory services.

Multiple Choice Questions (Bold means correct answer)

1. What is the primary characteristic of a systems approach" when applied to the soil-plant-atmosphere (SPA) continuum?
   a) Breaking down complex problems into isolated, individual parts for study.
   b) Focusing solely on the most productive component of the system.
   c) Viewing the soil, plant, and atmosphere as an integrated, dynamic whole with interconnected components.
   d) Prioritizing laboratory experiments over field observations.
2. Which of the following best describes an emergent property" in the context of the SPA system?
   a) A characteristic that is inherent to a single component, like soil texture.
   b) A quality or behavior that arises from the complex interactions between system components and cannot be predicted from individual parts.
   c) A constant value that defines a specific property of a component, like a crop coefficient.
   d) An external input that drives the system, such as rainfall.
3. What is the main purpose of defining system boundaries" in the SPA continuum?
   a) To physically separate the system from its environment.
   b) To simplify complexity and focus analysis on a manageable scope.
   c) To ensure that no external influences affect the system.
   d) To prove the absolute independence of system components.
4. In the SPA system, how does the atmosphere" primarily act as an environmental driver?
   a) By physically holding the soil particles together.
   b) By directly regulating nutrient uptake by plants.
   c) By providing energy (solar radiation) and demand for water (temperature, humidity, wind).
   d) By acting as the primary reservoir for soil moisture.

5. Which of the following is a key process facilitated by the soil" component in the SPA continuum?
   a) Transpiration of water vapor.
   b) Photosynthesis and biomass accumulation.
   c) Infiltration and storage of water.
   d) Regulation of atmospheric CO_2 levels.
6. The plant" component in the SPA system primarily acts as a:
   a) Static reservoir for water.
   b) Passive absorber of solar radiation.
   c) Water transporter and consumer, releasing water to the atmosphere.
   d) Geomorphological agent shaping the landscape.
7. Which of the following represents a flow" of water within the SPA system?
   a) Soil moisture content.
   b) Plant biomass.
   c) Root uptake.
   d) Atmospheric temperature.
8. What is the significance of feedback loops" in the SPA system?
   a) They prevent any interaction between the system's components.
   b) They cause the system to always return to its initial state regardless of inputs.
   c) They describe how outputs from one process influence subsequent inputs, creating dynamic responses.
   d) They simplify the system by eliminating complex interactions.
9. When modeling the SPA continuum, state variables" typically represent:
   a) External environmental conditions that drive the system.
   b) Constant values specific to soil or crop properties.
   c) Quantities that change over time and define the system's current condition.
   d) Rates of processes, like infiltration or transpiration.
10. The process where water vapor is released from plant leaves into the atmosphere is known as:
    a) Infiltration.
    b) Percolation.
    c) Evaporation.
    d) Transpiration.

11. What is the combined process of water loss from both the soil surface and plant leaves called?
    a) Runoff.
    b) Deep percolation.
    c) Evapotranspiration.
    d) Capillary rise.
12. A parameter" in a simulation model of the SPA continuum would typically be:
    a) Daily rainfall amount.
    b) Soil moisture content at a specific depth.
    c) The saturated hydraulic conductivity of a soil type.
    d) The current leaf area index of the crop.
13. Which simulation paradigm is best suited for modeling the continuous accumulation of plant biomass and the continuous flow of water through different soil layers over time?
    a) Discrete Event Simulation.
    b) Agent-Based Modeling.
    c) Continuous Simulation (System Dynamics).
    d) Monte Carlo Simulation (as a standalone paradigm).
14. When would a field-level" system boundary be most appropriate for an SPA study?
    a) Analyzing global climate patterns.
    b) Optimizing irrigation scheduling for a specific crop in a single field.
    c) Studying the long-term impacts of deforestation on regional rainfall.
    d) Modeling the intricate chemical reactions within a single soil particle.
15. What is the role of driving variables" in SPA simulation models?
    a) They are outputs generated by the model.
    b) They are internal system properties that change over time.
    c) They are constant values that remain fixed throughout the simulation.
    d) They are external environmental inputs that compel system behavior.
16. The phenomenon where excess water flows over the land surface, often carrying away topsoil, is known as:
    a) Deep percolation.
    b) Runoff.
    c) Capillary rise.
    d) Infiltration.

17. If a plant closes its stomata to conserve water during a drought, this is an example of what kind of feedback loop?
    a) Positive feedback (amplifying change).
    b) Negative feedback (dampening change/promoting stability).
    c) Open loop.
    d) Linear feedback.
18. What aspect of the soil" component is most directly related to its ability to release water to plant roots?
    a) Soil temperature.
    b) Organic matter content.
    c) Water holding capacity (field capacity, wilting point).
    d) Soil color.
19. In the context of the SPA system, why are system boundaries often described as permeable"?
    a) Because they are physically invisible.
    b) Because they allow for continuous exchange of matter, energy, and information with the environment.
    c) Because they change randomly during a simulation.
    d) Because they prevent any influence from external factors.
20. Which of the following is an example of a flux" in the SPA system?
    a) Soil moisture content.
    b) Plant leaf area.
    c) Photosynthesis rate.
    d) Amount of rainfall in a month.
21. The statement The behavior of the whole system is more than the sum of its individual parts" directly relates to which principle of a systems approach?
    a) Reductionism.
    b) Isolation.
    c) Emergence.
    d) Linearity.
22. Which of the following is not typically a primary output of the SPA system when defined at the field level?
    a) Crop yield.
    b) Evapotranspiration.
    c) Deep percolation.
    d) Global atmospheric CO_2 concentration.

23. Which simulation paradigm is most effective for modeling the spread of a pest or disease through a field by tracking the movement and interactions of individual organisms?

a) Continuous Simulation. b) Discrete Event Simulation.

c) Agent-Based Modeling. d) System Dynamics.

24. The principle that system boundaries are arbitrary but logical" means:

a) Boundaries can be drawn randomly without justification.

b) Boundaries are physically fixed but conceptually flexible.

c) Boundaries are chosen by the analyst based on study objectives, even if not physically impenetrable.

d) Boundaries must always encompass the entire planet.

25. In the context of agroadvisory, how does a systems approach improve predictive capabilities?

a) By simplifying complex interactions into single-factor effects.

b) By focusing only on past events and observed data.

c) By forecasting future outcomes based on dynamic interactions and potential scenarios.

d) By eliminating the need for any form of data collection.

26. Why is solar radiation" a crucial atmospheric input to the SPA system?

a) It directly causes soil compaction.

b) It fuels both evaporation/transpiration and photosynthesis.

c) It increases the plant's wilting point.

d) It primarily determines wind speed.

27. The ability of a systems-informed agroadvisory to provide optimal irrigation scheduling is a direct benefit of understanding:

a) The chemical composition of fertilizers.

b) The specific market prices of crops.

c) The dynamic water balance within the SPA continuum.

d) The historical yield data of a single crop.

28. What is the main difference between state variables" and parameters" in a simulation model?

a) State variables are external inputs, while parameters are outputs.

b) State variables change over time, while parameters are constant values defining properties.

c) State variables are always measured, while parameters are always estimated.

d) State variables describe processes, while parameters describe entities.

29. If an agroadvisory aims to analyze the overall water quality impact of agriculture across an entire river basin, what scale of system boundary would be most appropriate?

a) Individual plant scale.

b) Field/farm scale.

c) Watershed/regional scale.

d) Global atmospheric scale.

30. A systems approach helps agroadvisory in risk assessment by:

a) Eliminating all uncertainties from agricultural production.

b) Guaranteeing a specific yield outcome regardless of conditions.

c) Quantifying the probability and severity of risks under uncertain conditions using techniques like Monte Carlo simulation.

d) Recommending a single, fixed solution for all potential problems.

2

Crop Models - A Dynamic Representation of Agricultural Systems

2.1 Introduction

The preceding chapter introduced the foundational concept of the systems approach, emphasizing the interconnectedness of components within complex entities and the power of simulation to understand their dynamic behavior. Building upon this understanding, This chapter delves into one of the most vital applications of system simulation in agriculture: crop models. These sophisticated computational tools stand as the cornerstone for quantitatively representing the intricate interplay between the soil, the plant, and the atmosphere, providing a dynamic lens through which to examine agricultural productivity and resource use efficiency.

Crop models are essentially mathematical descriptions that mimic the growth, development, and yield of agricultural crops under various environmental conditions and management practices. They encapsulate our scientific understanding of how plants absorb light, water, and nutrients, how they convert these resources into biomass, and how they respond to stresses like drought, heat, or nutrient deficiency. Far from being mere statistical regressions, these are often process-based models, meaning they attempt to simulate the underlying biological, physical, and chemical mechanisms that govern crop performance. By integrating knowledge from disciplines such as plant physiology, soil science, meteorology, and agronomy, crop models provide a holistic and dynamic representation of the entire agricultural system.

The evolution of crop modeling has mirrored advancements in computational power and scientific understanding. From early empirical models that relied heavily on statistical relationships between yield and a few environmental factors, the field has progressed to complex mechanistic models that simulate detailed physiological processes on a daily or even hourly basis. This evolution has transformed crop models from purely research-oriented tools into powerful instruments for real-world decision-making. Their ability to conduct virtual experiments—simulating hundreds or thousands of growing seasons under different scenarios without the time, cost, or risk of actual field trials—makes them indispensable for both scientific inquiry and practical agricultural management.

In the context of agroadvisory, crop models are truly revolutionary. They provide a quantitative basis for precise recommendations, moving beyond generalized advice to offer tailored strategies for specific locations, soil types, and climate conditions. They enable the prediction of crop yields, the optimization of irrigation schedules, the fine-tuning of fertilizer applications, and the assessment of climate change impacts, all of which are critical for enhancing farm profitability and promoting sustainable practices.

This chapter will systematically unpack the world of crop models. It will begin by exploring their fundamental structure and components, detailing how biological processes are translated into mathematical functions. Subsequently, it will delve into the types of crop models, differentiating between empirical, mechanistic, and hybrid approaches, and discussing the strengths and limitations of each. A significant portion will be dedicated to the crucial steps of data requirements, calibration, and validation, emphasizing how models are made relevant and reliable for specific contexts. Furthermore, the chapter will showcase various applications of crop models in diverse agricultural contexts, from tactical farm management to strategic policy planning. By the end of this chapter, readers will possess a comprehensive understanding of crop models as indispensable tools for both scientific research and modern agroadvisory.

2.2 Concept of Crop Models

At its most fundamental level, a crop model is a simplified, mathematical representation of the complex biological, physical, and chemical processes that govern the growth, development, and yield formation of a specific agricultural crop (Fig.2.1).

It serves as a quantitative mimicry of how a plant interacts with its surrounding environment—the soil and the atmosphere—and how it responds to human management interventions. The core concept behind crop models lies in translating our scientific understanding of these intricate interactions into a set of logical rules and mathematical equations that can be executed by a computer.

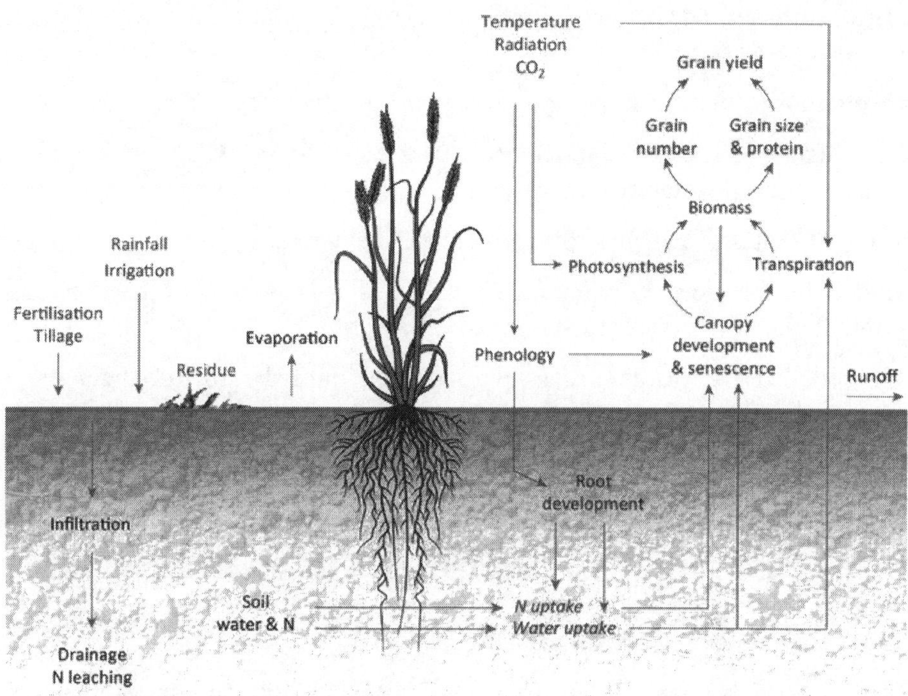

Fig. 2.1: Crop model schematic and processes in crop growth and development

The concept of a crop model is rooted in the system approach discussed in the preceding chapter. Instead of studying isolated plant organs or single environmental factors, a crop model treats the entire soil-plant-atmosphere (SPA) continuum as an integrated system. It recognizes that a crop‘s performance is an emergent property arising from the continuous interplay of numerous components and processes. For instance, the rate of photosynthesis is not only dependent on light availability but also on the plant‘s water status (influenced by soil moisture) and atmospheric carbon dioxide concentration. A crop model, therefore, captures these interdependencies and feedback loops.

Fundamentally, a crop model operates by taking a set of initial conditions (e.g., planting date, initial soil moisture, initial crop biomass), driving variables (daily or hourly weather data like temperature, rainfall, solar radiation), and parameters (crop-specific genetic characteristics, soil hydraulic properties). Using these inputs, it simulates the rates of various physiological processes over time. These processes include:

Photosynthesis: The conversion of light energy into chemical energy, driving biomass production.

Respiration: The consumption of energy for maintenance and growth.

Transpiration: The loss of water vapor from leaves, influenced by atmospheric demand and soil water availability.

Nutrient Uptake: The absorption of essential elements from the soil.

Biomass Allocation: How the newly produced biomass is distributed among different plant organs (roots, stems, leaves, grains).

Phenological Development: The progression through growth stages (e.g., germination, flowering, maturity), often driven by temperature (thermal time).

As these processes unfold in the simulation, the model continuously updates its state variables, which represent the current condition of the system. Examples of state variables include total dry biomass, leaf area index, root depth, and soil water content at different depths. By tracking these state variables over the entire growing season, the model can ultimately predict key outputs such as final yield, total water use, or nutrient uptake.

The power of this concept lies in its predictive capability and its ability to act as a virtual laboratory. Once a crop model is calibrated and validated for a specific environment and crop, it can be used to:

Project future scenarios: How might a crop perform under different climate change scenarios, or with altered rainfall patterns?

Evaluate management strategies: What is the optimal planting date, irrigation schedule, or fertilizer application rate for a given location?

Analyze sensitivities: Which environmental factors or crop characteristics have the most significant impact on yield?

Identify knowledge gaps: Where does our scientific understanding of a process need improvement, as indicated by model discrepancies with observed data?

In essence, a crop model is more than just a predictive tool; it is a dynamic hypothesis, a formalized expression of our understanding of how crops grow (Fig.2.2). It provides a quantitative framework for integrating diverse scientific knowledge, testing hypotheses, and ultimately generating actionable insights for agroadvisory and sustainable agricultural management.

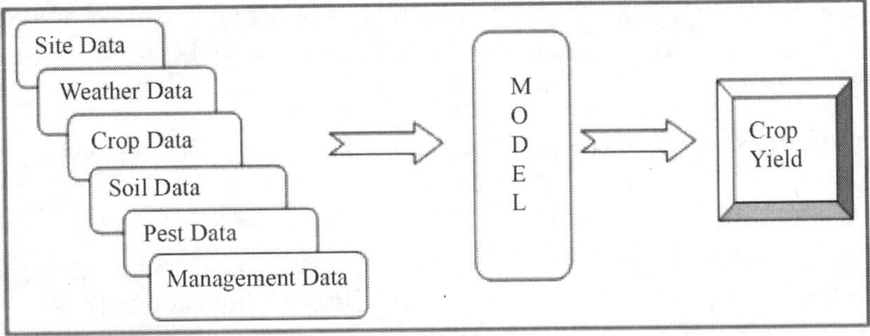

Fig. 2.2: Inputs and system approach of crop growth models

2.3. Types of Crop Models: An Elaborated Explanation

The field of crop modeling is characterized by a rich diversity of approaches, each designed to capture different facets of crop growth and environmental interaction with varying degrees of detail and predictive power. Understanding these distinctions is paramount for anyone seeking to apply crop models effectively in research, management, or agroadvisory. These models can be categorized based on their underlying conceptualization, their treatment of uncertainty, their temporal representation, and the factors they consider.

2.3.1 Empirical vs. Mechanistic vs. Hybrid Models -The Core Distinction

This classification defines the fundamental approach a model takes to represent the processes governing crop performance:

- **Empirical Models (or Statistical Models)**

Mechanism: These models function much like a black box. They establish direct statistical relationships between observed crop outputs (like final yield) and a few key environmental inputs or management practices (e.g., total seasonal rainfall, average temperature, fertilizer application rates). They are built purely from historical data, often through regression analysis. They do not attempt to simulate the biological or physical processes that cause the observed outcomes; they simply describe the correlation.

Analogy: Imagine trying to predict the boiling time of water simply by observing how long it takes at different room temperatures, without understanding heat transfer, specific heat capacity, or latent heat. You might find a correlation, but you wouldn't know *why* it boils.

Advantages: They are relatively simple to develop, requiring less detailed scientific knowledge of plant physiology or soil physics. They are also computationally efficient, making them fast to run. For localized predictions within the range of conditions from which they were derived, they can be quite accurate.

Disadvantages: Their primary limitation is their lack of explanatory power; they tell you what might happen, but not why. This makes them notoriously poor at extrapolation—predicting outcomes outside the specific conditions (e.g., different soil types, climate change scenarios, or novel crop varieties) from which their relationships were established. They are inherently site-specific and difficult to transfer globally.

- **Mechanistic Models (or Process-Based Models)**

Mechanism: These models operate as a "transparent box" or a detailed engineering diagram. They attempt to simulate the fundamental biological, chemical, and physical processes that *cause* crop growth and yield, based on our current scientific understanding. They use mathematical equations to describe rates of processes like:

Photosynthesis: How light energy and CO_2 are converted into sugars.

Respiration: Energy consumption for plant maintenance and growth.

Transpiration: Water movement from roots through the plant and its release as vapor, influenced by atmospheric demand and stomatal regulation.

Nutrient Uptake: Absorption of essential elements from the soil, dependent on concentration and root activity.

Biomass Partitioning: How assimilated sugars are allocated to different plant organs (leaves, stems, roots, fruits/grains).

Phenological Development: Progression through growth stages (e.g., germination, flowering, maturity) often driven by thermal time (accumulated temperature).

These processes interact dynamically, updating state variables (e.g., leaf area index, total biomass, soil water content) over small time steps (daily or hourly).

Analogy: This is akin to designing a car engine from scratch, understanding how fuel burns, pistons move, and gears engage to produce motion.

Advantages: High explanatory power, providing insights into the why behind crop performance. They are robust for extrapolation to new environments, management practices, or even novel genetic traits, making them invaluable for climate change impact assessment and breeding programs. They provide detailed information on intermediate processes, not just final yield.

Disadvantages: They are significantly more complex to develop, requiring extensive scientific knowledge and a large volume of detailed input data (e.g., hourly weather, multi-layered soil characteristics, specific crop parameters). Their calibration and validation are rigorous and computationally intensive.

• Hybrid Models

Mechanism: These models seek a pragmatic balance by combining the strengths of both mechanistic and empirical approaches. They typically use a mechanistic framework for well-understood, core physiological processes but integrate empirical relationships or statistical components for processes that are either too complex, poorly understood, or where detailed mechanistic data is scarce (e.g., specific pest damage functions, or highly localized soil interactions). Increasingly, they incorporate **Machine Learning (ML)** or **Artificial Intelligence (AI)** algorithms to learn complex patterns from data and refine relationships within the mechanistic structure.

Analogy: This is like building a custom car where the engine is meticulously designed from first principles, but certain comfort features or external aerodynamics are based on empirical testing and user preferences.

Advantages: Can leverage the best of both worlds, offering improved predictive accuracy and robustness compared to purely empirical models, while potentially being less data-intensive or complex than purely mechanistic models for specific applications. They are often more adaptable to real-world data availability.

Disadvantages: Can inherit some limitations from both parent types if the integration is not carefully managed. Designing optimal hybrid structures and integrating diverse modeling components can be challenging.

2.3.2 Deterministic vs. Stochastic Models: Handling Uncertainty

This classification distinguishes how models account for the inherent randomness and variability in natural systems:

Deterministic Models

Mechanism: These models operate with absolute certainty. Given a specific set of input values and parameters, a deterministic model will *always* produce one exact, singular output. There is no element of chance or randomness built into the model's internal calculations.

Analogy: A calculator: if you input 2 + 2, it will always deterministically output 4.

Use Cases: Ideal for simulating average or potential scenarios, or when the focus is on understanding the fundamental cause-and-effect relationships without considering variability. Most mechanistic crop models, when run with a single weather file, are deterministic.

Limitations: They provide a single "best guess prediction, which can be misleading in agriculture where natural variability (e.g., daily weather fluctuations, soil heterogeneity) is high. They do not quantify the uncertainty or risk associated with the predicted outcome.

- **Stochastic Models**

Mechanism: These models explicitly incorporate randomness and probability distributions for uncertain input variables or parameters. Instead of using a single fixed value, they draw values from a defined range based on a probability distribution. By running the model many times (often using **Monte Carlo simulation**), they generate an ensemble of possible outcomes, creating a probability distribution of the system's response.

Analogy: Rolling a die: you know the possible outcomes (1-6) and their probabilities, but the exact outcome of a single roll is random. Running many rolls gives you a distribution of results.

Use Cases: Essential for **risk assessment** and **uncertainty quantification** in agriculture. They help understand the variability of outcomes (e.g., the likelihood of achieving a certain yield under variable weather conditions, or the probability of crop failure due to drought).

Limitations: Computationally more demanding due to the need for multiple runs. Their outputs require statistical interpretation, which can be more complex than single-point predictions.

2.3.3 Static vs. Dynamic Models: Representing Time

This distinction relates to how models represent processes over time:

- **Static Models**

Mechanism: These models do not explicitly account for the passage of time in their internal calculations. They typically represent a single snapshot or an equilibrium state, focusing on overall input-output relationships over an entire period (e.g., a full growing season) rather than daily changes. Many simple empirical models are static.

Analogy: A photograph capturing a single moment, without showing the progression of events.

Use Cases: Simple long-term averages or yield potentials based on seasonal totals of environmental factors.

Limitations: Unable to capture the daily progression of crop growth, phenological stages, or the timing and severity of stresses (like a critical drought period during flowering), which are crucial for realistic agricultural decision-making.

- **Dynamic Models**

Mechanism: These models explicitly incorporate time as a variable. They simulate processes and update state variables at discrete time steps (e.g., daily, hourly, or even sub-hourly) throughout the crop's life cycle. Most mechanistic crop models are dynamic.

Analogy: A video recording that shows the continuous progression of events over time.

Use Cases: Detailed analysis of crop growth, phenological development, real-time stress monitoring, optimal timing for irrigation and fertilizer applications, and understanding the impact of short-term weather events.

Limitations: More complex to build and require high-resolution time-series data for inputs.

2.3.4 Production Level Models: Accounting for Limiting Factors

Within mechanistic modeling frameworks, crop models are often classified by the level of environmental constraints they incorporate:

Potential Production Models: These models simulate the maximum possible crop growth and yield determined solely by genotype, solar radiation, and temperature. They assume that water, nutrients, and pest/disease pressures are *non-limiting* and always optimally available. This represents an upper ceiling of productivity.

Water-Limited Production Models: These models extend potential models by introducing water availability as a limiting factor. They simulate the soil water balance, root water uptake, and how water deficits impact key physiological processes like photosynthesis and transpiration, thereby reducing potential growth and yield. Nutrient supply and pest pressure are still assumed optimal.

Nutrient-Limited Production Models: Building upon water-limited models, these further incorporate **nutrient dynamics** (e.g., nitrogen, phosphorus, potassium) into the simulation. They account for nutrient availability in the soil, uptake by roots, and how nutrient deficiencies can limit growth and yield, even if water is sufficient. Both water and nutrient limitations are considered.

Pest/Disease-Limited Production Models (or Yield-Reducing Factors Models): These are the most comprehensive, incorporating the impact of biotic stresses (pests, diseases, weeds) on crop growth and yield. They often do so by reducing photosynthetic efficiency, causing biomass loss, or altering nutrient uptake, building on the water- and nutrient-limited foundations.

This hierarchical approach helps in understanding *what* factors are primarily limiting production in a given environment and *how* management interventions (e.g., irrigation, fertilization, pest control) can alleviate these constraints.

2.3.5 Spatial vs. Non-Spatial Models: Geographical Representation

This classification addresses how models handle geographical variability:

- **Non-Spatial Models (Point-Based)**

Mechanism: These models assume uniformity within the area they simulate. They typically use a single set of environmental inputs (e.g., from a single weather station) and soil parameters to represent conditions at a specific "point or a homogeneous field. Outputs are generated for that single point.

Use Cases: Research plots, studies focusing on average conditions over a uniform area, or initial assessments where spatial variability is not the primary concern.

Limitations: Cannot capture within-field variability in soil properties, topography, or microclimates, which can lead to significant discrepancies when scaling up to large areas.

- **Spatial Models (Distributed)**

Mechanism: These models explicitly incorporate geographical variability by dividing a landscape into discrete spatial units (e.g., grid cells or polygons). They use Geographic Information Systems (GIS) to manage and process spatially explicit data on soil types, topography, climate, and management practices. The crop model is then run for each individual spatial unit, considering its unique local conditions, and the results are aggregated or visualized spatially.

Use Cases: Precision agriculture (e.g., variable rate irrigation or fertilizer application zones), regional yield mapping, watershed-level water quality modeling, and assessing the spatial impact of climate change.

Limitations: Extremely data-intensive, requiring high-resolution spatial input layers. They are also significantly more computationally demanding due to running the model for numerous individual units.

By understanding these diverse classifications, readers can appreciate the versatility and complexity of crop modeling, laying the groundwork for how these tools are applied to advance agroadvisory services effectively. The choice of model type is always a trade-off between the desired level of detail, the specific research question, and the availability of data and computational resources.

2.4 Techniques of Crop Modelling: Building, Calibrating, and Validating Models

The journey of creating and effectively utilizing a crop model is a rigorous, systematic, and often iterative process that transforms scientific understanding into a powerful predictive tool. It's an engineering feat for biological systems, demanding precision at every step from conceptual design to real-world application. Each technique involved plays a pivotal role in ensuring the model is not only scientifically robust but also practically valuable for informing agroadvisory and agricultural research.

The process typically begins with a thorough conceptual model design. Before a single line of code is written, a clear and precise understanding of the problem and the system is paramount. This initial phase involves explicitly defining the objectives the model should achieve, whether it's predicting final crop yield, optimizing irrigation schedules, or assessing nutrient uptake efficiency. This objective dictates the necessary level of detail for the model. Simultaneously, system boundaries are meticulously drawn to delineate what elements and processes are considered inside the model (e.g., the soil profile down to a specific depth, the plant canopy, the immediate atmospheric layer) versus those that are external environmental drivers. Within these boundaries, all key components and processes relevant to the objectives—from photosynthesis and transpiration to nutrient cycling and phenological development—are identified and their intricate relationships mapped out, often through diagrams illustrating the flow of water, energy, and biomass and highlighting critical feedback loops. This collaborative phase typically involves agronomists, plant physiologists, soil scientists, and meteorologists pooling their expertise.

Once the conceptual framework is solid, the next step is mathematical formulation. Here, the qualitative understanding gained in the previous stage is translated into quantitative terms. Every identified process and interaction is expressed as a set of mathematical equations, which could range from simple linear relationships to complex differential equations describing dynamic rates. For instance, the rate of photosynthesis might be formulated as a function of incoming solar radiation, ambient temperature, carbon dioxide concentration, and the plant's water status. This stage also rigorously defines all state variables, which represent the quantities that change over time within the system (e.g., daily plant biomass, hourly soil moisture content at various depths, leaf area index). Crucially, the equations for fluxes—the rates at which matter or energy move or transform (e.g., transpiration rate, water uptake rate, nutrient mineralization rate)—are specified. Furthermore, all parameters are identified; these are the constant values that define the specific characteristics of the crop (e.g., its maximum photosynthetic efficiency, thermal time requirements for flowering) or the soil (e.g., its water holding capacity, hydraulic conductivity). A suitable time step (e.g., daily or hourly) for the model's calculations is also chosen, balancing computational efficiency with the need to capture rapid dynamic changes.

No model can function without data, making data collection and preparation a critical, continuous effort. High-quality, comprehensive datasets are essential for both building and evaluating crop models. This includes historical weather data(daily or hourly records of solar radiation, temperature, rainfall, humidity, and wind speed), which serve as the primary external drivers for most models. Detailed soil data is also crucial, encompassing physical properties like texture (sand, silt, clay percentages), bulk density, organic matter content, and hydraulic characteristics across different soil horizons. For the biological component, crop data includes genetic coefficients specific to particular varieties (e.g., photoperiod sensitivity, biomass partitioning ratios) and extensive field experimental data documenting growth stages, biomass accumulation, water use, nutrient uptake, and final yield under a range of controlled management conditions. Information on management practices like planting dates, seeding densities, irrigation schedules, and fertilizer applications is also indispensable. A significant part of this stage involves meticulous data cleaning and formatting to resolve errors, fill gaps, and ensure consistency, preparing the data for seamless input into the model.

With the mathematical framework established and data prepared, the model is then brought to life through implementation. This involves writing the computer code in a chosen programming language (historically Fortran or C++, but increasingly Python or R for their flexibility). Good programming practice dictates a modular design, where distinct processes (like soil water dynamics, plant photosynthesis, or nutrient cycling) are encapsulated in separate code modules. This makes the model easier to understand, maintain, debug, and update. Robust routines for input/output handling are also developed to efficiently read the diverse datasets and generate simulation results in clear, accessible formats. For dynamic models, appropriate numerical solvers are incorporated to accurately integrate the rate equations over the chosen time steps, allowing the model to simulate the continuous progression of the crop's growth and interaction with its environment.

Once implemented, a crop model must be calibrated to align its simulations with real-world observations. Calibration is a fine-tuning process where a select subset of the model's parameters—typically those that are most sensitive to change and have the greatest impact on output, or those that are less well-known for a specific cultivar or soil—are adjusted. This adjustment aims to minimize the discrepancy between the model's predicted values and actual field data from specific experiments. Calibration can be done manually, through iterative trial-and-error adjustments while observing the model's response, which relies heavily on expert judgment but can be time-consuming. Alternatively, automated calibration techniques employ sophisticated optimization algorithms (such as genetic algorithms or shuffled complex evolution) that systematically search for the best parameter values by minimizing a predefined statistical objective function (e.g., the Root Mean Square Error between simulated and observed yield). It's crucial that the data used for calibration is distinct from the data used for subsequent validation to ensure a robust evaluation.

Following calibration, validation is the critical step that assesses the model's reliability and general applicability. This involves evaluating the model's performance against an independent dataset—data that was *not* used during the model's development or calibration. This independent testing helps confirm that the model can accurately predict outcomes under new, unseen conditions for the same crop and environment. Validation relies heavily on statistical metrics to quantify the agreement between simulated and observed values. Common measures include the Root Mean Square Error (RMSE), which indicates the average magnitude of prediction errors; the Coefficient of Determination (R^2), which explains how much of the observed variability is accounted for by the model; and various bias indicators. Beyond statistical analysis, graphical analysis (e.g., scatter plots of observed vs. simulated values, time-series plots) is essential for visually identifying systematic biases or deviations in the model's predictions. Furthermore, a qualitative reality check by domain experts ensures that the model's internal behavior and logic align with scientific understanding, fostering confidence in its predictive capabilities.

To thoroughly understand a model's behavior and its inherent uncertainties, sensitivity analysis is an indispensable technique. This involves systematically investigating how variations in the model's input parameters or driving variables influence its outputs. Simple approaches might involve varying one input at a time (OAT) while holding others constant, offering initial insights into individual factor impacts. However, more advanced global sensitivity analysis (GSA)techniques (e.g., using Monte Carlo simulations or Sobol' indices) simultaneously vary multiple parameters over their plausible ranges. This comprehensive approach reveals not only which parameters are most influential individually but also how they interact to affect the model's predictions. The insights gained from sensitivity analysis are invaluable for several reasons: they identify the most critical parameters that require precise measurement or calibration; they help understand the model's robustness to input uncertainties; and they guide future research by highlighting which aspects of the agricultural system have the greatest leverage on desired outcomes.

Once a crop model has undergone these rigorous development, calibration, validation, and sensitivity analysis phases, it becomes a powerful tool ready for diverse applications. These range from yield forecasting at local, regional, or national scales to inform market and food security decisions, to identifying optimal management strategies for farmers (e.g., precise irrigation amounts and timing, efficient fertilizer application rates, optimal planting dates for specific cultivars). They are also crucial for conducting climate change impact assessments, predicting how shifts in temperature and rainfall patterns will affect crop production, and for evaluating the performance of new crop breeding lines under a variety of future scenarios. Furthermore, crop models form the backbone of advanced decision support systems that deliver actionable agroadvisory, empowering stakeholders at all levels to make data-driven choices for more productive, resilient, and sustainable agriculture. The techniques outlined here underscore that crop modeling is not a static endeavor but a dynamic, evolving discipline committed to refining our understanding and management of complex agricultural systems.

2.5 Conclusion: The Transformative Power of Crop Models

This chapter has served as a foundational exploration into the intricate world of crop models, firmly establishing their role as vital mathematical representations of the dynamic interactions within the soil-plant-atmosphere (SPA) continuum. Readers have journeyed from understanding the fundamental concept of a crop model as a formalized scientific hypothesis, to appreciating the nuanced distinctions between empirical, mechanistic, and hybrid approaches, each offering unique insights and applications. Furthermore, the discussion delved into the critical considerations of deterministic versus stochastic modeling, highlighting how models can account for inherent variability and provide a spectrum of possible outcomes, rather than just a single prediction. The contrast between static and dynamic modelsilluminated how these tools capture either a snapshot or the continuous, evolving processes of crop growth over time.

Beyond their theoretical underpinnings, a significant portion of this chapter was dedicated to demystifying the practical techniques for model development, rigorous calibration, and robust validation. This detailed explanation underscored that building a reliable crop model is far from a simplistic endeavor; it is an iterative process demanding meticulous data collection, precise mathematical formulation, and thorough evaluation against real-world observations. The importance of sensitivity analysis was also highlighted, revealing how models can pinpoint the most influential factors impacting crop performance, thereby guiding both scientific inquiry and practical management decisions.

Ultimately, crop models emerge as far more than mere computational tools; they are sophisticated virtual laboratories. They provide an unparalleled environment to conduct countless what-if scenarios, simulating the impacts of diverse environmental conditions, management practices, and even novel genetic traits without the prohibitive costs, time, and risks associated with traditional field experimentation. This profound capability is precisely what makes them indispensable for advancing agroadvisory services. By enabling precision, prediction, and proactive guidance, crop models empower agricultural stakeholders to move beyond generalized advice. They facilitate the delivery of tailored, data-driven solutions for optimizing irrigation, fine-tuning nutrient applications, forecasting yields with greater accuracy, and mitigating the growing risks posed by an unpredictable climate.

The comprehensive understanding of crop models developed in this chapter is a cornerstone for the remainder of this book. It lays the essential groundwork for delving into the more advanced applications of system simulation and agroadvisory, exploring how these powerful tools are integrated with cutting-edge technologies to address the complex challenges and opportunities facing agriculture in the 21st century.

Multiple Choice Questions (Bold means correct answer)

1. What is the primary role of a crop model as a mathematical representation?
   a) To statistically analyze historical weather data.
   b) To mimic the growth, development, and yield of crops under various conditions.
   c) To visually represent soil properties in a field.
   d) To manage agricultural machinery.
2. A crop model is described as a dynamic hypothesis because it:
   a) Uses only empirical data.
   b) Provides a formalized expression of our understanding of how crops grow.
   c) Always produces the same results.
   d) Is only used for short-term predictions.
3. Which of the following is considered an initial condition for a crop model simulation?
   a) Daily rainfall.
   b) Final crop yield.
   c) Initial soil moisture.
   d) Photosynthesis rate.
4. The progression of a crop through stages like germination, flowering, and maturity is referred to as:
   a) Biomass allocation.
   b) Nutrient uptake.
   c) Phenological development.
   d) Respiration.
5. Which type of crop model relies purely on observed statistical correlations between inputs and outputs, without simulating underlying biological processes?
   a) Mechanistic model.
   b) Hybrid model.
   c) Empirical model.
   d) Dynamic model.
6. A key advantage of mechanistic crop models is their strong:
   a) Simplicity in development.
   b) Ability to extrapolate to novel environmental conditions.
   c) Reliance solely on historical yield data.
   d) Low computational demands.
7. When a crop model consistently produces the exact same output for the same set of inputs, it is classified as a:
   a) Stochastic model.
   b) Hybrid model.
   c) Deterministic model.
   d) Spatial model.

8. Which type of model is best suited for quantifying the uncertainty and risk associated with crop yield predictions due to variable weather?
   a) Deterministic model.
   b) Static model.
   c) Empirical model.
   d) Stochastic model.
9. Models that explicitly incorporate time as a variable and simulate processes at discrete time steps (e.g., daily or hourly) are known as:
   a) Static models.
   b) Empirical models.
   c) Dynamic models.
   d) Non-spatial models.
10. A crop model simulating growth under ideal conditions where water, nutrients, and pest pressures are non-limiting is classified as a:
    a) Water-limited production model.
    b) Nutrient-limited production model.
    c) Potential production model.
    d) Pest-limited production model.
11. What is the primary purpose of the calibration technique in crop modeling?
    a) To physically validate the model in a new field.
    b) To identify the most influential input parameters.
    c) To adjust model parameters to better match observed data from specific experiments.
    d) To generate random outputs for risk assessment.
12. Using an independent dataset to evaluate a model's ability to predict outcomes in new, unseen conditions is part of which technique?
    a) Conceptual model design.
    b) Calibration.
    c) Validation.
    d) Sensitivity analysis.
13. The process of translating a qualitative understanding of crop growth into mathematical equations is known as:
    a) Data collection.
    b) Implementation.
    c) Mathematical formulation.
    d) Application.
14. Which type of data is crucial for driving variables in most crop models?
    a) Soil texture analysis.
    b) Long-term historical weather records.
    c) Economic market prices.
    d) Genetic sequences of the crop.

15. What does a flux represent in the context of a crop model?
    a) A constant value for a soil property.
    b) The current amount of biomass accumulated.
    c) The rate of movement or transformation of matter/energy (e.g., transpiration rate).
    d) An external input that does not change over time.
16. What is the main disadvantage of a purely empirical crop model?
    a) It requires too much detailed physiological data.
    b) It has poor transferability to conditions outside its development range.
    c) It is too computationally intensive.
    d) It cannot predict yield.
17. If a crop model uses GIS to divide a landscape into grids and simulates growth for each unique spatial unit, it is:
    a) Non-spatial model.
    b) Point-based model.
    c) Spatial model.
    d) Static model.
18. Which of the following is a key reason for conducting sensitivity analysis on a crop model?
    a) To physically test the model in a laboratory.
    b) To fine-tune all model parameters simultaneously.
    c) To identify which input parameters or variables have the most significant influence on outputs.
    d) To simplify the model by removing all equations.
19. In model conceptualization, clearly defining system boundaries helps to:
    a) Physically isolate the farm.
    b) Determine what inputs and outputs cross into and out of the system.
    c) Eliminate all external influences.
    d) Ensure perfectly linear relationships between components.
20. A model that combines mechanistic principles for core processes with empirical relationships for less understood aspects is called a:
    a) Deterministic model.
    b) Hybrid model.
    c) Static model.
    d) Statistical model.

21. What is the primary role of state variables in a crop model?
    a) To provide constant environmental data.
    b) To define the rates of processes.
    c) To represent the quantities that change over time and define the system's current condition.
    d) To calibrate the model against observed data.
22. Which step in crop model development involves structuring the code into distinct modules and handling input/output files?
    a) Conceptual model design.
    b) Mathematical formulation.
    c) Implementation.
    d) Validation.
23. When evaluating a model, what does an R^2 value typically indicate?
    a) The average magnitude of error.
    b) The number of parameters in the model.
    c) The proportion of observed variance explained by the model.
    d) The complexity of the model's equations.
24. A farmer using a crop model to decide the optimal planting date for maximum yield on their specific field is an example of which stage of crop model usage?
    a) Calibration.
    b) Validation.
    c) Conceptualization.
    d) Application.
25. The principle of emergence in crop modeling implies that:
    a) All properties of a crop can be understood by studying individual cells.
    b) Crop yield is an unpredictable result of random events.
    c) The behavior of the entire crop system (e.g., yield) is greater than the sum of its isolated components.
    d) Models should only focus on short-term predictions.

3

Data Requirements for Crop Modelling

3.1 Introduction

The preceding chapters have firmly established the conceptual foundations of the systems approach and introduced the diverse world of crop models, highlighting their intricate mechanisms and their transformative potential for agroadvisory. However, even the most elegantly formulated and sophisticated crop model is ultimately a framework; its predictive power and reliability are inherently tethered to the quality and availability of the information it receives. As the adage in data science goes, Garbage in, garbage out– a truth that resonates profoundly within the realm of agricultural system simulation.

The chapter shifts focus from the theoretical constructs of models to the vital practical considerations of data requirements. This chapter will delve into the indispensable role that accurate, comprehensive, and well-structured data plays in every stage of crop modeling, from initial parameterization and rigorous calibration to robust validation and reliable application. Without the right data, even a theoretically perfect model cannot accurately mimic the complex dynamics of the soil-plant-atmosphere continuum or provide trustworthy insights for agricultural decision-making.

This chapter will systematically unpack the various categories of data essential for crop modeling. It will meticulously detail the specific characteristics and sources of weather data, which serve as the primary atmospheric drivers. It will then explore the crucial properties and measurement techniques for soil data, underpinning the understanding of water and nutrient availability. The discussion will extend to crop genetic and physiological data, which define the inherent capabilities and responses of specific plant varieties. Furthermore, the chapter will cover management data, illustrating how human interventions are integrated into the modeling framework. Beyond merely listing data types, this section will critically examine the challenges associated with data acquisition, including issues of quality, resolution, consistency, and spatial representativeness.

By the end of this chapter, readers will possess a comprehensive understanding of the diverse data landscapes necessary for crop modeling. They will learn to appreciate the paramount importance of data quality, recognize common data limitations, and grasp the methodologies involved in preparing and utilizing datasets to unlock the full predictive power of crop simulation models for effective agroadvisory.

3.2 Data: Essential Inputs for Crop Modelling

Weather data stands as one of the most critical categories of inputs for crop simulation models, acting as the primary atmospheric drivers that dictate the energy and water dynamics within the soil-plant-atmosphere continuum. The intricate processes of photosynthesis, transpiration, and phenological development are profoundly influenced by the immediate atmospheric conditions, making accurate and comprehensive weather data indispensable for reliable model predictions. Without this vital information, even the most sophisticated crop model remains a theoretical construct, unable to reflect the real-world variability that shapes agricultural outcomes (Fig.3.1).

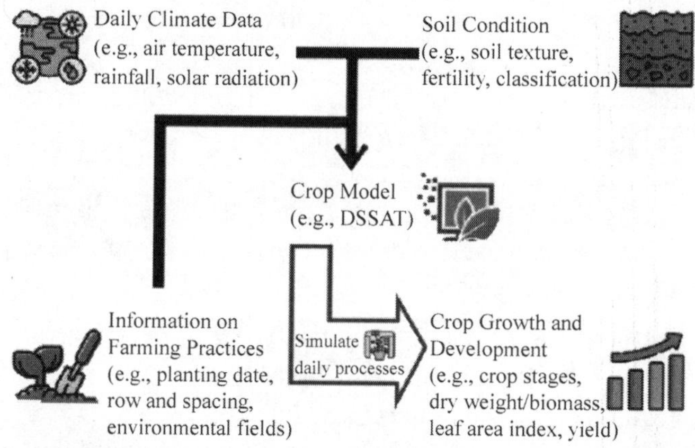

Fig. 3.1: Inputs and outputs for crop model simulation

3.2.1 Data Required

To accurately simulate crop growth and its interactions with the environment, crop models require several key meteorological variables. These typically include:

Solar Radiation (or Shortwave Radiation): This is arguably the most crucial energy input for plant growth, directly driving photosynthesis and evapotranspiration. It is often measured in Megajoules per square meter per day (MJ/m²/day) or Watts per square meter (W/m²). It determines the amount of energy available for biomass production and water evaporation.

Maximum and Minimum Air Temperature: Temperature directly influences the rates of most biological processes, including photosynthesis, respiration, and the progression of phenological stages (e.g., germination, flowering, maturity). Crop models often use these values to calculate daily average temperatures or growing degree days (GDD), which quantify the accumulation of heat required for development. Extreme temperatures (heat stress or frost) can also be critical.

Precipitation (Rainfall): The primary input of water into the agricultural system, directly affecting soil moisture content. It is typically measured as total daily rainfall in millimeters (mm). Accurate precipitation data is fundamental for simulating the water balance and predicting drought stress or waterlogging.

Relative Humidity (or Vapor Pressure Deficit - VPD): Humidity indicates the amount of moisture in the air. Vapor Pressure Deficit (VPD), which measures the drying power of the atmosphere, is often derived from temperature and humidity. High VPD drives higher rates of transpiration, influencing the plant's water demand and potential for stress. This affects how much water a plant needs to cool itself and transport nutrients.

Wind Speed: Wind influences the rate of evapotranspiration by removing humid air from around plant leaves and the soil surface. It is typically measured at a standard height (e.g., 2 or 10 meters) in meters per second (m/s) or kilometers per hour (km/h).

Reference Evapotranspiration (ETo): While sometimes calculated internally by models from the above variables, ETo (the evapotranspiration from a hypothetical reference crop under ideal conditions) is sometimes provided as a direct input, especially for simpler water balance models.

Atmospheric CO_2 Concentration: Increasingly relevant for modern crop models, especially when assessing climate change impacts. Elevated CO_2 levels can enhance photosynthesis, particularly for C3 crops, impacting biomass production and water use efficiency.

3.2.2 Frequency of Observation

The appropriate frequency of weather observations depends significantly on the complexity of the crop model and the specific processes being simulated:

Daily Data: For many general-purpose crop growth and yield models (e.g., DSSAT, APSIM), daily resolution is the most common and often sufficient.

Daily maximum and minimum temperatures, total daily rainfall, and total daily solar radiation are standard inputs. This frequency allows models to capture the overall daily progression of growth and water balance.

Hourly Data: More detailed mechanistic models, especially those focusing on rapid physiological responses, soil water dynamics, or microclimatic effects, often benefit from hourly or even sub-hourly data. For instance, hourly temperature and radiation allow for more accurate calculation of diurnal photosynthesis and transpiration patterns. Hourly rainfall data is crucial for precise simulation of infiltration and runoff events, as a large amount of rain over a short period behaves differently from the same amount spread throughout the day. Frost events, which can occur over a few critical hours, also necessitate hourly temperature data.

Sub-Daily (e.g., 30-minute, 15-minute): Highly detailed biophysical models or those integrated with real-time sensing for precision agriculture might utilize even finer temporal resolutions to capture rapid fluctuations in environmental conditions that impact immediate plant responses or irrigation scheduling.

Multi-Day/Seasonal Aggregates: While not direct inputs for process-based models, aggregated data (e.g., weekly averages, monthly totals) from historical records are often used for statistical analysis, regional climate characterization, or for driving empirical crop models.

3.2.3 Collection and Tabulation

The methods for collecting and tabulating weather data have evolved significantly with technological advancements, moving from manual observations to highly automated and remote sensing approaches:

- **Ground-Based Weather Stations**

Automated Weather Stations (AWS): These are the most common source of high-quality, continuous weather data. AWS are equipped with sensors that automatically measure and log parameters like air temperature, humidity, solar radiation, wind speed and direction, and rainfall. They can record data at various frequencies (e.g., 15-minute, hourly, daily) and transmit it wirelessly for immediate access. Installation and maintenance are crucial for data quality.

Manual Stations: Traditional meteorological observatories or farm-level rain gauges often involve manual readings at fixed times daily. While reliable, they provide lower temporal resolution and can be labor-intensive.

Network Density: The density of weather stations is critical. For accurate field-level crop modeling, data from a station as close as possible to the farm is ideal, as local microclimates can differ significantly from regional averages. National meteorological departments (like IMD in India) maintain extensive networks that provide data at district or sub-district levels.

- **Remote Sensing and Satellite-Derived Products**

Satellite Imagery: Satellites equipped with radiometers can infer various meteorological parameters across large geographical areas, overcoming the limitations of sparse ground station networks. They can estimate solar radiation, land surface temperature, cloud cover, and even precipitation (e.g., from radar or microwave sensors).

Reanalysis Datasets: These datasets combine historical weather observations from various sources (ground stations, balloons, satellites) with atmospheric models to produce a consistent, gridded, and globally complete dataset of meteorological variables. Examples include AgMERRA (Agricultural Model Intercomparison and Improvement Project's Modern-Era Retrospective analysis for Research and Applications) and NASA POWER. These are invaluable for historical simulations and regional studies, providing spatially explicit data even in areas with few ground stations.

Downscaling: Global or regional climate model outputs, often at coarse resolutions, need to be "downscaled" to local scales relevant for crop models. This involves statistical or dynamic methods to translate large-scale climate projections into localized weather inputs.

Weather Generators

Mechanism: When historical weather data is incomplete or unavailable for future climate scenarios, statistical weather generators are used. These programs simulate synthetic weather series based on the statistical properties (means, variances, correlations, probabilities of wet/dry days) of observed historical weather data for a given location.

Use Cases: Filling data gaps, extending short records, or generating multiple possible future weather sequences for stochastic simulations (e.g., Monte Carlo analyses) to assess risk.

Tabulation and Data Formats

Standardized Formats: Weather data for crop models are typically organized in structured formats for easy parsing by software. Common formats include simple text files (e.g., .WTH files for DSSAT, comma-separated values - CSV), where each row represents a date and columns contain the required meteorological variables.

Metadata: Crucial accompanying information (metadata) includes the weather station's latitude, longitude, and elevation, sensor heights, and units of measurement. This ensures proper model interpretation and calculation (e.g., daylength calculations based on latitude).

Quality Control: Before use, weather data undergoes rigorous quality control. This involves checking for extreme outliers (sensor errors), missing values, and consistency (e.g., minimum temperature cannot be higher than maximum temperature). Automated scripts and visual inspections are used for this purpose.

The meticulous acquisition, quality control, and proper tabulation of weather data are foundational steps in ensuring the accuracy and reliability of crop model simulations. The choice of data source and resolution depends heavily on the model's complexity, the spatial scale of the study, and the specific questions being addressed, ultimately impacting the utility of the crop model for agroadvisory.

3.3 Soil Data: Understanding the Subterranean Engine of Growth

While weather data dictates the atmospheric conditions above ground, **soil data** provides crucial insights into the subterranean environment where plant roots anchor, absorb water, and acquire essential nutrients. The soil acts as a complex, living matrix, serving as the primary reservoir for water and nutrients, and influencing the plant's ability to access these vital resources. For crop models to accurately simulate water and nutrient cycling, root growth, and overall crop performance, comprehensive and precise soil data are absolutely essential. This section will delve into the critical soil parameters required by crop models and explain the importance of each.

3.3.1 Essential Soil Data Parameters and Their Importance

- **Soil Profile Description (Horizon Depths)**

Description: The soil profile is a vertical cross-section of the soil, revealing distinct layers or horizons (e.g., O, A, B, C horizons). For modeling purposes, the depth to the bottom of each significant horizon is typically recorded. These horizons differ in their physical, chemical, and biological properties.

Importance: Crop models simulate processes in discrete layers, or nodes, down the soil profile. Knowing the depths of these horizons allows the model to correctly attribute properties (like texture, organic matter, bulk density) to the appropriate soil depths. This is crucial for accurately simulating root growth and distribution, water movement, and nutrient availability at different depths, as roots explore specific horizons for resources. For example, a shallow restrictive layer might limit root penetration and water uptake.

- **Soil Texture (Sand, Silt, and Clay Percentages)**

Description: Soil texture refers to the relative proportions of sand (largest particles), silt (medium particles), and clay (smallest particles) in a given soil sample (Fig.3.2). These percentages determine the soil's textural class (e.g., sandy loam, clay, silt loam).

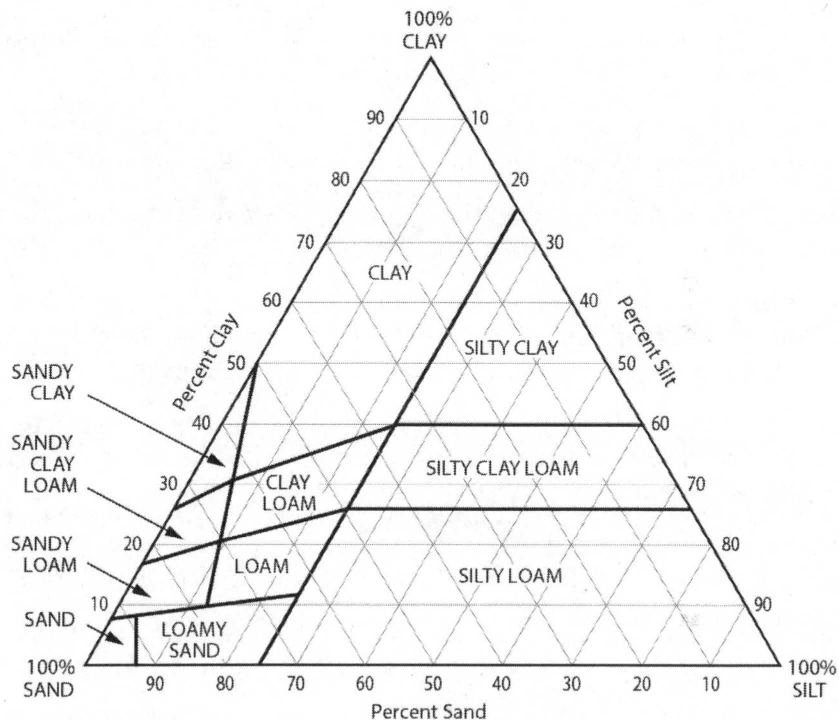

Fig. 3.2: A soil texture triangle determined by the proportion of sand, silt and clay

Importance: Soil texture is a fundamental property that profoundly influences virtually all soil physical and hydrological processes simulated in crop models:

Water Retention: Clayey soils, with their large surface area and small pores, have a high water-holding capacity, retaining more water against gravity. Sandy soils, with larger pores, drain quickly and have lower water-holding capacity. Silt falls in between. This directly impacts the model's calculation of available water for plants.

Water Movement (Hydraulic Properties): Texture dictates how easily water infiltrates and moves through the soil. Water moves faster through sandy soils and slower through clayey soils.

Nutrient Retention: Finer textured soils (more clay) generally have a higher capacity to retain nutrients due to their larger surface area and higher cation exchange capacity.

Aeration: Sandy soils are typically well-aerated, while very clayey soils can become waterlogged and poorly aerated, impacting root function.

- **Soil Water Retention Characteristics (Water Content at Different Potentials)**

Description: This describes the relationship between soil water content and soil water potential (or tension/suction). Key points on this curve include:

Saturated Water Content (θs): The maximum water content when all pores are filled with water (often close to total porosity).

Field Capacity (θFC): The water content remaining in the soil after gravitational water has drained away (typically 2-3 days after saturation). This is considered the upper limit of readily available water for plants.

Permanent Wilting Point (θPWP): The water content at which plants can no longer extract water from the soil and permanently wilt. This is the lower limit of available water.

Available Water Capacity (AWC): The difference between field capacity and permanent wilting point (θFC−θPWP), representing the total amount of water stored in the soil that is accessible to plants.

- **Importance:** These parameters are critical for accurate water balance simulations. Crop models use these values to:

Calculate the amount of water available for plant uptake.

Determine when a plant experiences water stress.

Simulate infiltration, drainage, and capillary rise based on the energy status of water in the soil.

Inform irrigation scheduling recommendations: knowing AWC helps define when and how much water to apply.

- **Saturated Hydraulic Conductivity (Ksat)**

Description: This parameter measures the ease with which water moves through saturated soil. It represents the maximum rate of water flow through the soil.

Importance: Ksat is vital for simulating infiltration rates (how quickly rainwater or irrigation water enters the soil) and deep percolation (how quickly water drains below the root zone). A high Ksat indicates rapid drainage, while a low Ksat can lead to ponding and surface runoff. It directly influences how much water actually enters the soil profile to be available to plants.

- **Bulk Density**

Description: Bulk density is the mass of dry soil per unit volume, including both soil particles and pore space. It is inversely related to porosity.

Importance: Bulk density directly impacts:

Root Penetration: High bulk density (compacted soil) impedes root growth and proliferation, restricting the plant's access to water and nutrients in deeper layers. Crop models use this to simulate root growth limitations.

Water Infiltration and Movement: Compacted soils with high bulk density have reduced macroporosity, slowing down water infiltration and drainage.

Aeration: High bulk density reduces air-filled pore space, leading to poor soil aeration, which can stress roots and limit nutrient uptake.

Volumetric Water Content: Bulk density is used to convert gravimetric water content (mass of water per mass of dry soil) to volumetric water content (volume of water per volume of soil), which is the standard unit for water balance calculations in models.

- **Organic Carbon / Organic Matter Content**

Description: The percentage of organic carbon or total organic matter in the soil. Organic matter is composed of decomposed plant and animal residues.

Importance: Soil organic matter (SOM) is a critical component influencing almost all soil properties:

Nutrient Source: SOM is a significant reservoir of essential nutrients, particularly nitrogen, phosphorus, and sulfur, which are released through mineralization. Models simulate nutrient cycling based on SOM dynamics.

Water Retention: SOM improves soil structure and increases the soil's water-holding capacity, especially in sandy soils.

Soil Structure and Aggregation: SOM helps bind soil particles into stable aggregates, improving aeration, water infiltration, and root growth.

Cation Exchange Capacity (CEC): SOM contributes significantly to CEC, enhancing the soil's ability to retain positively charged nutrients.

- **Soil pH**

Description: A measure of the acidity or alkalinity of the soil solution, expressed on a scale from 0 to 14.

Importance: Soil pH is a master variable controlling nutrient availability and microbial activity:

Nutrient Availability: Most essential plant nutrients are optimally available within a specific pH range (typically 6.0-7.5 for most crops). Extreme pH values (too acidic or too alkaline) can make nutrients unavailable, even if present in the soil, or can lead to toxic concentrations of elements like aluminum. Crop models often use pH-dependent functions for nutrient uptake.

Microbial Activity: Soil pH influences the activity of beneficial microorganisms that play vital roles in nutrient cycling (e.g., nitrogen fixation, organic matter decomposition).

- **Cation Exchange Capacity (CEC)**

Description: CEC is the total capacity of a soil to hold exchangeable cations (positively charged ions like K^+,Ca^{2+}, Mg^{2+}, NH^{4+}). It is a measure of the soil's nutrient-holding capacity.

Importance: CEC directly relates to the soil's ability to retain essential plant nutrients against leaching. Soils with higher CEC (typically clayey or organic-rich soils) are more fertile as they can store more nutrients, making them available for plant uptake over time. Models use CEC to simulate nutrient retention and release dynamics.

3.3.2 Data Collection and Tabulation for Soil Data

Sampling Strategy: Soil samples are typically collected from different depths within each identified soil horizon at representative locations across a field. For spatial models, a denser sampling grid might be required.

Laboratory Analysis: Samples are sent to specialized laboratories for detailed analysis of physical properties (texture, bulk density, water retention curves) and chemical properties (pH, organic carbon, nutrient content, CEC).

Field Measurements: Some parameters, like bulk density, can also be measured directly in the field.

Data Organization: Soil data is often tabulated by horizon, with each row representing a specific soil layer and columns containing the measured parameters. This structured format allows crop models to easily access the properties for each depth.

Georeferencing: For spatial applications, the precise geographic coordinates of each sampling point are recorded to link the soil data to other spatial datasets (e.g., satellite imagery, weather grids).

The accuracy and detail of soil data directly translate into the reliability of crop model predictions regarding water and nutrient stress, ultimately influencing the precision and effectiveness of agroadvisory services. Without a clear understanding of the soil's characteristics, a crop model's ability to simulate the subterranean engine of growth is severely compromised.

3.4 Crop Genetic and Physiological Data: The Plant's Blueprint

While weather data provides the environmental stimuli and soil data describes the medium for growth, crop genetic and **physiological data** represent the biological potential and internal blueprint of a plant species or cultivar. These parameters include genotype-specific traits such as growth duration, canopy structure, leaf area development, rooting depth, thermal time requirements, photosynthetic efficiency, and yield components.

3.4.1 Weather Data: Essential Inputs for Crop Modelling

Weather data stands as one of the most critical categories of inputs for crop simulation models, acting as the primary atmospheric drivers that dictate the energy and water dynamics within the soil-plant-atmosphere continuum. The intricate processes of photosynthesis, transpiration, and phenological development are profoundly influenced by the immediate atmospheric conditions, making accurate and comprehensive weather data indispensable for reliable model predictions. Without this vital information, even the most sophisticated crop model remains a theoretical construct, unable to reflect the real-world variability that shapes agricultural outcomes.

- **Data Required**

To accurately simulate crop growth and its interactions with the environment, crop models require several key meteorological variables. These typically include:

Solar Radiation (or Shortwave Radiation): This is arguably the most crucial energy input for plant growth, directly driving photosynthesis and evapotranspiration. It is often measured in Megajoules per square meter per day ($MJ/m^2/day$) or Watts per square meter (W/m^2). It determines the amount of energy available for biomass production and water evaporation.

Maximum and Minimum Air Temperature: Temperature directly influences the rates of most biological processes, including photosynthesis, respiration, and the progression of phenological stages (e.g., germination, flowering, maturity). Crop models often use these values to calculate daily average temperatures or growing degree days (GDD), which quantify the accumulation of heat required for development. Extreme temperatures (heat stress or frost) can also be critical.

Precipitation (Rainfall): The primary input of water into the agricultural system, directly affecting soil moisture content. It is typically measured as total daily rainfall in millimeters (mm). Accurate precipitation data is fundamental for simulating the water balance and predicting drought stress or waterlogging.

Relative Humidity (or Vapor Pressure Deficit - VPD): Humidity indicates the amount of moisture in the air. Vapor Pressure Deficit (VPD), which measures the drying power of the atmosphere, is often derived from temperature and humidity. High VPD drives higher rates of transpiration, influencing the plant's water demand and potential for stress. This affects how much water a plant needs to cool itself and transport nutrients.

Wind Speed: Wind influences the rate of evapotranspiration by removing humid air from around plant leaves and the soil surface. It is typically measured at a standard height (e.g., 2 or 10 meters) in meters per second (m/s) or kilometers per hour (km/h).

Reference Evapotranspiration (ETo): While sometimes calculated internally by models from the above variables, ETo (the evapotranspiration from a hypothetical reference crop under ideal conditions) is sometimes provided as a direct input, especially for simpler water balance models.

Atmospheric CO_2 Concentration: Increasingly relevant for modern crop models, especially when assessing climate change impacts. Elevated CO_2 levels can enhance photosynthesis, particularly for C3 crops, impacting biomass production and water use efficiency.

- **Frequency of Observation**

The appropriate frequency of weather observations depends significantly on the complexity of the crop model and the specific processes being simulated:

Daily Data: For many general-purpose crop growth and yield models (e.g., DSSAT, APSIM), daily resolution is the most common and often sufficient. Daily maximum and minimum temperatures, total daily rainfall, and total daily solar radiation are standard inputs. This frequency allows models to capture the overall daily progression of growth and water balance.

Hourly Data: More detailed mechanistic models, especially those focusing on rapid physiological responses, soil water dynamics, or microclimatic effects, often benefit from hourly or even sub-hourly data. For instance, hourly temperature and radiation allow for more accurate calculation of diurnal photosynthesis and transpiration patterns. Hourly rainfall data is crucial for precise simulation of infiltration and runoff events, as a large amount of rain over a short period behaves differently from the same amount spread throughout the day. Frost events, which can occur over a few critical hours, also necessitate hourly temperature data.

Sub-Daily (e.g., 30-minute, 15-minute): Highly detailed biophysical models or those integrated with real-time sensing for precision agriculture might utilize even finer temporal resolutions to capture rapid fluctuations in environmental conditions that impact immediate plant responses or irrigation scheduling.

Multi-Day/Seasonal Aggregates: While not direct inputs for process-based models, aggregated data (e.g., weekly averages, monthly totals) from historical records are often used for statistical analysis, regional climate characterization, or for driving empirical crop models.

- **Collection and Tabulation**

The methods for collecting and tabulating weather data have evolved significantly with technological advancements, moving from manual observations to highly automated and remote sensing approaches:

Ground-Based Weather Stations

Automated Weather Stations (AWS): These are the most common source of high-quality, continuous weather data. AWS are equipped with sensors that automatically measure and log parameters like air temperature, humidity, solar radiation, wind speed and direction, and rainfall. They can record data at various frequencies (e.g., 15-minute, hourly, daily) and transmit it wirelessly for immediate access. Installation and maintenance are crucial for data quality.

Manual Stations: Traditional meteorological observatories or farm-level rain gauges often involve manual readings at fixed times daily. While reliable, they provide lower temporal resolution and can be labor-intensive.

Network Density: The density of weather stations is critical. For accurate field-level crop modeling, data from a station as close as possible to the farm is ideal, as local microclimates can differ significantly from regional averages. National meteorological departments (like IMD in India) maintain extensive networks that provide data at district or sub-district levels.

Remote Sensing and Satellite-Derived Products

Satellite Imagery: Satellites equipped with radiometers can infer various meteorological parameters across large geographical areas, overcoming the limitations of sparse ground station networks. They can estimate solar radiation, land surface temperature, cloud cover, and even precipitation (e.g., from radar or microwave sensors).

Reanalysis Datasets: These datasets combine historical weather observations from various sources (ground stations, balloons, satellites) with atmospheric models to produce a consistent, gridded, and globally complete dataset of meteorological variables. Examples include AgMERRA (Agricultural Model Intercomparison and Improvement Project's Modern-Era Retrospective analysis for Research and Applications) and NASA POWER. These are invaluable for historical simulations and regional studies, providing spatially explicit data even in areas with few ground stations.

Downscaling: Global or regional climate model outputs, often at coarse resolutions, need to be downscaled to local scales relevant for crop models. This involves statistical or dynamic methods to translate large-scale climate projections into localized weather inputs.

Weather Generators

Mechanism: When historical weather data is incomplete or unavailable for future climate scenarios, statistical weather generators are used. These programs simulate synthetic weather series based on the statistical properties (means, variances, correlations, probabilities of wet/dry days) of observed historical weather data for a given location.

Use Cases: Filling data gaps, extending short records, or generating multiple possible future weather sequences for stochastic simulations (e.g., Monte Carlo analyses) to assess risk.

• Tabulation and Data Formats

Standardized Formats: Weather data for crop models are typically organized in structured formats for easy parsing by software. Common formats include simple text files (e.g., .WTH files for DSSAT, comma-separated values - CSV), where each row represents a date and columns contain the required meteorological variables.

Metadata: Crucial accompanying information (metadata) includes the weather station's latitude, longitude, and elevation, sensor heights, and units of measurement. This ensures proper model interpretation and calculation (e.g., daylength calculations based on latitude).

Quality Control: Before use, weather data undergoes rigorous quality control. This involves checking for extreme outliers (sensor errors), missing values, and consistency (e.g., minimum temperature cannot be higher than maximum temperature). Automated scripts and visual inspections are used for this purpose.

The meticulous acquisition, quality control, and proper tabulation of weather data are foundational steps in ensuring the accuracy and reliability of crop model simulations. The choice of data source and resolution depends heavily on the model's complexity, the spatial scale of the study, and the specific questions being addressed, ultimately impacting the utility of the crop model for agroadvisory. The biological data represent the inherent biological blueprint and dynamic responses of the plant itself. This category of data is absolutely fundamental for crop models, as it defines how a specific crop species or cultivar will interpret and respond to the environmental conditions and management practices it encounters. It dictates the plant's potential for growth, its developmental timeline, its efficiency in resource use, and its tolerance to stress. Without precise genetic and physiological parameters, a crop model cannot accurately mimic the unique characteristics of different crops or even different varieties of the same

3.5 Conclusion: The Foundation of Data-Driven Agricultural Simulation

This chapter has meticulously detailed the indispensable nature of data in the intricate world of crop modeling. It has become abundantly clear that crop models, regardless of their inherent sophistication, are only as reliable and insightful as the data fed into them. We have systematically explored the critical categories of information: from the dynamic atmospheric forces captured by weather data (solar radiation, temperature, precipitation, humidity, wind speed) to the static and dynamic properties of the soil (texture, water retention, hydraulic conductivity, bulk density, organic matter, pH, CEC) that define the root environment. Furthermore, we delved into the intrinsic characteristics of the plant itself, elucidated by crop genetic and physiological data (phenological timings, RUE, biomass partitioning, stress tolerances), and critically, the pervasive influence of human decisions reflected in management data (planting, irrigation, fertilization, pest control, tillage, residue management). The emphasis throughout has been on the precision required for each parameter, the appropriate frequency of observation, and the meticulous processes of collection and tabulation. Ultimately, this chapter solidifies the understanding that high-quality, comprehensive, and accurately collected data forms the bedrock upon which reliable crop model simulations are built. It is this foundational data that empowers models to move beyond theoretical constructs, enabling them to provide the accurate, context-specific, and actionable insights that are absolutely essential for effective agromet advisory, ensuring a more productive, resilient, and sustainable future for agriculture.

Multiple Choice Questions (Bold means correct answer)

1. What is the fundamental principle regarding data quality in crop modeling, often summarized as Garbage in, garbage out?
   a) Only garbage data should be used for initial model runs.
   b) Low-quality data is acceptable if the model is complex enough.
   c) The reliability and predictive power of a crop model are directly dependent on the quality of its input data.
   d) Data collection should be minimized to avoid errors.
2. Which of the following is considered the most crucial energy input for plant growth, directly driving photosynthesis and evapotranspiration in crop models?
   a) Wind speed. b) Relative humidity.
   c) Solar radiation. d) Minimum air temperature.
3. What is the typical unit for measuring daily rainfall data required by most crop models?
   a) Megajoules per square meter per day (MJ/m²/day).
   b) Degrees Celsius (°C).
   c) Millimeters (mm).
   d) Meters per second (m/s).

4. For simulating rapid physiological responses and precise infiltration events, which frequency of weather data is often preferred by detailed mechanistic models?
   a) Monthly averages.
   b) Annual totals.
   c) Daily data.
   d) Hourly or sub-hourly data.
5. What primary atmospheric phenomenon is directly influenced by wind speed and relative humidity in a crop model's water balance calculations?
   a) Photosynthesis rate.
   b) Nutrient uptake.
   c) Evapotranspiration.
   d) Soil temperature.
6. Which type of data collection method for weather involves combining historical observations with atmospheric models to produce consistent, gridded datasets globally?
   a) Manual weather stations.
   b) Automated Weather Stations (AWS).
   c) Remote sensing/Satellite imagery.
   d) Reanalysis datasets.
7. What is the importance of knowing the latitude, longitude, and elevation of a weather station for crop modeling?
   a) It determines the cost of data collection.
   b) It allows for accurate calculation of daylength and solar geometry for the site.
   c) It helps in identifying the type of crop grown in the area.
   d) It defines the soil type of the region.
8. What is the primary function of weather generators in crop modeling?
   a) To directly measure real-time weather conditions.
   b) To simulate synthetic weather series based on statistical properties for gap-filling or future scenarios.
   c) To forecast short-term weather events for irrigation.
   d) To calibrate weather sensors for accuracy.
9. Which soil parameter primarily determines the soil's capacity to hold water against gravity and influence nutrient retention due to its particle size and surface area?
   a) Bulk density.
   b) Soil pH.
   c) Soil texture.
   d) Saturated hydraulic conductivity.

10.The concept that describes the maximum amount of water remaining in the soil after gravitational water has drained away (e.g., 2-3 days after saturation) is known as:

a) Saturated water content.

b) Permanent wilting point.

c) Available water capacity.

d) Field capacity.

11. Why is soil bulk density a critical parameter for crop models, particularly concerning root growth?

a) It determines the soil's nutrient content.

b) High bulk density (compaction) can impede root penetration and access to resources.

c) It indicates the soil's pH level.

d) It directly controls solar radiation absorption by the soil.

12. Which soil parameter is a measure of the ease with which water moves through saturated soil, directly affecting infiltration and drainage rates?

a) Organic carbon content.

b) Cation Exchange Capacity (CEC).

c) Saturated hydraulic conductivity.

d) Soil profile depth.

13. How does soil organic matter (SOM) primarily benefit soil properties in the context of crop modeling?

a) It makes the soil impermeable to water.

b) It decreases soil aeration and increases bulk density.

c) It acts as a reservoir of nutrients, improves water retention, and enhances soil structure.

d) It is a direct input for photosynthesis.

14. What is the main reason soil pH is considered a master variable in crop models regarding nutrient availability?

a) It indicates the precise amount of available water in the soil.

b) It directly controls the chemical forms and solubility of most essential plant nutrients.

c) It determines the physical structure of the soil aggregates.

d) It dictates the rate of photosynthesis.

15. A soil with a high Cation Exchange Capacity (CEC) implies:
    a) Poor drainage and aeration.
    b) A strong ability to hold and retain positively charged nutrients.
    c) High acidity.
    d) Low water-holding capacity.
16. What does a crop model use phenological parameters (like Growing Degree Days) to primarily simulate?
    a) The plant's water use efficiency.
    b) The timing of key developmental stages (e.g., flowering, maturity).
    c) The rate of nutrient uptake.
    d) The maximum potential biomass production.
17. Radiation Use Efficiency (RUE) quantifies a crop's ability to convert:
    a) Absorbed water into biomass.
    b) Intercepted solar radiation into dry biomass.
    c) Soil nutrients into protein.
    d) Wind energy into structural growth.
18. What is the significance of biomass partitioning coefficients in crop models?
    a) They define the plant's sensitivity to drought.
    b) They determine how total biomass is allocated among different plant organs (leaves, stems, roots, grains).
    c) They measure the efficiency of water use by the plant.
    d) They control the soil's water retention capacity.
19. Why is accurately simulating Leaf Area Index (LAI) development crucial in crop models?
    a) LAI directly influences the plant's root depth.
    b) It determines the amount of solar radiation intercepted for photosynthesis and the rate of transpiration.
    c) It indicates the soil's nutrient content.
    d) It is a direct measure of final grain yield.
20. A crop model's ability to predict yield reductions under water scarcity is directly linked to its Water Use Efficiency (WUE) parameters and:
    a) The amount of fertilizer applied.
    b) Its phenological development.

c) Its ability to simulate the soil water balance and plant water stress.

d) The maximum potential biomass.

21. Data on critical nutrient concentrations and maximum uptake rates primarily fall under which category of crop data?

    a) Weather data.

    b) Soil data.

    c) Nutrient uptake and use efficiency parameters.

    d) Management data.

22. What do stress tolerance parameters in a crop model define?

    a) The ideal environmental conditions for maximum yield.

    b) The plant's ability to resist pest attacks only.

    c) The plant's response to various abiotic (e.g., heat, drought) and biotic stresses.

    d) The optimal harvesting time.

23. Harvest Index (HI) is a key parameter that represents the ratio of:

    a) Total biomass to intercepted radiation.

    b) Economic yield to total above-ground biological biomass.

    c) Water used to nutrient uptake.

    d) Leaf area to stem biomass.

24. What is a common method for collecting phenological and biomass accumulation data for crop model calibration and validation?

    a) Analyzing satellite imagery without ground truth.

    b) Conducting controlled field experiments with meticulous measurements.

    c) Relying solely on historical climate averages.

    d) Estimating values based on soil texture.

25. The specific date of planting is a critical management input because it dictates the crop's:

    a) Response to soil pH.

    b) Exposure to prevailing weather patterns throughout its growth cycle.

    c) Maximum potential yield under ideal conditions.

    d) Rooting depth.

26. Which management data directly influences inter-plant competition for resources and overall canopy development?
    a) Irrigation amount.
    b) Fertilizer type.
    c) Planting density.
    d) Tillage depth.
27. When modeling a crop, specifying the cultivar or variety name links the model to its:
    a) Water Use Efficiency.
    b) Genetic and physiological parameters.
    c) Solar radiation input.
    d) Soil organic matter content.
28. Records of date and amount of water applied fall under which category of management data?
    a) Harvest information.
    b) Fertilizer application data.
    c) Irrigation data.
    d) Tillage practices.
29. Why is the timing and amount of fertilizer application crucial for crop models?
    a) It defines the plant‘s photoperiod sensitivity.
    b) It directly influences nutrient availability in the soil and subsequent uptake by the crop.
    c) It determines the soil‘s bulk density.
    d) It controls the maximum solar radiation received by the crop.
30. Which type of management data directly impacts soil physical properties like bulk density, porosity, and surface residue cover?
    a) Pest control measures.
    b) Irrigation events.
    c) Tillage practices.
    d) Harvest date.

31. Leaving crop residues on the soil surface (mulch) is a management practice that primarily impacts:
    a) Plant phenological development.
    b) Soil temperature, soil water evaporation, and erosion.
    c) Crop genetic potential.
    d) Atmospheric CO_2 concentration.
32. For models simulating multi-cropping systems, what is the importance of the harvest date of a preceding crop?
    a) It determines the future rainfall patterns.
    b) It dictates the start of the next crop's simulation period and influences initial soil conditions.
    c) It affects the cultivar choice for the subsequent crop.
    d) It changes the genetic parameters of the next crop.
33. What is the common term for the cumulative heat units (e.g., °C-days) required for a crop to progress through its developmental stages?
    a) Radiation Use Efficiency (RUE).
    b) Vapor Pressure Deficit (VPD).
    c) Growing Degree Days (GDD).
    d) Cation Exchange Capacity (CEC).
34. Which of the following is a primary source for obtaining highly detailed crop genetic and physiological data?
    a) Public weather archives.
    b) Global soil databases.
    c) Controlled environment studies and meticulous field phenotyping.
    d) Aggregated national agricultural statistics.
35. The method of downscaling in weather data refers to:
    a) Reducing the number of weather parameters used.
    b) Converting local weather data to global averages.
    c) Translating large-scale climate model outputs into localized weather inputs.
    d) Simplifying complex weather equations for faster computation.

36. Which soil water retention point represents the upper limit of readily available water for plants after gravitational drainage?
    a) Permanent Wilting Point (θPWP).
    b) Saturated Water Content (θs).
    c) Field Capacity (θFC).
    d) Residual Water Content (θr).
37. A key challenge in data acquisition for crop modeling, as mentioned in the introduction, includes issues of:
    a) Excessive data availability.
    b) Consistency, resolution, and spatial representativeness.
    c) Lack of model complexity.
    d) Over-reliance on theoretical constructs.
38. Which type of data collection method for management data is gaining prominence due to GPS-enabled machinery and integrated software?
    a) Manual field logs.
    b) Laboratory soil analysis.
    c) Precision agriculture platforms.
    d) Scientific literature reviews.
39. The process of adjusting a subset of a model's parameters to make its simulations better match observed data is called:
    a) Validation.
    b) Sensitivity analysis.
    c) Calibration.
    d) Conceptualization.
40. Why is atmospheric CO_2 concentration becoming increasingly relevant as an input for modern crop models?
    a) It is a key indicator of soil organic matter.
    b) It directly influences soil pH.
    c) It can enhance photosynthesis, especially for C3 crops, impacting biomass and WUE.
    d) It primarily drives wind speed

41. Which of the following is a direct impact of planting depth in crop models?
    a) Final grain yield.
    b) Soil water retention curve.
    c) Emergence rates and initial root establishment.
    d) Radiation use efficiency.
42. What does the term Available Water Capacity (AWC) of soil refer to?
    a) The total water content at saturation.
    b) The water content at permanent wilting point.
    c) The difference between field capacity and permanent wilting point.
    d) The amount of water lost due to evaporation.
43. Which of the following is an example of an output obtained from a well-calibrated and validated crop model?
    a) Daily maximum temperature.
    b) Soil clay percentage.
    c) Predicted crop yield.
    d) Planting density.
44. The data indicating nitrogen content in the soil (mg/kg) falls under which category?
    a) Weather data.
    b) Management data (if applied).
    c) Crop genetic data.
    d) Soil data (nutrient status).
45. Why is it crucial to use an independent dataset for model validation that was not used during calibration?
    a) To increase the speed of simulation.
    b) To prove that the model can accurately predict outcomes in new, unseen conditions.
    c) To simplify the model's structure.
    d) To reduce the number of required input parameters.
46. What is the role of High-Throughput Phenotyping (HTP) in collecting crop genetic and physiological data?
    a) It is used to measure soil hydraulic conductivity.
    b) It allows for rapid, non-destructive measurement of plant traits over large populations.

c) It records daily weather variables.

d) It tabulates farmer management records.

47. What is the main reason crop models benefit from hourly rainfall data rather than just daily totals?

a) To determine the soil's organic matter content.

b) To more precisely simulate infiltration rates and runoff events.

c) To calculate the total solar radiation received.

d) To predict the exact planting date.

48. When choosing functional forms for mathematical formulation, what does this primarily involve?

a) Deciding on the programming language for the model.

b) Selecting the type of statistical analysis for outputs.

c) Defining the mathematical equations that represent biological/ physical relationships.

d) Determining the number of soil horizons.

49. Which of the following is a common statistical metric used to quantify the agreement between simulated and observed values during model validation?

a) Thermal Time (GDD).

b) Vapor Pressure Deficit (VPD).

c) Root Mean Square Error (RMSE).

d) Cation Exchange Capacity (CEC).

50. The concept of data requirements in crop modeling ultimately connects to the overall goal of agroadvisory by:

a) Limiting the scope of advice to general recommendations.

b) Enabling models to provide accurate, context-specific, and actionable insights for sustainable agriculture.

c) Reducing the need for field observations.

d) Simplifying complex biological processes into single variables.

4

Relationship Diagrams in Crop-Weather Models: Visualizing Complexity

4.1 Introduction

Crop-weather models are sophisticated constructs, meticulously designed to replicate the dynamic and intricate interplay among atmospheric conditions, complex soil processes, nuanced plant physiology, and the decisive influence of human management interventions. The agricultural system is a prime example of a highly interconnected and non-linear system, where changes in one component can ripple through and affect many others, often with time lags and feedback loops. To truly comprehend, effectively design, rigorously develop, and clearly communicate the inner workings of such multifaceted simulations, relying solely on dense textual descriptions or abstract mathematical equations can prove daunting and often insufficient. This is precisely where relationship diagrams emerge as indispensable tools.

At their core, relationship diagrams are powerful visual representations that translate the abstract scientific concepts and mathematical formulations embedded within crop-weather models into intuitive graphical formats. They are essentially the architectural blueprints and flowcharts of these complex systems, allowing us to see, rather than just read, how various components of the agricultural ecosystem interact. Imagine trying to understand the intricate wiring of a skyscraper or the precise movements of a symphony orchestra purely from a written description; diagrams provide the essential visual mapping that brings clarity and coherence to complexity.

These diagrams are not mere aesthetic additions; they are fundamentally required because of the inherent complexity and interdisciplinary nature of crop modeling. They bridge critical gaps:

Bridging Disciplinary Gaps: Agricultural system modeling brings together experts from diverse fields – agronomists who understand plant responses, meteorologists who provide weather insights, soil scientists who grasp soil physics and chemistry, and computer scientists who implement the models.

Each discipline speaks its own technical language. Relationship diagrams offer a universal, visual lexicon that transcends these disciplinary boundaries, fostering mutual understanding and collaborative model development.

Enhancing Conceptual Clarity: When modelers embark on designing a new model or refining an existing one, the sheer number of interacting variables (e.g., how temperature affects photosynthesis, which affects biomass, which affects leaf area, influencing light interception, and so on) can be overwhelming. Diagrams provide a structured way to visualize system boundaries, identify key state variables (quantities that accumulate or deplete), define processes (the rates of change), and map out the cause-and-effect relationships, including vital feedback loops. This clarity ensures that the conceptual model is robust, logical, and comprehensive before computational implementation.

Guiding Systematic Development: As the model moves from concept to mathematical formulation and then to computer code, these diagrams serve as invaluable guides. They ensure that the logical flow of calculations is accurately translated into modular code structures, helping developers organize functions and data interactions efficiently. This systematic approach minimizes errors and ensures that the implemented model truly reflects the intended scientific hypotheses.

Facilitating Debugging and Analysis: When a crop model produces unexpected or illogical results, relationship diagrams become powerful diagnostic tools. By tracing the flow of information and material through the visualized pathways, modelers can pinpoint where errors in logic, parameterization, or data input might be occurring. This dramatically streamlines the debugging process, turning a potentially frustrating endeavor into a more manageable analytical task.

Serving as Comprehensive Documentation: Finally, these diagrams provide clear, concise, and accessible documentation of the model's structure and underlying assumptions. This is vital not only for current users but also for future model developers who might need to understand, update, or extend the model. Good documentation, supported by clear diagrams, ensures model transparency, reproducibility, and transferability across different contexts and research teams.

In essence, relationship diagrams are the essential blueprints that guide the entire lifecycle of crop-weather models. They are not merely illustrations, but dynamic tools that enable the complex process of simulating agricultural systems to be managed effectively, leading to more robust, understandable, and ultimately, more useful models for informed agroadvisory and scientific advancement.

4.2 Relationship Diagrams Used in Crop-Weather Models

Crop-weather models are, by nature, complex systems that simulate intricate interactions among environmental factors, plant physiology, and management practices. To effectively conceptualize, design, develop, and communicate these models, scientists and developers rely heavily on relationship diagrams. These visual tools serve as blueprints, translating complex scientific hypotheses and mathematical formulations into intuitive graphical representations. They are essential for understanding how different components of the agricultural system influence one another, how material and information flow, and how feedback loops drive the system's dynamics.

4.2.1 Purpose and Significance of Relationship Diagrams

The primary purposes of employing relationship diagrams in crop modeling are multifaceted:

- **Conceptual Clarity and Design:** Before writing any code, diagrams help modelers visualize the system boundaries, identify key state variables, define processes, and map out the interactions between them. This structured approach ensures a comprehensive and logical design.
- **Communication and Collaboration:** Crop modeling is inherently interdisciplinary, involving agronomists, meteorologists, soil scientists, computer scientists, and economists. Diagrams provide a universal language that facilitates understanding and collaboration among these diverse stakeholders, regardless of their specific technical backgrounds. They make complex ideas accessible.
- **Model Development and Implementation:** Diagrams serve as a guide during the programming phase, ensuring that the mathematical equations and algorithms are correctly translated into code modules and that the flow of data and logic is maintained.
- **Debugging and Analysis:** When model outputs are unexpected, relationship diagrams can help trace the flow of calculations and identify potential errors in logic or parameterization, making the debugging process more efficient.
- **Documentation:** They provide clear and concise documentation of the model's structure and underlying assumptions, which is crucial for future updates, transferability, and reproducibility.

4.2.2 Types of Relationship Diagrams in Crop-Weather Models

Several types of diagrams are commonly employed, each offering a different perspective on the model's structure and dynamics:

- **Flowcharts (Process Flow Diagrams)**

Description: Flowcharts are perhaps the most universally recognized diagrams. They depict the sequential steps and decision points within a process or algorithm. In crop models, they illustrate the flow of calculations from input data through various physiological processes to output predictions. Standard symbols (e.g., rectangles for processes, diamonds for decisions, arrows for flow) are used.

Elements and Interpretation

Rectangles (Process/Activity): Represent a specific calculation or a module where a set of processes occurs (e.g., "Calculate Photosynthesis," "Update Soil Water Balance").

Diamonds (Decision): Indicate points where a condition is evaluated, leading to different pathways (e.g., "Is water stress severe?").

Arrows (Flow/Sequence): Show the direction of data or control flow between processes.

Ovals/Terminators (Start/End): Mark the beginning and end of the simulation or a specific sub-process.

Data Symbols (e.g., parallelogram): Can represent input or output data.

Application in Crop Models: A flowchart might show the daily simulation loop: reading weather data, then sequentially calculating light interception, photosynthesis, respiration, biomass partitioning, water uptake, evapotranspiration, and updating state variables (e.g., biomass, LAI, soil moisture) before repeating for the next day. They are excellent for illustrating the operational sequence of the model.

4.2.3 Stock and Flow Diagrams (System Dynamics Diagrams)

Description: These diagrams are central to System Dynamics modeling, focusing on the accumulation of quantities (stocks) and the rates (flows) that change these quantities over time. They are particularly effective at visualizing feedback loops and delays.

Elements and Interpretation

Rectangles (Stocks/Levels): Represent accumulations within the system (e.g., "Total Biomass," "Soil Water Content," "Leaf Area Index"). These are quantities that can increase or decrease.

Valves/Arrows (Flows/Rates): Represent the rates at which material or information flows into or out of stocks (e.g., "Photosynthesis Rate" adding to "Total Biomass," "Transpiration Rate" draining "Soil Water Content").

Circles/Clouds (Auxiliary Variables/Sources/Sinks)

Auxiliary Variables: Intermediate calculations that influence flows but are not accumulations themselves (e.g., Vapor Pressure Deficit,Nutrient Uptake Efficiency).

Sources/Sinks: Infinite reservoirs from which flows originate or terminate (e.g., "Atmosphere" as a source of CO_2, "Deep Percolation" as a sink for soil water).

Information Links (Dashed Arrows): Show that the value of one variable influences another, but no material or energy flows.

Application in Crop Models: A stock and flow diagram for a crop model would explicitly show "Total Biomass" as a stock, influenced by the"Photosynthesis Rate" (an inflow) and "Respiration Rate" (an outflow). Photosynthesis Rate might be influenced by "Solar Radiation" (an auxiliary variable) and "Leaf Area Index" (another stock), which in turn influences "Light Interception." This visualizes the interconnected dynamic processes.

4.2.4 Causal Loop Diagrams (CLDs)

Description: CLDs illustrate the cause-and-effect relationships between variables in a system, explicitly highlighting feedback loops (reinforcing or balancing) and their direction. They are qualitative diagrams used for conceptualizing system behavior and identifying key drivers and leverage points.

Elements and Interpretation

Variables (Words/Phrases): Represent key elements or conditions within the system (e.g., "Soil Moisture," "Crop Growth Rate," "Yield").

Arrows (Causal Links): Connect variables to show cause-and-effect relationships.

Polarity (+ or -): A "+" sign at the arrowhead indicates that an increase in the cause leads to an increase in the effect (or a decrease leads to a decrease). A "-" sign indicates that an increase in the cause leads to a decrease in the effect (or vice-versa).

Loops (R or B)

Reinforcing (R) Loop: A loop where a change in one variable propagates around the loop, returning to reinforce the initial change (positive feedback, often leading to exponential growth or decline).

Balancing (B) Loop: A loop where a change in one variable propagates around the loop, returning to counteract the initial change (negative feedback, often leading to equilibrium or goal-seeking behavior).

Application in Crop Models: A CLD might show: "Increased Soil Moisture" (+) leads to "Increased Crop Growth Rate" (+), which leads to "Increased Water Uptake" (+), which (-) leads to "Decreased Soil Moisture." This forms a balancing loop (B) illustrating how soil moisture depletion regulates growth. Another loop might show "Increased Leaf Area" (+) leading to "Increased Light Interception" (+), leading to "Increased Photosynthesis" (+), which leads to "Increased Biomass" (+), and thus "Increased Leaf Area" (R), showing a reinforcing growth process.

4.2.5 Unified Modeling Language (UML) Diagrams

Description: UML is a standardized visual language used in software engineering for specifying, visualizing, constructing, and documenting the artifacts of a software system. In crop modeling, particularly for object-oriented mechanistic models, UML diagrams (like Class Diagrams or Sequence Diagrams) can illustrate the software architecture.

Elements and Interpretation

Class Diagrams: Show the classes (e.g., "Plant," "Soil," "WeatherStation"), their attributes (data they hold), and their methods (functions they perform), along with relationships between them (e.g., "Plant" uses "Soil" data).

Sequence Diagrams: Illustrate the order of messages passed between objects over time, useful for depicting how different model components interact during a simulation step.

Application in Crop Models: A UML class diagram might show a Crop class interacting with a Soil class and a Weather class. The Crop class might have attributes like leafAreaIndex and methods like calculatePhotosynthesis(). It would interact with the Soil class to get soilMoisture and tell the Soil class about waterUptake. This is highly detailed and more relevant for software developers.

4.2.6 Conceptual Diagrams (Simplified Block Diagrams)

Description: These are often high-level, simplified diagrams that use blocks or boxes to represent major components or modules of the crop system or model, with arrows indicating general relationships or flows. They abstract away the fine details to provide an overarching view.

Elements and Interpretation: Simple boxes for components (e.g., "Weather," "Soil," "Crop," "Management"), and arrows showing inputs, outputs, and influences.

Application in Crop Models: Often used in introductions or overviews to broadly show how environmental factors (weather, soil), genetic traits, and management influence crop growth and yield. They provide a quick and easy way to grasp the overall structure without getting lost in specifics.

In conclusion, relationship diagrams are far more than mere illustrations; they are integral tools in the entire lifecycle of crop-weather models. From clarifying initial scientific concepts and guiding the complex process of mathematical formulation and coding, to aiding in debugging and serving as invaluable communication aids across diverse disciplines, these visual representations ensure that the complexity inherent in simulating agricultural systems is managed effectively, leading to more robust, understandable, and ultimately, more useful models for agroadvisory and research.

4.3 Simple Relationship Diagrams in Crop-Weather Models

When trying to understand or explain how a crop model works, especially to a diverse audience, simple relationship diagrams are incredibly helpful. They strip away the complex equations and programming details, offering a clear visual representation of the key components and how they interact. Here are some of the most commonly used simple diagrams:

4.3.1 Conceptual Diagram (or Block Diagram)

What it is: This is the most basic type of diagram, showing the major components of the system as "blocks": and the general influences or flows between them as arrows. It provides a high-level overview, illustrating the primary inputs, the core processes, and the main outputs.

Why it's used: It's excellent for introducing the model's scope and its main interacting elements without getting bogged down in detail. It emphasizes the "what" and "who" (which components interact) rather than the "how" or "when."

How it applies to crop models: You would typically see blocks (Fig.4.1) for "Weather," "Soil," "Crop," and "Management," with arrows showing how weather and soil influence the crop, and how management influences all three.

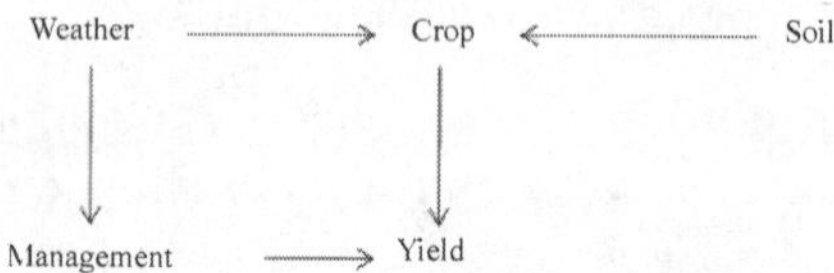

Fig. 4.1: Simplified flow chart with blocks

Explanation: This simplified flowchart shows the main steps within a single day of a crop simulation. The model reads weather, calculates energy capture, allocates growth, updates water status, and then decides whether the crop has reached maturity. If not, the loop continues to the next day.

4.3.2 Basic Stock and Flow Diagram

What it is: This diagram type focuses on showing accumulations (stocks) and the rates (flows) that change these accumulations. It's particularly useful for visualizing how quantities build up or deplete over time, and what drives those changes (Fig.4.2).

Why it's used: It explicitly highlights the dynamic nature of a system, making it clear what is accumulating (e.g., biomass, water) and what processes are driving that accumulation or depletion. It's a foundational tool in "system dynamics" thinking.

How it applies to crop models: Crop growth is fundamentally about the accumulation of biomass and water in the soil.

Explanation: This simplified flowchart shows the main steps within a single day of a crop simulation. The model reads weather, calculates energy capture, allocates growth, updates water status, and then decides whether the crop has reached maturity. If not, the loop continues to the next day.

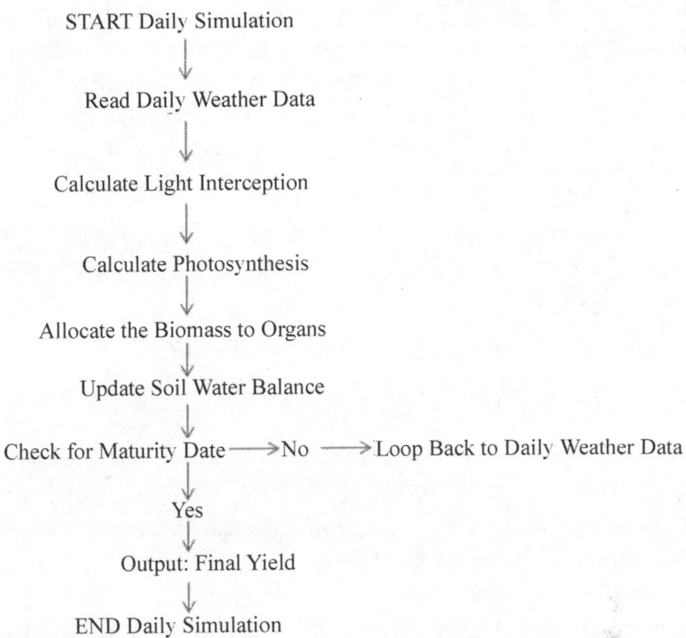

Fig. 4.2: Flow chart showing steps within a single day of crop simulation

4.3.3 Basic Stock and Flow Diagram

What it is: This diagram type focuses on showing accumulations (stocks) and the rates (flows) that change these accumulations. It's particularly useful for visualizing how quantities build up or deplete over time, and what drives those changes.

Why it's used: It explicitly highlights the dynamic nature of a system, making it clear what is accumulating (e.g., biomass, water) and what processes are driving that accumulation or depletion. It's a foundational tool in "system dynamics" thinking.

How it applies to crop models: Crop growth is fundamentally about the accumulation of biomass and water in the soil.

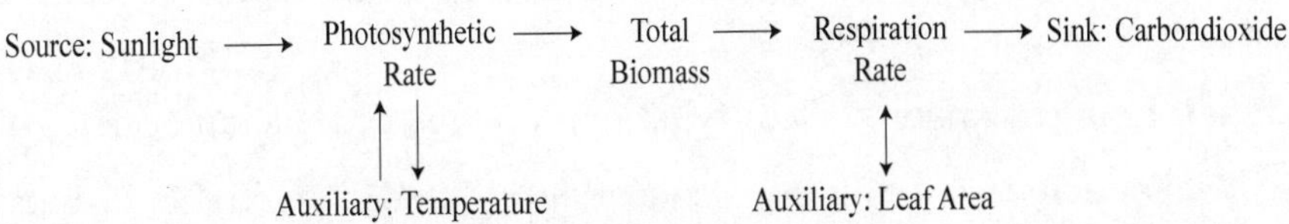

Fig. 4.3: Flow chart adding to total biomass

Explanation: Here, [Total Biomass] is a stock (a rectangle, representing an accumulation). [Photosynthesis Rate] is a flow (a valve-like arrow) that adds to Total Biomass. [Respiration Rate] is a flow that depletes Total Biomass. Circles like Sunlight, Temperature, and Leaf Area are auxiliary variables or sources/sinks that influence the rates. This diagram clearly shows how biomass accumulates based on the balance of photosynthetic production and respiratory consumption, driven by light and temperature.

These simple diagrams are powerful communication tools that help demystify the complex internal logic of crop-weather models, making them more accessible for design, discussion, and practical application.

4.4 Composite Relationship Diagrams in Crop-Weather Models

While simple conceptual diagrams, flowcharts, and basic stock-and-flow illustrations are excellent for introducing specific aspects of a crop model, real-world agricultural systems are characterized by a high degree of interdependency and feedback loops. To fully capture this complexity, modelers often employ **composite relationship diagrams**. These diagrams integrate elements and concepts from multiple simpler diagramming techniques into a single, more comprehensive visualization.

4.4.1 What are Composite Relationship Diagrams?

Composite relationship diagrams are designed to illustrate a more holistic view of the crop-weather system and the model that simulates it. Instead of focusing on just sequences, or just accumulations, or just causal links, they attempt to weave together several layers of information onto one visual. This allows for a richer understanding of how different sub-systems within the model interact and influence each other dynamically. They often combine:

Modular Organization: Showing distinct sub-models or components (like in a conceptual diagram).

Material/Energy Flows: Indicating physical transfers (e.g., water, carbon, nutrients).

Information Flows/Influences: Showing how one component‘s state or output affects another‘s process or rate.

Feedback Loops: Explicitly identifying reinforcing or balancing cycles.

Time Dependencies: Implying sequential processes within a dynamic framework.

4.4.2 Why are They Required?

Composite diagrams are necessary when:

- **Complexity Demands Integration:** The real agricultural system doesn't operate in isolated boxes. Photosynthesis affects leaf area, which affects light interception, which affects photosynthesis – a reinforcing loop. Soil water influences nutrient uptake, and root growth influences water uptake. These interdependencies are difficult to show fully with just one type of simple diagram.
- **Holistic Understanding:** They provide a "system-level" perspective, helping stakeholders understand not just individual processes but how they contribute to the emergent properties of the whole system (like yield or water use efficiency).
- **Advanced Model Design:** For complex mechanistic and hybrid models, composite diagrams serve as vital blueprints for designing the interactions between different modules (e.g., a "plant module" interacting with a soil module and a weather module).
- **Identifying Leverage Points:** By visualizing feedback loops and interconnections, composite diagrams can help identify critical points where interventions (e.g., management practices) can have the most significant impact on the system's behavior.

4.4.3 Example of a Composite Relationship Diagram for a Crop-Weather Model: The Soil-Plant-Atmosphere Continuum

This example attempts to integrate key concepts from conceptual, stock-and-flow, and causal loop diagrams to illustrate the core dynamics within a crop-weather model (Fig.4.4).

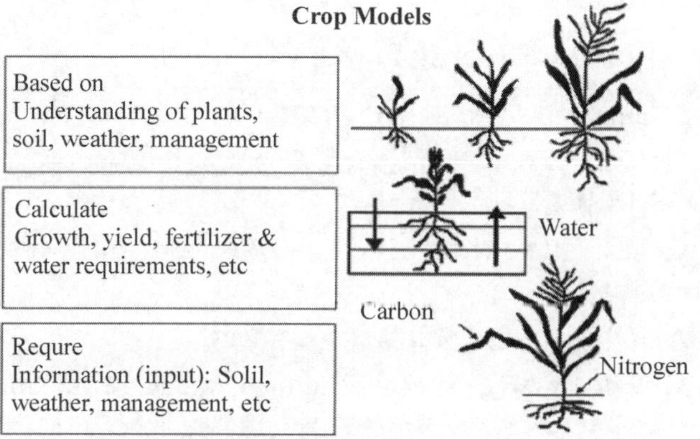

Fig. 4.4: Components in crop weather modelling

- **Explanation of Elements and Interactions**
- **Boxes (Major Components/Modules)**
    - [Weather]: External driving forces for the atmosphere.
    - [SOIL PROFILE]: Represents the physical and chemical characteristics of the soil layers.
    - [Atmosphere-Canopy Interactions]: A module that calculates physiological rates based on weather and plant state.
    - [Canopy (LAI)]: Represents the leaf area index, a key state variable for the plant.
    - [Total Biomass]: A primary state variable, representing the total accumulated dry matter of the plant.
    - [Plant Organs]: Represents the partitioning and accumulation of biomass in different parts of the plant.
    - [Root Biomass]: Specifically highlights the root component, crucial for soil interactions.
    - [Soil Water Content]: A state variable representing the amount of water in the soil.
    - [Nutrient Cycling]: A module simulating the transformations and availability of nutrients in the soil.
    - [Atmosphere (CO_2, Humidity)]: Represents the atmospheric reservoir of CO_2 and water vapor.
    - [Irrigation]: A management input.
    - (Sink: Groundwater): Represents water leaving the system.
    - (Source: Sunlight): (Implicit, originating from Weather) the energy source for photosynthesis.

Solid Arrows (Material/Energy Flows)

- Weather components like Rain directly flow into [Soil Water Content].
- "Water Vapor" (from Atmosphere-Canopy Interactions) flows into [ATMOSPHERE].
- [Photosynthesis Rate] (from Atmosphere-Canopy Interactions) adds to [Total Biomass].
- [IRRIGATION] adds to [Soil Water Content].
- [SOIL WATER CONTENT] can lead to "Drainage, Leaching" (sink).

Dashed Arrows (Information/Influence Flows)

- [Weather] (Temperature, Radiation) influences [Atmosphere-Canopy Interactions].
- [Atmosphere] (CO_2, Humidity) influences [Atmosphere-Canopy Interactions].
- [CANOPY (LAI)] influences [Atmosphere-Canopy Interactions] (e.g., how much light is intercepted).
- [SOIL PROFILE] properties (Texture, BD, OM, pH, CEC) influence [Soil Water Content] and [Nutrient Cycling].
- [Soil Water Content] influences [Atmosphere-Canopy Interactions] (via water stress) and [Root Biomass] (via water uptake).
- [Root Biomass] influences [Soil Water Content] (via uptake) and [Nutrient Cycling] (via nutrient uptake).
- [NUTRIENT CYCLING] influences [Plant Organs] (nutrient supply) and [Root Biomass].
- [Total Biomass] influences [Plant Organs] through partitioning.
- [Plant Organs] feedback to [Canopy (LAI)] (leaf growth) and [Root Biomass] (root growth).

Feedback Loops (Implied)

Growth Loop: [Total Biomass] increases [Canopy (LAI)], which increases light interception, increasing [Photosynthesis Rate], which increases [Total Biomass] (a reinforcing loop).

Water Balance Loop: [Soil Water Content] influences Transpiration, leading to a decrease in [Soil Water Content], which in turn can reduce Transpiration (a balancing loop).

Nutrient Cycling Loop: [Nutrient Cycling] provides nutrients for [Plant Organs], which contribute to Total Biomass, which upon senescence, contributes to Organic Matter in [SOIL PROFILE], fueling [Nutrient Cycling] (a reinforcing loop).

This composite diagram, while still simplified, offers a far more integrated view than any single diagram type. It visually demonstrates the dynamic, interconnected nature of the soil plant atmosphere system and how a crop model attempts to simulate these complex relationships to provide a holistic understanding of crop performance.

4.5 Complex Relationship Diagrams in Crop Weather Models

While simple conceptual diagrams and flowcharts provide foundational understanding, the true sophistication and mechanistic depth of modern crop models often necessitate complex relationship diagrams. These are not merely expanded versions of simpler diagrams; they represent a higher level of granularity, integrate multiple layers of information, and delve into the intricate feedback loops, mathematical formulations, and even software architecture that define a sophisticated simulation. They move beyond the what to illuminate the "how" and "why" at a much finer resolution.

4.5.1 What Makes a Diagram Complex"?

A relationship diagram becomes complex when it incorporates:

- **High Granularity:** Breaking down major processes into their constituent sub-processes and showing more detailed state variables and parameters.
- **Explicit Mathematical Relationships:** Often implying or directly showing where specific equations or functions are applied.
- **Multi-level Feedback Loops:** Detailing intricate positive (reinforcing) and negative (balancing) feedback mechanisms that regulate system behavior over time.
- **Integration of Diverse Sub-systems:** Clearly illustrating how seemingly disparate modules (e.g., carbon assimilation, nitrogen dynamics, water transport) are mechanistically linked.
- **Software Architecture:** For object-oriented models, these diagrams might represent classes, objects, and their methods and interactions, providing a blueprint for the code itself.
- **Uncertainty and Probabilistic Relationships:** In some advanced cases, they can even hint at or explicitly model probabilistic dependencies.

4.5.2. Why are Complex Diagrams Required?

These advanced diagrams are indispensable for several reasons:

Precise Mechanistic Understanding: They allow scientists to articulate and verify their understanding of detailed physiological and biophysical processes, ensuring that the model accurately reflects the underlying science.

Rigorous Software Engineering: For developing robust, modular, and maintainable crop model software, detailed diagrams (like UML) are critical for guiding programmers and ensuring code consistency.

Identifying Critical Linkages and Bottlenecks: By exposing fine-grained interactions, these diagrams can reveal subtle dependencies or potential bottlenecks in resource flows that simpler diagrams might miss.

Facilitating Model Refinement and Extension: When adding new components or improving existing ones (e.g., integrating a detailed pest damage module), complex diagrams provide the necessary context to ensure seamless integration without introducing unintended side effects.

Interdisciplinary Communication (Advanced Level): While challenging for novices, these diagrams become the common language for experienced modelers and domain experts to discuss specific model behaviors, assumptions, and limitations at a deep mechanistic level.

- **Examples of**

- **Detailed Stock and Flow Diagrams with Multiple Feedback Loops**

Building on the basic stock and flow concept, a complex version might focus on a specific critical process like carbon allocation or water dynamics, breaking down each stock and flow into multiple influencing factors and explicitly drawing every feedback loop.

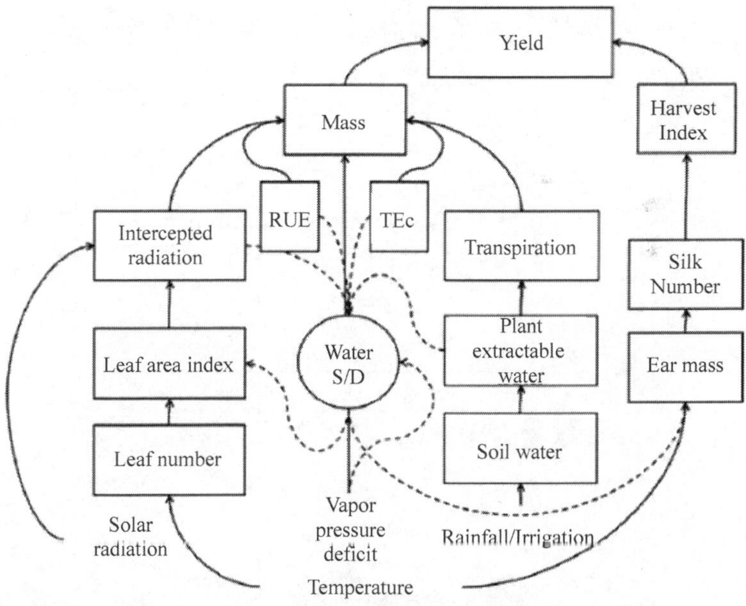

Fig. 4.5: Schematic representation of the structure of the crop growth model

Explanation of Complexity: While this DFD example is at Level 1 (high-level), more complex DFDs (Level 2, 3) would decompose each "Process" block (e.g., "Crop Simulation") into sub-processes, showing the flow of specific data elements (e.g., "Daily Temperature", "Soil Moisture Content") between internal computational steps and local data stores. They clarify the flow of information through the system's architecture.

4.5.3 Integrated Modeling Framework Diagrams

These diagrams illustrate how a crop model might be coupled with other models (e.g., climate models, economic models, hydrological models) to form a larger, more complex system for comprehensive analysis (Fig.4.6).

- **Example: Crop Model Integrated into a Climate Change Impact Assessment Framework**

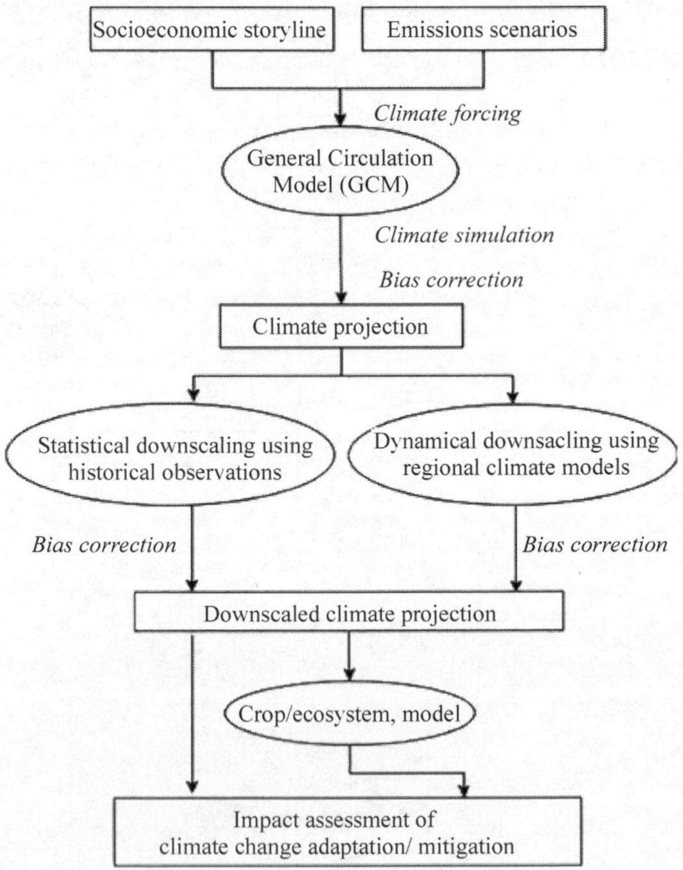

Fig. 4.6: Crop model integrated in a climate change impact assessment framework

Explanation of Complexity: This diagram illustrates complexity by showing:

Multiple Models: Explicitly linking different types of models (GCM, RCM, Crop Model, Economic Model).

Data Exchange Interfaces: Implying the specific data formats and protocols needed for models to communicate (e.g., GCM outputs becoming RCM inputs).

Multi-scale Integration: Bridging global climate projections down to regional and farm-level crop simulations, then up to economic and policy implications.

End-to-End Workflow: Presenting a complete analytical pipeline from climate science to agricultural outcomes and policy recommendations.

In summary, complex relationship diagrams are essential tools for anyone deeply involved in the design, development, and advanced application of crop models. They provide the necessary detail to understand the intricate workings, manage the software engineering challenges, and effectively communicate the comprehensive insights derived from these powerful simulation systems.

4.6 Conclusion: The Indispensable Role of Relationship Diagrams in Crop Modelling

This chapter has provided a focused exploration into the critical function of **relationship diagrams** within the intricate domain of crop-weather models. We have traversed from the fundamental premise of these visual tools as essential aids for conceptual clarity to a detailed examination of various diagrammatic types—ranging from simple conceptual block diagrams and flowcharts that illustrate basic system components and sequential processes, to more advanced stock and flow diagrams that explicitly represent accumulations and dynamic rates, and finally, to complex diagrams like those from UML that map out sophisticated software architectures.

The overarching message is clear: relationship diagrams are not merely supplementary illustrations; they are integral and indispensable components in the entire lifecycle of crop model development and application. They empower modelers to meticulously articulate scientific hypotheses, refine mathematical formulations, and guide the precise implementation of algorithms. By visually untangling the complex web of interactions among weather, soil, plant, and management factors, these diagrams facilitate profound understanding, enable effective interdisciplinary communication, streamline the debugging process, and foster the development of more robust and transparent models. Ultimately, the skillful use of relationship diagrams is paramount for translating the inherent complexity of agricultural systems into actionable insights, ensuring that crop-weather models can truly deliver reliable and context-specific agroadvisory for sustainable food production.

Multple Choice Questions (Bold means correct answer)

1. What is the primary purpose of using relationship diagrams in crop-weather modeling?
   a) To complicate the model's understanding.
   b) To visually represent the interactions between system components.
   c) To replace mathematical equations entirely.
   d) To store historical weather data.
2. Relationship diagrams are considered indispensable because they:
   a) Are only used by software developers.
   b) Simplify complex scientific hypotheses and mathematical formulations into visual formats.
   c) Eliminate the need for data collection.
   d) Automatically run model simulations.
3. One key benefit of relationship diagrams in crop modeling is their ability to bridge disciplinary gaps among experts. This means they:
   a) Force all experts to learn the same programming language.
   b) Provide a universal visual language for communication and mutual understanding.
   c) Limit the number of experts involved in model development.
   d) Only show relationships within a single scientific discipline.
4. During the conceptual clarity and design phase of model development, relationship diagrams primarily help modelers to:
   a) Calibrate model parameters.
   b) Visualize system boundaries, state variables, and processes.
   c) Validate model outputs with observed data.
   d) Collect field measurements.
5. Which of the following is NOT a general purpose or significance of relationship diagrams in crop modeling?
   a) Facilitating debugging and analysis.
   b) Guiding systematic development.
   c) Performing automated statistical calculations.
   d) Providing comprehensive documentation.

6. Which type of diagram provides the most basic, high-level overview of a crop model's scope, showing major components as blocks and general influences?
   a) Flowchart.
   b) Conceptual Diagram (or Block Diagram).
   c) Stock and Flow Diagram.
   d) Causal Loop Diagram.
7. In a simple Conceptual Diagram, if you see a block labeled WEATHER pointing an arrow towards a block labeled CROP, what does this typically represent?
   a) The crop controls the weather.
   b) Weather data is an input that directly influences the crop's behavior.
   c) The crop consumes the weather.
   d) Weather and crop are unrelated.
8. A simplified Conceptual Diagram showing Weather,Soil, and Management influencing Crop and Yield is primarily used for:
   a) Detailing specific mathematical equations.
   b) Illustrating the main interacting elements and inputs/outputs of the system.
   c) Debugging coding errors.
   d) Validating model performance.
9. What is a common limitation of a purely simple Conceptual Diagram?
   a) It is too complex for general understanding.
   b) It does not explicitly show the sequence of events or detailed mechanisms.
   c) It cannot represent external inputs.
   d) It only applies to statistical models.
10. The simplicity of a Conceptual Diagram makes it particularly useful for:
    a) Performing complex numerical simulations.
    b) Introducing the model's scope to a broad audience.
    c) Debugging specific logical errors in the code.
    d) Quantifying uncertainty in predictions.

11. What is the main strength of a Flowchart in crop modeling?
    a) It highlights feedback loops.
    b) It depicts the accumulation of quantities over time.
    c) It shows the sequential steps and decision points in a process or algorithm.
    d) It represents the software architecture of the model.
12. In a flowchart for a daily crop simulation, what does a diamond symbol typically represent?
    a) A data input.
    b) A processing step.
    c) A decision point or conditional logic.
    d) An output result.
13. If a flowchart for a crop model shows a loop from Calculate Photosynthesis back to Read Daily Weather Data, this indicates:
    a) An error in the model's logic.
    b) That the simulation is performed iteratively over time steps.
    c) That photosynthesis is the only process that matters.
    d) That weather data is only read once.

14 Flowcharts are particularly effective for illustrating the how of a crop model, referring to its:
    a) Overall goals.
    b) Operational sequence.
    c) Genetic parameters.
    d) Economic impacts.

15 What would be a common Process/Activity represented by a rectangle in a crop model flowchart?
    a) Solar Radiation (an input).
    b) Is crop mature? (a decision).
    c) Calculate Evapotranspiration."
    d) Yield (an output).

16 Which type of diagram is central to System Dynamics modeling and explicitly focuses on accumulations and rates of change?
    a) Flowchart.
    b) Causal Loop Diagram.

c) Stock and Flow Diagram.

d) Conceptual Diagram.

17 In a Stock and Flow Diagram, what do rectangles typically represent?

a) Sources or sinks of material.

b) Rates of change.

c) Accumulations or levels within the system.

d) External influences.

18. In a Stock and Flow Diagram, if an arrow (flow) labeled Photosynthesis Rate points towards a rectangle (stock) labeled Total Biomass, what does this indicate?

a) Photosynthesis depletes biomass.

b) Photosynthesis is an inflow that adds to total biomass.

c) Biomass controls photosynthesis.

d) Photosynthesis is a fixed constant.

19. What do dashed arrows typically represent in a Stock and Flow Diagram?

a) Material or energy flows.

b) Information or influence links (no material transfer).

c) Errors in the model.

d) External inputs that are not used.

20. A Stock and Flow Diagram for a crop model would be particularly useful for visualizing:

a) The specific algorithm for calculating a single parameter.

b) How quantities like biomass or soil water accumulate and deplete over time.

c) The genetic code of the crop.

d) The social impact of agricultural policies.

21. What is the primary focus of Causal Loop Diagrams (CLDs)?

a) To detail software components and their methods.

b) To explicitly show cause-and-effect relationships and feedback loops.

c) To illustrate sequential computational steps.

d) To quantify the exact amount of material flowing between components.

22. In a CLD, what does a + sign on a causal link signify?
    a) A negative correlation.
    b) An increase in the cause leads to an increase in the effect (or decrease to decrease).
    c) The relationship is inversely proportional.
    d) The variable is constant.

23 A Reinforcing (R) loop in a CLD typically represents:
    a) A system returning to equilibrium.
    b) Positive feedback, leading to exponential growth or decline.
    c) A stable, self-correcting mechanism.
    d) A linear relationship.

24. Which type of diagram is a formal notation used in software engineering to specify, visualize, and document software systems, often used for object-oriented crop models?
    a) Flowchart.
    b) Causal Loop Diagram.
    c) Unified Modeling Language (UML) Class Diagram.
    d) Conceptual Diagram.

25. What is a key advantage of complex relationship diagrams (like a detailed Stock and Flow or UML) compared to simpler ones?
    a) They are easier for non-experts to understand.
    b) They explicitly incorporate probabilistic dependencies for all parameters.
    c) They provide higher granularity and integrate multiple layers of information, revealing intricate feedback loops and software architecture.
    d) They eliminate the need for model calibration.

5

Evaluation of Crop Responses to Weather

5.1 Introduction

The preceding chapters have laid a robust foundation, introducing the concept of agricultural systems, delving into the architecture and types of crop models, and meticulously detailing the essential data requirements for their operation. We have established that crop models are powerful tools for simulating the complex interactions within the soil-plant-atmosphere continuum, driven by a rich tapestry of environmental and management inputs. Among these inputs, weather stands out as arguably the most dynamic and influential factor, wielding profound control over every aspect of crop life, from the precise timing of germination to the final quantity and quality of the harvest.

Chapter 5 shifts our focus to the direct and indirect consequences of this critical environmental driver: the evaluation of crop responses to weather. This chapter is dedicated to understanding, quantifying, and interpreting how variations in meteorological conditions—such as temperature, solar radiation, precipitation, humidity, and wind—directly influence the physiological processes, developmental progression, biomass accumulation, and ultimate yield of cultivated crops. It‘s about moving from simply providing weather data to actually *assessing* its impact on the plant‘s performance.

The ability to accurately evaluate these responses is paramount for several interconnected reasons:

Firstly, from a fundamental scientific perspective, it deepens our understanding of plant ecophysiology. By analyzing how crops react to specific weather patterns, we gain invaluable insights into the underlying biological mechanisms that govern growth and development under diverse environmental pressures. This knowledge is crucial for advancing agricultural science itself.

Secondly, for agricultural management and decision-making, this evaluation is directly actionable. Understanding optimal weather conditions for specific growth stages, or predicting the impact of adverse events, enables farmers and agronomists to refine planting dates, optimize irrigation schedules, adjust fertilizer applications, and implement timely pest and disease control measures. It transforms reactive farming into proactive, weather-smart agriculture.

Thirdly, in the face of increasing climate variability and change, evaluating crop responses becomes a critical component of risk assessment and adaptation strategies. It allows us to predict the vulnerability of current cropping systems to future climate scenarios, identify regions and crops most at risk from extreme heat, prolonged drought, or unseasonal frosts, and develop resilient farming practices and crop varieties capable of withstanding these challenges.

Finally, for crop breeding programs, understanding the weather-response traits of different genotypes can accelerate the development of new, high-yielding, and climate-resilient varieties. By identifying which physiological processes are most sensitive to specific weather variables, breeders can target traits that enhance performance under future climatic conditions.

This chapter will systematically explore various methodologies for evaluating crop responses. It will detail how phenological events (e.g., emergence, flowering, maturity) are driven by thermal time and photoperiod, and how deviations from optimal conditions impact these critical timelines. We will examine the influence of weather on biomass accumulation throughout the season, dissecting how solar radiation and temperature dictate photosynthetic rates and carbon allocation. Furthermore, the chapter will delve into the complex relationship between weather and yield components, such as grain number, grain size, and harvest index, illustrating how stress during specific developmental windows can disproportionately affect final production. Special attention will be given to the impact of extreme weather events (e.g., heatwaves, severe droughts, hailstorms) and methods for quantifying their damaging effects. By the conclusion of this chapter, readers will be equipped with a comprehensive understanding of the vital approaches and considerations involved in evaluating how crops respond to the ever-changing canvas of weather, forming a critical link in the chain of data-driven agricultural intelligence.

5.2 Seasonality of Crops: Adapting to the Rhythms of Weather and Thermal Regimes

The concept of seasonality in crops describes the natural, cyclical pattern of growth, development, and production of agricultural plants in response to the annual progression of weather conditions. It fundamentally dictates when crops are typically planted, when they reach critical developmental stages, and when they are ultimately harvested (Table.5.1).

Table 5.1: Crop seasons in India

Cropping Season	Major Crops Cultivated	
	Northern States	**Southern States**
Kharif June-September	Rice. Cotton. Bajra. Maize, Jowar. Tur	Rice. Maize, Ragi. Jowar. Groundnut
Rabi October - March	Wheat, Gram, Rapeseeds and Mustard. Barley	Rice. Maize. Ragi, Groundnut, Jowar
Zaid April-June	Vegetables. Fruits. Fodder	Rice. Vegetables. Fodder

This inherent rhythm is a cornerstone of traditional agriculture and remains a critical factor in modern crop production and modeling. Understanding seasonality is not merely about calendar dates; it's about comprehending the intricate biological adaptations of crops to specific environmental cues, with the thermal regime playing a paramount role.

5.2.1 What is Crop Seasonality?

At its most fundamental level, crop seasonality describes the suitability of particular crops to grow and thrive during specific periods of the year due to the unique combination of prevailing weather conditions. Each crop species, and even different cultivars within a species, has evolved or been bred to optimize its life cycle under a defined range of temperature, light, and water availability. This adaptation leads to distinct growing seasons that are optimal for a given crop in a particular geographical region.

5.2.2 The Dominant Influence of the Thermal Regime

While solar radiation, precipitation, humidity, and wind all play significant roles, the thermal regime (temperature patterns) is arguably the most dominant factor influencing crop seasonality and performance. Crops are biological organisms whose physiological and developmental rates are highly sensitive to temperature.

- **Cardinal Temperatures for Growth and Development**

Every crop species and cultivar has specific cardinal temperatures that define its thermal requirements for various biological processes:

Minimum (Base) Temperature (Tbase)**:** The lowest temperature at which a physiological process (e.g., germination, leaf expansion, photosynthesis) can occur. Below this, development effectively ceases.

Optimum Temperature (Topt)**:** The temperature at which a physiological process occurs at its fastest or most efficient rate. Growth and development are maximized within or near this range.

Maximum Temperature (Tmax)**:** The highest temperature at which a physiological process can occur. Above this, the process rate declines sharply, leading to stress, damage, or even plant death.

These cardinal temperatures vary significantly among crop types. For example:

Cool-season crops (e.g., wheat, barley, spinach) have lower Tbase, Topt, and Tmax ranges (e.g., wheat Tbase $\approx 0-5^0$C, Topt$\approx 20-25^0$C). They thrive in cooler springs and milder summers.

Warm-season crops (e.g., maize, rice, cotton, soybean) have higher Tbase, Topt, and Tmax ranges (e.g., maize Tbase$\approx 8-10^0$C, Topt $\approx 28-32^0$C). They require warmer temperatures for successful growth.

Beyond the Tmax, high temperatures can cause heat stress, leading to reduced photosynthesis, accelerated senescence, pollen sterility, and significant yield losses. Similarly, temperatures below Tbase can cause chilling or freezing injury.

- **Accumulated Heat (Growing Degree Days - GDD)**

Beyond instantaneous temperatures, the accumulation of heat over time is a fundamental driver of phenological progression. This is quantified using Growing Degree Days (GDD), also known as Thermal Time.

Calculation: GDD is typically calculated daily by averaging the daily maximum and minimum temperatures, subtracting the crop's Tbase , and accumulating these values over time. For example: GDD=$\sum$((Tmax +Tmin)/2$-$Tbase), with specific rules for when temperatures fall outside the effective range.

Phenological Triggers: Each developmental stage (e.g., emergence, leaf initiation, flowering, grain filling, physiological maturity) for a given crop cultivar requires a specific, accumulated amount of GDD to be completed. Crop models extensively use these GDD requirements to simulate the timing of these critical events.

Importance: The seasonal march of temperature determines when these GDD thresholds are met, thus dictating the timing of these key phenological transitions. For example, an early warming trend in spring will lead to faster accumulation of GDD and thus earlier planting and maturity.

- **Specific Thermal Cues**

Vernalization: Certain crops, particularly winter annuals (e.g., winter wheat, winter barley), require a specific period of exposure to cold temperatures (vernalization) to induce or accelerate flowering. This evolutionary adaptation prevents premature flowering in autumn and ensures reproductive development only after the winter chill has passed, leading to optimal conditions in the following spring.

Chilling Requirement: For some perennial fruit crops, a period of cold is also necessary to break dormancy and ensure uniform bud break and flowering.

- **Other Key Weather Drivers (Interacting with Thermal Regime)**

Solar Radiation (Light): The intensity and duration of solar radiation are crucial for photosynthesis. Seasonal variations in day length and cloud cover directly impact the energy available for biomass production. Critically, temperature influences the efficiency with which this light is converted.

Photoperiod: For many crops, the length of the day is a critical cue for triggering flowering. Short-day plants flower when days are shorter than a critical length, while long-day plants require longer days. Day-neutral plants are less sensitive. The interplay with temperature ensures reproductive development occurs at an optimal time.

Precipitation (Rainfall): The quantity and seasonal distribution of rainfall are crucial, especially in rain-fed agricultural systems. Crops are adapted to utilize monsoon rains, winter precipitation, or rely on stored soil moisture. Seasonal thermal regimes influence evaporation and transpiration rates, thereby affecting the effectiveness of rainfall.

Humidity and Wind: Seasonal variations in humidity (influencing vapor pressure deficit) and wind speed affect evapotranspiration rates. High temperatures combined with low humidity and strong winds create high evaporative demand, exacerbating water stress even if soil moisture is adequate.

- **Phenological Implications of Seasonality and Thermal Regime**

The interplay of these factors, particularly the thermal regime, profoundly impacts the phenological progression of crops. The rhythm of the seasons, driven by temperature cycles, ensures that:

Germination and Emergence occur when soil temperatures and moisture are favorable.

Vegetative Growth aligns with periods of ample light and moderate temperatures for maximum biomass accumulation.

Flowering and Pollination happen during specific thermal windows conducive to successful reproduction, avoiding extreme heat or cold that can lead to pollen sterility or poor fruit set.

Grain Filling and Maturity occur under conditions that promote optimal quality and allow for efficient harvesting (e.g., sufficiently warm for rapid filling but not excessively hot to prematurely halt the process, followed by dry periods for grain drying).

Crop models extensively use genetic parameters (thermal time coefficients, photoperiod sensitivities, cardinal temperatures) to simulate this phenological clock, driven by the weather inputs.

• Ecological and Agronomic Importance

Understanding crop seasonality, particularly the nuances of the thermal regime, is fundamental for practical agriculture and scientific research:

- **Optimizing Planting and Harvesting:** Farmers traditionally time their operations to align with predictable seasonal thermal patterns to maximize yields and minimize risks.
- **Resource Management:** It informs decisions on irrigation scheduling (when is water demand highest due to temperature?), fertilizer application (when is nutrient uptake most active given optimal temperatures?), and pest/disease management (many pests and diseases have temperature-dependent life cycles).
- **Crop and Cultivar Selection:** Matching specific crop varieties or species to the local seasonal thermal regime is essential for successful cultivation and selecting heat/cold tolerant varieties.
- **Risk Mitigation:** Anticipating adverse thermal events (e.g., early frost, heatwaves during flowering) based on seasonal forecasts allows for proactive measures.
- **Market Dynamics:** Seasonality, driven by thermal conditions, influences the supply and pricing of agricultural commodities.

• Classification of Crops by Seasonality (Examples)

Different regions classify crops based on their predominant growing season, largely defined by the thermal regime:

Temperate Zones

Spring Crops: Sown in spring when soil temperatures rise above Tbase , harvested in late summer/early autumn (e.g., spring wheat, maize, soybeans).

They require a frost-free period for their entire life cycle.

Winter Crops: Sown in autumn, requiring sufficient cold for vernalization, and harvested in late spring/early summer (e.g., winter wheat, winter barley). This strategy allows them to utilize winter moisture and escape intense summer heat by maturing earlier.

Indian Subcontinent (Monsoon-Driven, but thermally distinct)

Kharif Crops (Monsoon Crops): Sown with the onset of the southwest monsoon (June-July) when temperatures are high and humidity is increasing. Harvested in autumn (September-October) as temperatures begin to cool. Highly dependent on monsoon rains. Examples: Rice (requires warm, humid conditions), Maize, Sorghum, Cotton, Soybean.

Rabi Crops (Winter Crops): Sown after the monsoon (October-November) as temperatures decline, entering the cooler winter period. Harvested in spring (March-April) as temperatures begin to rise. They rely on residual moisture from the monsoon or winter irrigation. Examples: Wheat, Barley, Chickpea, Mustard – all benefiting from cooler growing conditions.

Zaid Crops (Summer Crops): A short season between Rabi and Kharif (March-June), grown with irrigation during very hot conditions. Examples: Cucumbers, Melons, some pulses – crops tolerant to high temperatures.

- **Implications of Climate Change on Seasonality and Thermal Regimes**

Climate change is fundamentally disrupting established seasonal thermal patterns, presenting major challenges:

Shifting Growing Seasons: Rising global temperatures are leading to earlier springs and extended frost-free periods in many regions. This accelerates GDD accumulation, potentially shifting optimal planting and harvesting dates. While this might allow for new crops or multiple cropping in some areas, it can also lead to phenological mismatches where, for instance, a sensitive flowering stage occurs during an unexpected heatwave.

Increased Heat Stress: More frequent and intense heatwaves, particularly during sensitive reproductive stages (flowering, grain filling), can cause significant yield reductions even if total GDD is met, as temperatures exceed Tmax for critical processes.

Altered Chilling/Vernalization: Warmer winters can reduce the effectiveness of vernalization for winter crops, leading to delayed or irregular flowering, or even preventing yield in regions reliant on this mechanism.

Changed Water Demand: Higher temperatures increase potential evapotranspiration, demanding more water from the soil, thus intensifying drought risk even if rainfall amounts don‘t change.

Pest and Disease Dynamics: Warmer winters and extended growing seasons can allow pests and diseases to overwinter more successfully, complete more life cycles, or expand their geographic range, altering their seasonal occurrence and impact on crops.

Yield Variability: The increasing unpredictability of seasonal thermal regimes leads to greater yield variability and increased risk for farmers.

Crop models play a crucial role in evaluating these shifting seasonal impacts, helping to identify adaptation strategies (e.g., new heat/cold tolerant varieties, adjusted planting dates, altered irrigation schemes) to maintain agricultural productivity and resilience in a changing climate. By explicitly simulating crop responses to the temporal dynamics of the thermal regime, these models provide the insights needed to navigate the evolving rhythms of agricultural seasonality.

5.3 Crop Growth and Development

Within the complex tapestry of the agricultural system, the concepts of crop growth and crop development are central to understanding how plants respond to their environment and ultimately produce yield (Fig.5.1).

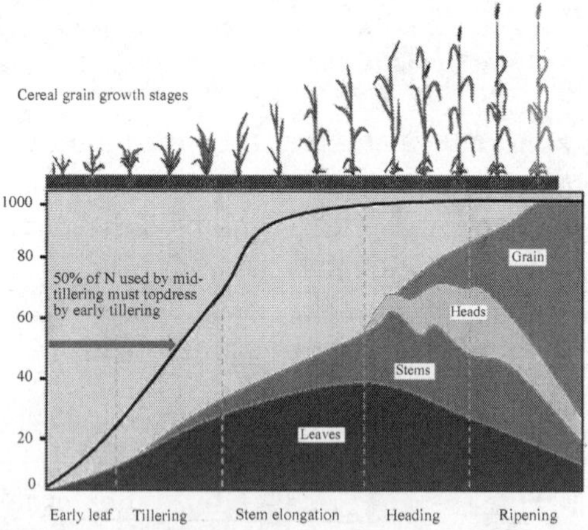

Fig. 5.1: Different stages of growth and development in cereal crops

While often used interchangeably in common parlance, in the realm of crop modeling and plant science, these two terms refer to distinct yet intricately linked processes that unfold throughout the plant‘s life cycle. A thorough understanding of their mechanisms and interactions is paramount for accurate crop simulation and effective agroadvisory.

5.3.1 Distinguishing Growth and Development

- **Crop Development (Phenology)**

Definition: Crop development refers to the irreversible progression of a plant through distinct morphological and physiological stages of its life cycle. It's about changes in form and function, marking qualitative shifts. These stages are often referred to as phenological stages.

Nature: It's primarily a qualitative change (e.g., from seedling to tillering, from vegetative to flowering, from flowering to maturity). It dictates the timing and sequencing of events.

Importance: The timing of these developmental stages is critical because:

Resource Demands Shift: Water and nutrient requirements change dramatically at different growth stages.

Susceptibility to Stress Varies: A crop's vulnerability to heat, cold, or drought stress is often highest during specific phenological windows (e.g., flowering is highly sensitive to heat stress).

Yield Determination: Key yield components (e.g., number of grains, size of grains) are largely determined during specific developmental phases.

Management Timing: Agricultural practices (e.g., fertilization, irrigation, pest control) are often timed according to these stages.

- **Drivers of Development:** Phenological progression is primarily driven by:

Thermal Time (Growing Degree Days - GDD): This is the most common driver. Plants integrate temperature over time. Each developmental phase requires a specific accumulation of heat units (often expressed in °C-days or °F-days) to complete.

Cardinal Temperatures: For each stage, there's a base temperature (Tbase) below which development ceases, an optimum temperature (Topt) where development is fastest, and a maximum temperature (Tmax) above which development slows or stops. The GDD concept usually calculates heat accumulation within the Tbase and Tmax range, favoring Topt . For example, wheat needs accumulated GDD to transition from tillering to jointing, then to heading, and finally to physiological maturity. Different cultivars have different GDD requirements.

Photoperiodism: The plant's response to the relative lengths of day and night. For many crops, the onset of flowering is triggered by specific day lengths.

Short-Day Plants: Flower when day length is shorter than a critical duration (e.g., soybean, maize in certain regions).

Long-Day Plants: Flower when day length is longer than a critical duration (e.g., wheat, barley, spinach).

Day-Neutral Plants: Flower independently of day length (e.g., tomato).

This ensures reproduction occurs at a time of year conducive to seed set and maturation.

Vernalization: A cold temperature requirement for some plants to induce or accelerate flowering. Many winter cereals (e.g., winter wheat) need exposure to prolonged cold periods to switch from vegetative to reproductive growth, preventing premature flowering in autumn.

Interaction of Drivers: It's crucial to note that these drivers often interact. For instance, photoperiod sensitivity might modify the GDD requirement for flowering.

- **Key Developmental Stages and Their Significance**

Germination & Emergence: The initial phase where the seed imbibes water and the radicle and plumule emerge. Critical for stand establishment and vigor.

Vegetative Growth (e.g., Tillering, Leaf Development): Focuses on the production of leaves, stems, and roots. This phase builds the factory for photosynthesis. Leaf Area Index (LAI) rapidly increases during this period.

Reproductive Development

Flowering/Anthesis: The period of pollen shed and fertilization. Extremely sensitive to environmental stresses (e.g., heat stress can cause pollen sterility, reducing grain number).

Grain/Fruit/Tuber Filling: The period of biomass accumulation in the economic yield portion. Duration and rate are critical for final yield.

Maturity (Physiological Maturity): The point at which maximum dry matter accumulation is reached, and the plant begins senescence. Water and nutrient uptake largely cease.

Harvest Maturity: The stage when the economic product has reached an optimal moisture content or quality for harvest.

5.3.2 Crop Growth (Biomass Accumulation)

Definition: Crop growth refers to the **irreversible increase in the dry mass (biomass) of the plant** over time. It's a quantitative increase in the physical size and weight of the plant.

Nature: It's a quantitative change, typically measured in units of dry weight per unit area (e.g., grams per square meter (g/m^2) or kilograms per hectare (kg/ha)).

Core Process: Photosynthesis: The engine of growth. Plants convert light energy, carbon dioxide (CO_2), and water into carbohydrates (sugars), which form the building blocks of plant tissue.

Light Interception: The amount of Photosynthetically Active Radiation (PAR) intercepted by the canopy is crucial. This is heavily influenced by the Leaf Area Index (LAI) – the total green leaf area per unit ground area. Models simulate LAI development and its impact on light interception.

CO_2 Assimilation: The rate at which the plant takes up CO_2 from the atmosphere through its stomata. This is influenced by CO_2 concentration, temperature, and water status (stomatal closure under stress).

Radiation Use Efficiency (RUE): The efficiency with which intercepted PAR is converted into dry biomass (g biomass / MJ intercepted PAR). This is a key genetic parameter for a crop type.

Respiration: The process by which plants break down carbohydrates to release energy for metabolic processes. Respiration consumes some of the carbohydrates produced by photosynthesis.

Maintenance Respiration: Energy expended to maintain existing plant tissues (e.g., protein turnover). This rate is highly temperature-dependent, increasing with higher temperatures.

Growth Respiration: Energy expended to synthesize new plant tissue from assimilated carbohydrates.

Net Photosynthesis: The actual amount of carbohydrates available for growth, calculated as Gross Photosynthesis minus Total Respiration.

Biomass Partitioning (Allocation): The process by which the newly assimilated dry matter is distributed or allocated among different plant organs (leaves, stems, roots, and reproductive structures like grain, fruit, or tubers).

This is a highly dynamic process, changing significantly with the plant's developmental stage. Early in the season, a higher proportion of biomass might be allocated to leaves and roots to maximize light interception and resource acquisition. During reproductive development, a larger fraction is directed towards the harvestable organs.

Environmental stresses (e.g., water deficit, nutrient deficiency) can significantly alter partitioning patterns, often favoring root growth under drought, or reducing allocation to reproductive organs under severe stress.

Yield Formation: The ultimate goal of crop production.

Harvest Index (HI): The ratio of economic yield (e.g., dry grain weight) to the total above-ground biological biomass at maturity. It reflects the efficiency with which the plant converts total biomass into the harvestable product.

Yield Components: Final yield is a product of several components determined at different developmental stages:

Grain Number/Fruit Number: Often determined around flowering and early reproductive stages (e.g., number of ears per unit area, number of grains per ear). Highly sensitive to stress during this period.

Grain Weight/Fruit Size: Determined during the grain/fruit filling period. Influenced by the duration of filling and the availability of assimilates.

5.3.3 The Interdependence of Growth and Development in Crop Models

Crop models are designed to capture the intricate feedback loops between growth and development:

Development Influences Growth: As a plant moves through developmental stages, its physiological capacities change, and resources are re-prioritized. For example, the transition to flowering triggers a redirection of assimilates towards reproductive organs, potentially reducing vegetative growth. The onset of senescence (a developmental process) marks the end of active growth.

Growth Influences Development: The amount of accumulated biomass (growth) can influence the rate of development. For instance, sufficient leaf area (LAI) is crucial for intercepting enough light to produce the energy required for phenological progression.

Stress as a Modifier: Environmental stresses (e.g., water, heat, nutrient deficiencies) impact both the *rate* of development (often accelerating phenological progression, leading to shorter growing seasons) and the *amount* of growth (reducing biomass accumulation). The timing of stress is critical; for example, heat stress during flowering can drastically reduce grain number (a developmental impact), while heat stress during grain filling can reduce grain weight (a growth impact). Crop models explicitly simulate these stress factors to adjust both developmental rates and growth efficiencies.

By mechanistically simulating these intertwined processes of growth and development, driven by environmental and management inputs, crop models can accurately predict crop performance under a wide range of conditions, offering invaluable insights for optimizing agricultural practices and responding to global food security challenges.

5.4 Effect of Weather Parameters on Crop Development and Monitoring Criteria

Crop development, often referred to as phenology, describes the progression of a plant through its distinct growth stages, from germination to maturity. Unlike growth (which is the accumulation of biomass), development is a qualitative change in the plant's form and physiological state. The timing and rate of these developmental transitions are exquisitely sensitive to environmental cues, with weather parameters being the primary drivers. Understanding these relationships is crucial for accurate crop modeling, predicting yield, and optimizing management practices.

5.4.1 Effect of Weather Parameters on Crop Development

- **Temperature (The Foremost Driver):** Temperature is the single most important environmental factor governing the rate of crop development. Plant metabolic and enzymatic activities are highly temperature-dependent.

Cardinal Temperatures: For every crop species and cultivar, there are specific cardinal temperatures:

Base Temperature (Tbase)**:** The minimum temperature below which no development or growth occurs.

Optimum Temperature (Topt)**:** The temperature at which the rate of development is maximal.

Maximum Temperature (Tmax)**:** The temperature above which the rate of development declines rapidly, leading to stress, damage, or even cessation of development.

Growing Degree Days (GDD)/Thermal Time: Crop development is largely a function of accumulated heat. GDD, or thermal time, quantifies this accumulation.

Calculation: GDD is typically calculated daily by averaging the daily maximum and minimum temperatures and subtracting the Tbase for the specific crop.

$$GDD = \frac{\text{Maximum temp} + \text{Minimum temp}}{2} - \text{Base temp}$$

By adding the heat units used by the crop on daily basis, the total number heat units used by the crop during a particular phenophase or its growth cycle can be calculated .The base temperature is generally considered as 8-10^{0}C for Kharif (summer) crops and 5^{0}C for rabi (winter) crops .

Significance: Each phenological stage (e.g., emergence, flowering, physiological maturity) requires a specific accumulated amount of GDD to be completed. This allows crop models to predict the timing of these stages based solely on temperature inputs.

Vernalization: For certain temperate crops (e.g., winter wheat, some fruit trees), exposure to a period of chilling temperatures (usually above freezing but below a certain threshold) is required to induce or accelerate flowering. This prevents premature flowering in autumn and ensures synchronized reproduction in spring.

Extreme Temperatures

Heat Stress: Temperatures consistently above Topt or exceeding Tmax (even for short periods, especially during reproductive stages like flowering) can accelerate development prematurely, shorten critical grain-filling periods, and cause sterility, leading to significant yield losses.

Chilling/Freezing Stress: Temperatures below Tbase or freezing points can delay development, cause tissue damage, or outright kill plants, halting development.

- **Photoperiod (Day Length):** Photoperiod, the duration of light in a 24-hour cycle, acts as a crucial signal for the initiation of certain developmental stages, particularly flowering.

Short-Day Plants: Require day lengths shorter than a critical maximum to flower (e.g., soybean, maize in tropical regions).

Long-Day Plants: Require day lengths longer than a critical minimum to flower (e.g., wheat, barley, spinach).

Day-Neutral Plants: Flowering is not primarily controlled by photoperiod (e.g., tomato, cucumber).

Significance: Photoperiodism ensures that a crop‘s reproductive phase occurs at an optimal time of the year, maximizing successful pollination and seed/

fruit development when conditions (e.g., temperature, water) are likely to be favorable. Its interaction with temperature is complex; for example, a crop might be a short-day plant, but also have a high GDD requirement, delaying its flowering until both conditions are met.

• **Water Availability (Moisture Stress):** While temperature and photoperiod are direct drivers of developmental *rate*and *timing*, water availability can significantly modify these processes, especially under stress conditions.

Water Deficit (Drought Stress): Can accelerate development (e.g., force premature flowering or maturity, a drought escape mechanism) or, paradoxically, delay it if severe enough to halt metabolic activity. It often shortens critical yield-determining phases like grain filling.

Waterlogging (Excess Water): Can lead to oxygen deprivation in the root zone, stressing the plant and potentially delaying development or causing plant death.

Relative Humidity (Vapor Pressure Deficit - VPD): High VPD (low humidity) combined with high temperatures increases transpirational demand, leading to rapid water loss and inducing water stress more quickly, which then indirectly affects development.

• **Solar Radiation (Light Quantity and Quality):** While solar radiation is primarily critical for biomass accumulation (growth), its quantity and quality also have subtle effects on development:

Total Daily Radiation: Adequate light ensures sufficient energy for all metabolic processes, supporting healthy development. Low light conditions (e.g., persistent cloudiness) can slow down developmental processes, even if GDD is accumulating, by limiting the energy available for growth and maintenance.

Light Quality (Spectral Composition): The ratio of red to far-red light can influence morphological development, such as stem elongation or branching patterns, although its effect on phenological timing is usually less direct than photoperiod.

5.4.2 Criteria Used for Monitoring Crop Development

Monitoring crop development accurately is fundamental for field management, research, and validating crop models. Various criteria and scales are employed:

• **Phenological Scales/Codes (Standardized Visual Assessment):** These scales provide a universal language for describing plant development.

BBCH Scale (Biologische Bundesanstalt, Bundessortenamt und CHemische Industrie): This is the most widely adopted and comprehensive phenological scale. It uses a decimal code system (00-99) to describe the principal (first digit) and secondary (second digit) growth stages for a wide range of mono- and dicotyledonous crops.

Principal Growth Stages (0-9)

0: Germination, sprouting, bud development

1: Leaf development

2: Formation of side shoots, tillering

3: Stem elongation or rosette growth, shoot development

4: Development of harvestable vegetative plant parts

5: Inflorescence emergence, heading

6: Flowering

7: Development of fruit/seed

8: Ripening or maturity of fruit/seed

9: Senescence, beginning of dormancy

Secondary Growth Stages (0-9): Provide finer detail within each principal stage (e.g., for cereals, BBCH 13 means 3rd leaf unfolded; BBCH 65 means full flowering, 50% of flowers open).

Advantages: Universality, precision, facilitates data comparison across studies and regions, and helps time agricultural interventions.

Zadoks Scale: An older, specialized decimal code primarily used for cereals (e.g., wheat, barley), often still referenced but largely superseded by BBCH for broader applications.

Crop-Specific Scales: Some crops may have their own historical or localized scales (e.g., Fehr and Caviness for soybeans, Iowa Corn Staging Guide for maize), often mapping to or being translatable to BBCH.

5.4.3 Direct Field Observations

This involves trained observers visually inspecting plants in the field.

Regular Visits: Typically weekly, but more frequently (daily/bi-daily) during critical, rapidly changing stages (e.g., emergence, flowering onset).

Record Keeping: Documenting the date of occurrence for key phenological events for a representative sample of plants in a plot.

Criteria: Based on a clear visual definition of each stage as per the chosen phenological scale (e.g., first tiller visible,50% of plants showing first flower,grains milky ripe).

5.4.4 Quantitative Measurements

While developmental stages are qualitative, quantitative metrics can track their progress or serve as proxies.

Days After Planting (DAP) / Days After Emergence (DAE): A simple, but often insufficient, measure as development rate varies with weather.

Cumulative GDD: The most accurate *predictor* of development. Field observations are used to calibrate and validate the GDD requirements for specific stages and cultivars.

Leaf Number Counts: For some crops (e.g., maize, cereals), the number of fully developed leaves can be a good indicator of vegetative developmental progress.

Plant Height / Canopy Cover: While primarily growth indicators, significant changes can correlate with developmental phases (e.g., rapid height increase during stem elongation).

5.4.5 Remote Sensing and Proximal Sensing

Advanced technologies offer non-destructive, large-scale, and often continuous monitoring of crop development.

Vegetation Indices (VIs): Time-series analysis of indices like Normalized Difference Vegetation Index (NDVI), Enhanced Vegetation Index (EVI), or Green Chlorophyll Index (GCI) derived from satellite (e.g., Sentinel-2, Landsat), aerial (UAVs/drones), or ground-based sensors.

Changes in VI values over time reflect changes in canopy greenness, biomass, and structure, which are tightly linked to phenological stages (e.g., rapid increase during vegetative growth, peak at flowering/early grain fill, decline during senescence).

Algorithms can identify green-up dates (emergence/start of season), peak greenness (full canopy/flowering), and brown-down dates (senescence/ maturity).

PhenoCams: Automated, ground-based digital cameras that capture time-lapse images of specific plots. These provide continuous, high-temporal resolution visual data, allowing for precise determination of phenological events.

Lidar/Radar: Can provide information on plant height and canopy structure, which can be linked to developmental stages, especially in complex canopies.

Hyperspectral Imaging: Provides more detailed spectral information that can differentiate subtle physiological changes associated with developmental transitions or stress.

5.4.6 Crop Models (Predictive Tool)

Crop models themselves are powerful tools for monitoring and predicting development.

By inputting daily weather data, a calibrated crop model can simulate the progression of phenological stages.

Model predictions can then be compared with actual field observations (using BBCH scale, etc.) to evaluate the model‘s accuracy and refine its parameters.

In summary, the interplay of temperature, photoperiod, and water availability orchestrates the developmental rhythm of crops. Monitoring these qualitative shifts is crucial, utilizing a combination of standardized scales like BBCH, direct field observation, quantitative measurements, and increasingly, advanced remote and proximal sensing technologies. This integrated approach provides a comprehensive understanding of how crops navigate their life cycle under varying environmental conditions.

5.5 Effect of Weather Parameters on Crop Growth (Biomass Accumulation)

While crop development refers to the qualitative progression through distinct phenological stages, crop growth is the quantitative increase in the plant‘s dry matter or biomass. This accumulation of dry weight (carbon compounds) is the fundamental process that leads to the formation of leaves, stems, roots, and ultimately, the harvestable yield. Crop growth is the net result of photosynthesis (carbon gain) minus respiration (carbon loss), and it is profoundly influenced by various weather parameters. Understanding these intricate relationships in detail is essential for mechanistic crop modeling.

5.5.1 The Core Process: Photosynthesis and Respiration Balance

Crop growth is driven by the balance between two primary physiological processes:

- **Photosynthesis:** The process by which green plants convert light energy, carbon dioxide (CO_2), and water into carbohydrates (sugars) and oxygen. These carbohydrates are the building blocks for all plant tissues.

  $6CO_2 + 6H_2O + \text{Light Energy} \rightarrow C_6H_{12}O_6 \text{ (Glucose)} + 6O_2$
- **Respiration:** The process by which plants break down stored carbohydrates to release energy for metabolic activities, maintenance of existing tissues, and synthesis of new tissues. Respiration consumes carbohydrates and releases CO_2.

  $\text{Carbohydrates} + O_2 \rightarrow CO_2 + H_2O + \text{Energy}$
- **Net Growth = Gross Photosynthesis - Total Respiration**

Therefore, any weather factor that affects either the rate of photosynthesis or the rate of respiration will directly influence the overall rate of dry matter accumulation (growth).

5.5.2 Detailed Effects of Weather Parameters on Crop Growth

- **Solar Radiation (Light): The Energy Source** Solar radiation is the indispensable energy source that powers photosynthesis, making it the most fundamental determinant of potential crop growth.

Photosynthetically Active Radiation (PAR): Plants primarily utilize wavelengths between 400 and 700 nanometers for photosynthesis. This portion is called PAR. Crop models often convert total incoming solar radiation to PAR, assuming about 45-50% of global solar radiation is PAR.

Light Interception: Not all incoming PAR is used. Only the light intercepted by the plant canopy contributes to photosynthesis. The Leaf Area Index (LAI) – the total green leaf area per unit ground area – is critical here. A higher LAI (up to a certain point, usually LAI 3-5 depending on canopy structure) means more light is intercepted, leading to higher rates of canopy photosynthesis. Models simulate LAI development and then calculate intercepted PAR based on this LAI and incoming solar radiation.

Radiation Use Efficiency (RUE): This is a key crop-specific genetic parameter that quantifies the efficiency with which intercepted PAR is converted into total dry biomass (g biomass / MJ intercepted PAR). RUE varies among crop species (e.g., C4 crops generally have higher RUE than C3 crops under optimal conditions) and can be reduced by stresses.

Light Saturation and Photoinhibition: While more light generally means more photosynthesis, there's a point of light saturation where photosynthetic enzymes are working at their maximum capacity, and further increases in light intensity don't increase the rate. Beyond this, excessively high light intensity, especially when combined with other stresses (like water deficit), can lead to photoinhibition, where the photosynthetic machinery is damaged, reducing efficiency and growth.

Overall Impact: Higher daily sums of intercepted PAR (due to high incoming radiation and a well-developed canopy) directly lead to higher rates of carbon assimilation and, consequently, greater biomass accumulation, assuming other factors are not limiting.

- **Temperature:** The Regulator of Metabolic Rates - Temperature influences virtually all biochemical and physiological processes in plants, including photosynthesis and respiration, thus profoundly affecting growth.

Enzyme Kinetics: All plant metabolic reactions are catalyzed by enzymes, whose activity is highly sensitive to temperature. There's an optimal temperature range for these enzymes to function efficiently.

Photosynthesis Rate: Within a species-specific range, increasing temperature generally increases the rate of photosynthesis up to an optimum (Topt for photosynthesis). Beyond this optimum, the rate declines rapidly due to enzyme denaturation and damage to the photosynthetic apparatus.

Respiration Rate: Respiration rates generally increase exponentially with temperature (often approximated by a Q10 factor, where the rate doubles for every 10^0C increase) up to a certain point.

- **Net Photosynthesis and Growth**

Optimal Range: When temperatures are within the optimal range for both photosynthesis and respiration, net carbon gain is maximized, leading to rapid biomass accumulation.

Low Temperatures (Chilling/Freezing): Temperatures below the base temperature (Tbase) drastically slow down or halt metabolic processes, leading to very slow growth or no growth. Chilling or freezing temperatures can cause physical damage to cell membranes and proteins, inhibiting growth even after warming.

High Temperatures (Heat Stress): Temperatures above the optimum for photosynthesis, especially when they exceed the maximum (Tmax), can lead to:

Reduced Photosynthesis: Damage to photosystems, enzyme denaturation, and increased photorespiration (especially in C3 plants).

Increased Respiration: While photosynthesis declines, respiration continues to increase with temperature, leading to a negative carbon balance (more carbon lost than gained), resulting in reduced net growth or even biomass loss.

Accelerated Senescence: Premature aging and death of leaves, reducing the photosynthetic factory.

Altered Partitioning: Resources may be re-directed away from productive growth (e.g., grain filling) towards stress defense.

High Night Temperatures: Can be particularly detrimental as they drive high rates of maintenance respiration without concurrent photosynthesis, leading to a significant depletion of carbohydrate reserves accumulated during the day.

- **Water Availability (Precipitation & Soil Moisture)**

The Solvent and Reactant- Water is essential for nearly all physiological processes. Its availability directly impacts photosynthesis and overall growth.

Stomatal Conductance: Plants regulate water loss (transpiration) through tiny pores on their leaves called stomata. When water is abundant, stomata are fully open, allowing for maximum CO_2 uptake for photosynthesis.

Photosynthesis Reduction: Under **drought stress** (insufficient soil water):

Stomatal Closure: Plants close their stomata to conserve water. While this reduces water loss, it also limits CO_2 entry, directly reducing the rate of photosynthesis.

Reduced Enzyme Activity: Severe water stress can directly impair photosynthetic enzymes and machinery.

Reduced Leaf Expansion: Water deficits limit cell expansion, leading to smaller leaves and reduced LAI, which in turn reduces light interception and overall canopy photosynthesis.

Nutrient Transport: Water is the medium for nutrient transport from the soil to the roots and throughout the plant. Water stress reduces nutrient uptake, leading to nutrient deficiencies that stunt growth.

Cell Turgor: Water maintains cell turgor, which is essential for cell expansion and overall plant rigidity. Loss of turgor leads to wilting and reduced growth.

Waterlogging (Excess Water): Too much water in the soil can lead to anaerobic conditions (lack of oxygen) in the root zone. This damages roots, impairs nutrient and water uptake, and can lead to the production of toxic compounds, severely inhibiting growth.

Overall Impact: Optimal water availability is crucial for maintaining high photosynthetic rates and cell expansion, leading to robust biomass accumulation. Both deficit and excess water significantly reduce growth.

5.5.3 Carbon Dioxide (CO_2) Concentration

The Photosynthetic Substrate- CO_2 is a direct reactant in photosynthesis. Its concentration in the atmosphere influences the rate of carbon fixation.

C3 vs. C4 Plants

C3 Plants (e.g., wheat, rice, soybean, potato): Their primary photosynthetic enzyme (RuBisCO) can also react with oxygen (photorespiration), which is a wasteful process. Increasing atmospheric CO_2 directly increases photosynthesis and reduces photorespiration, leading to a significant CO_2 **fertilization effect** (enhanced growth).

C4 Plants (e.g., maize, sorghum, sugarcane, millet): Have a specialized mechanism that concentrates CO_2 around RuBisCO, making them more efficient at lower CO_2 concentrations and less prone to photorespiration. They show less direct photosynthetic response to elevated CO_2 compared to C3 plants, but can still benefit indirectly by improving water use efficiency (they don‘t need to open stomata as widely to get enough CO_2, thus losing less water).

Overall Impact: Rising global atmospheric CO_2 (due to climate change) is expected to have a positive fertilization effect on C3 crop growth, although the magnitude depends on other limiting factors (water, nutrients, temperature).

5.5.4 Wind Speed

Boundary Layer and Physical Stress: Wind influences growth through its effects on the microclimate around the plant and physical damage.

Boundary Layer: Wind affects the thickness of the boundary layer of still air immediately surrounding the leaf surface. A thinner boundary layer (due to higher wind) increases the rate of gas exchange (CO_2 uptake, water vapor loss).

Transpiration: Increased wind speed typically increases transpiration rates, potentially leading to faster onset of water stress if soil water supply is limited.

CO_2 Supply: Moderate wind can improve CO_2 supply to the canopy, slightly enhancing photosynthesis, especially in dense canopies where CO_2 near the leaves might be depleted.

Physical Damage: High wind speeds can cause physical damage to plants (e.g., leaf tearing, stem breakage, lodging—where stems bend over), reducing the photosynthetic area, altering plant architecture, and leading to yield losses. This mechanical stress can also trigger physiological responses that divert resources from growth to repair.

Overall Impact: Moderate wind is generally beneficial, but strong, persistent winds are detrimental to crop growth due to increased water stress and physical damage.

5.5.5 Humidity (Vapor Pressure Deficit - VPD)

The Driving Force for Transpiration- Humidity, especially when expressed as Vapor Pressure Deficit (VPD), is a critical component of the atmospheric demand for water, directly influencing transpiration and indirectly affecting photosynthesis and growth.

VPD Definition: VPD is the difference between the amount of moisture in the air at saturation (how much moisture the air *can* hold) and the actual amount of moisture in the air. A high VPD means the air is very dry and has a high pulling power for water.

Transpiration Rate: High VPD drives higher rates of transpiration (water loss from leaves).

Stomatal Response: To conserve water under high VPD, plants often partially close their stomata. As discussed, this reduces CO_2 uptake, thereby limiting photosynthesis and growth. Even if soil water is abundant, high VPD can induce atmospheric drought conditions.

Nutrient Uptake: Reduced transpiration due to low VPD (high humidity) can sometimes slow down the mass flow of water-borne nutrients to the roots, potentially leading to nutrient deficiencies (e.g., calcium deficiency in some crops).

Overall Impact: Optimal VPD allows for efficient transpiration without excessive water loss, supporting high rates of photosynthesis and growth. Extreme high VPD can severely limit growth, while very low VPD (very high humidity) can also be detrimental by reducing transpiration-driven nutrient transport and increasing the risk of foliar diseases.

In conclusion, crop growth, as measured by biomass accumulation, is a dynamic and complex process that is intricately controlled by the interplay of all major weather parameters. Solar radiation provides the energy, temperature regulates the rates of biochemical reactions, water enables the processes and transport,

CO_2 provides the carbon building blocks, and wind and humidity modify the exchange rates. Crop models integrate these individual and interacting effects to simulate the quantitative progression of plant dry matter throughout the growing season, providing essential insights for agricultural research and management.

5.6 Methods Employed to Evaluate Weather Effects on Crops: An In-Depth Analysis

The pervasive and often capricious nature of weather profoundly influences every facet of crop life, from the precise biochemical reactions occurring within plant cells to the ultimate economic yield realized across agricultural landscapes. Consequently, the accurate and comprehensive evaluation of how various weather parameters impact crops stands as a critical pillar of agricultural science. This endeavor provides invaluable insights necessary for enhancing productivity, bolstering global food security, and formulating robust adaptation strategies in the face of escalating climate variability. This evaluation is not a singular approach but rather a sophisticated, multi-faceted process that strategically deploys a diverse range of methodologies. Each method offers distinct advantages and inherent limitations, collectively painting a holistic picture of plant-environment interactions, from meticulously controlled laboratory settings to expansive real-world field observations and advanced computational simulations.

5.6.1 Observational and Empirical Methods

These approaches form the bedrock of our understanding, relying on direct observation and the statistical analysis of historical data to deduce relationships between prevailing weather conditions and subsequent crop performance.

- **Field Experiments (Controlled and On-farm):** The most foundational approach involves cultivating crops directly in the field, meticulously monitoring ambient weather conditions, soil characteristics, and implemented management practices. Researchers typically design these experiments to either capture the impact of natural inter-annual or spatial weather variability or to deliberately introduce specific weather-related stresses. For instance, comparing crop phenological progression, biomass accumulation, and final yield across multiple growing seasons with differing rainfall patterns, temperature regimes, or solar radiation levels allows for a direct assessment of how these natural variations influence crop performance. In more controlled field setups, specialized equipment might even simulate specific stresses, such as induced drought conditions or artificial misting to alter humidity, thereby isolating the effects of particular weather parameters on plant responses. The primary strength of field experiments lies in their real-world relevance, as the data are collected under actual farming conditions, inherently capturing the full complexity of interactions within the soil-plant-atmosphere continuum. They provide irrefutable ground truth, offering direct, measurable evidence of how crops perform under specific environmental contexts, which is indispensable for validating findings from other methods. Furthermore, these experiments offer highly site-specific information, revealing nuanced adaptations or sensitivities unique to a particular location. However, field experiments suffer from a fundamental limitation: the uncontrollable nature of natural weather events. Researchers cannot precisely dictate specific weather sequences, making it exceedingly challenging to isolate the effect of a single weather variable in isolation. Replicating rare or extreme weather events, such as a severe heatwave at a critical developmental stage, is practically impossible. Moreover, conducting multi-year, multi-location field trials is notoriously costly and time-consuming, and the site-specific nature of the findings means they may not always be directly transferable to other regions with different environmental conditions or management systems without further validation.

- **Historical Weather-Yield Correlations / Statistical Models:** This method entails the rigorous statistical analysis of extensive historical records. Crop yield data, often aggregated at regional or national scales over several decades, are correlated with corresponding historical weather data. This weather data might include metrics such as mean monthly temperature, seasonal precipitation totals, or accumulated growing degree days (GDD) during specific crop phases. Techniques like multiple linear regression, correlation analysis, or more advanced machine learning algorithms are then employed to identify empirical relationships and sensitivities. Through this process, statistical models can reveal which weather variables, and during which specific windows within the growing season, have historically exerted the most significant influence on yield fluctuations. For example, analysis might demonstrate a strong negative correlation between maize yield and the number of days exceeding a certain heat threshold during the flowering period, or a positive correlation with rainfall during critical vegetative phases. The main advantage of this approach is its reliance on readily available historical datasets, often spanning long periods and large geographical areas, making it a relatively cost-effective method for broad-scale analysis. It excels at identifying broad patterns and major climatic drivers of yield variability over extended timeframes. Nevertheless, a critical limitation is the distinction between correlation and causation. Statistical relationships, however strong, do not inherently prove direct cause-and-effect mechanisms; other unmeasured or confounding factors (e.g., changes in technology, pest outbreaks, policy shifts) could be influencing both weather and yield. These models inherently lack mechanistic detail, offering little explanation as to *why* a particular weather factor affects yield at a physiological level. Consequently, they often perform poorly when extrapolating beyond the range of historical data, struggling to reliably predict outcomes under novel climate conditions, with new crop varieties, or under altered management practices.

- **Phenological Networks / Citizen Science** This method involves the systematic and often long-term collection of observed phenological dates (e.g., planting, emergence, onset of tillering, flowering, physiological maturity, harvest) across a wide geographic distribution. These observations are frequently gathered through established research station networks or increasingly, through large-scale citizen science initiatives where volunteers contribute local observations. By comparing these observed phenological timings with corresponding historical weather data over decades, researchers can effectively identify subtle or significant shifts in crop calendars in response to evolving temperature regimes, changes in day length, or other long-term climatic trends. This provides crucial insights into how regional climate change might be altering the timing of agricultural seasons and impacting the synchronization between plant development and environmental conditions. The key strengths of phenological networks lie in their ability to provide extensive datasets across vast spatial and temporal scales, which are invaluable for analyzing regional and long-term trends in developmental timing. They offer direct observations of critical developmental milestones, capturing the actual moments when plants transition through their life stages. However, a potential limitation, particularly with citizen science initiatives, is variability in data quality and consistency if observation protocols are not rigorously standardized and followed. Furthermore, these networks primarily focus on developmental timing and typically provide limited direct detail on quantitative aspects of growth, such as biomass accumulation or final yield quantity.

5.6.2 Controlled Environment Methods

These methodologies offer the distinct advantage of allowing researchers to precisely manipulate individual environmental factors, thereby isolating their specific effects on crop growth and development.

- **Growth Chambers / Phytotrons:** These are highly sophisticated and controlled indoor facilities designed to create specific and reproducible environmental conditions for plant growth. Parameters such as air temperature, light intensity (including spectrum and photoperiod), CO_2 concentration, relative humidity, and often soil moisture can be set and precisely varied. Plants are cultivated in pots or small beds within these chambers, allowing for meticulous control over their immediate environment. This level of precision enables researchers to systematically change one weather parameter at a time while holding all others constant. For example, scientists can expose plants to a series of increasing temperatures to determine the cardinal temperatures for specific physiological processes (e.g., optimal temperature for photosynthesis) or vary light intensity to study the efficiency of light conversion into biomass. This controlled approach is instrumental in gaining deep mechanistic understanding of plant responses. The paramount advantage of growth chambers is their high precision and control, which allows for the unambiguous isolation of individual environmental factors and their direct impact on plant physiology and growth. This leads to high repeatability of experiments, minimizing experimental variability and increasing confidence in the results. They are ideal for deriving mechanistic insights at the leaf or whole-plant level, uncovering the fundamental biological principles governing growth. Moreover, these chambers allow for the testing of hypothetical future climate scenarios (e.g., sustained elevated CO_2 levels or higher average temperatures) without having to wait for actual climate change to unfold. However, a significant drawback is the artificiality of the environment, which may not fully replicate the complex interactions, soil conditions, or wind effects encountered in natural field settings. Experiments are also typically conducted on individual plants or very small populations, making it challenging to directly extrapolate findings to field-scale crop stands. Finally, the setup and operational costs of sophisticated growth chambers are considerably high, limiting their widespread accessibility.
- **Free Air CO_2 Enrichment (FACE) / Temperature Free Air Enrichment (T-FACE):** Representing a bridge between controlled environments and open-field conditions, FACE and T-FACE systems

are large-scale, open-field experimental setups designed to modify atmospheric conditions around a growing crop without enclosing it. In FACE experiments, CO_2 gas is released from a ring of vertical pipes to maintain an elevated concentration of CO_2 around the crop, simulating future atmospheric conditions. Similarly, T-FACE experiments use infra-red heaters to consistently warm the air surrounding the plants. These systems allow researchers to evaluate the direct impact of sustained changes in CO_2 or air temperature on crop growth and yield under otherwise natural, fluctuating field conditions. The primary advantage of FACE and T-FACE is their high level of realism, as they conduct experiments directly within the natural field environment, capturing the complex interactions between soil, plant, and atmosphere, as well as the influence of pests, diseases, and natural weather variability. They enable measurements at a larger scale (e.g., yield per hectare) that are directly relevant to agricultural production. However, these systems are extremely expensive to establish and operate, requiring significant logistical coordination and resources. They also offer limited factor control, as typically only one or two parameters (CO_2 or temperature) can be controlled, while all other weather variables remain natural and therefore fluctuate unpredictably, potentially confounding results.

5.6.3 Modeling Approaches

These methods leverage computational power to simulate crop responses based on input weather data, offering flexible and scalable tools for evaluation.

- **Process-Based Crop Models (Simulation Models)** These models (e.g., DSSAT, APSIM, WOFOST) are sophisticated mathematical representations that simulate the fundamental physiological and biophysical processes within a crop, typically on a daily or sub-daily time step. They mechanistically simulate processes such as photosynthesis, respiration, transpiration, nutrient uptake, and biomass partitioning, all driven by detailed inputs of weather (solar radiation, temperature, precipitation, humidity, wind), soil characteristics, and management practices. Process-based models are invaluable for evaluating weather effects through several mechanisms:

Scenario Analysis: By inputting different historical weather datasets, models can assess the impact of past climate variability on crop performance. More powerfully, they can be run with future climate change projections (derived from Global Climate Models (GCMs) or Regional Climate Models (RCMs)) to predict how crops might perform under novel future environmental conditions.

Sensitivity Analysis: This involves systematically varying one specific weather parameter (e.g., increasing daily maximum temperature by 1^0C or reducing seasonal rainfall by 20%) while keeping all other factors constant. This controlled perturbation allows researchers to quantify the crop‘s sensitivity to that particular weather factor and identify critical thresholds.

Yield Gap Analysis: Models can simulate the potential yield of a crop under ideal, non-limiting weather and management conditions. By comparing this potential with simulated yields under actual weather conditions (e.g., rainfed or under specific management), the model helps identify yield losses directly attributable to sub-optimal weather, thus quantifying the weather-induced yield gap.

The primary strengths of process-based models include providing mechanistic understanding by simulating the underlying biological processes, thus explaining *why* weather affects crop performance. They enable the exploration of an almost infinite number of what-if" scenarios, including those that are impractical or impossible to conduct in real-world field experiments (e.g., long-term climate change impacts, extreme event frequencies). Once accurately calibrated and validated, running numerous simulations is relatively cost-effective, making them ideal for large-scale assessments. However, these models are highly data-intensive, requiring comprehensive input data for both their setup (parameterization) and subsequent calibration and validation. They are inherently based on simplifications and assumptions of complex biological systems, leading to inherent uncertainty in their predictions. Ensuring their reliability for a specific region, crop, or management system demands significant effort in calibration and validationagainst high-quality observed field data.

- **Hybrid Models-** Hybrid models represent a pragmatic approach that combines the strengths of both statistical and mechanistic modeling techniques. For instance, a primarily process-based model might incorporate empirically derived statistical relationships for certain complex processes where detailed mechanistic understanding is limited or computationally too intensive (e.g., modeling the effect of a specific pest or disease outbreak on yield reduction based on observed weather conditions rather than simulating the entire life cycle of the pest). Alternatively, statistical downscaling methods might be used to provide localized weather inputs for a mechanistic model. The main advantage of hybrid models is their ability to improve accuracy and robustness in situations where purely mechanistic models might struggle due to complexity or data gaps. They can potentially be less data-intensive for certain components while still retaining the explanatory power of a mechanistic framework. However, they inherit the limitations of both statistical and mechanistic modeling, and their development and validation can be complex due to the need to integrate disparate approaches.

5.6.4 Remote Sensing and Geospatial Technologies

These cutting-edge methodologies offer powerful, non-destructive means to monitor crop responses to weather from a distance, providing spatial and temporal insights at various scales.

- **Satellite Imagery and Vegetation Indices (VIs)-** Satellites equipped with various sensors continuously capture spectral reflectance data from the Earth's surface. This raw data is then processed to calculate various Vegetation Indices (VIs), such as the Normalized Difference Vegetation Index (NDVI) or the Enhanced Vegetation Index (EVI). These VIs serve as proxies for vegetation greenness, density, and overall health. They are fundamentally linked to the amount of photosynthetically active biomass.

Evaluation of Weather Effects: Time-series analysis of VIs (tracking how the VI value for a given agricultural area changes over days, weeks, or months) is a powerful tool for phenological monitoring. Distinct patterns in VI changes can detect key phenological events like "green-up" (emergence/start of active growth), "peak greenness" (maximum vegetative growth/flowering), and "brown-down" (senescence/maturity). Shifts in the timing of these events can directly indicate the impact of a changing thermal regime or altered rainfall patterns. Furthermore, abrupt drops or sustained low VI values compared to historical averages for a specific time of year can effectively indicate crop stress (e.g., drought, heat stress, or disease outbreaks potentially exacerbated by weather) over large areas. VIs can also be correlated with actual yields (after calibration with ground truth data) to provide broad-scale yield forecasts, implicitly reflecting the cumulative impact of seasonal weather on overall productivity.

The primary benefits of satellite imagery include its vast spatial coverage, allowing for efficient and cost-effective monitoring of immense agricultural regions. Long-term historical archives of satellite data provide invaluable records for analyzing past weather impacts and long-term trends. The method is non-destructive, requiring no direct interaction with the crop. However, optical sensors are significantly hindered by cloud cover, which can lead to data gaps, especially in humid or monsoon-dominated regions. The spatial and temporal resolution of satellite data, while improving, may still not be fine enough for specific field-level management decisions. Crucially, VIs are proxies; they do not directly measure biomass or physiological rates and therefore require robust ground truth data for accurate calibration and interpretation. Additionally, VIs can saturate in very dense canopies, meaning they may not accurately distinguish further increases in biomass once a certain level of greenness is reached.

5.6.5 Unmanned Aerial Vehicles (UAVs / Drones)

Drones, equipped with diverse payloads of sensors (e.g., high-resolution RGB cameras, multispectral, thermal, or hyperspectral imagers), capture imagery at very low altitudes. This allows for the acquisition of ultra-high-resolution data at the farm or even sub-field scale.

Evaluation of Weather Effects: UAVs enable the precise monitoring of detailed phenological events and canopy development, such as quantifying emergence rates, assessing plant counts, and tracking early-stage canopy cover within a field with unprecedented spatial detail. Thermal cameras can effectively detect localized water stress by identifying areas where plant canopy temperatures are elevated due to reduced transpiration. Multispectral cameras can reveal subtle changes in plant health indicative of early nutrient deficiencies or disease outbreaks that might be linked to specific weather conditions (e.g., prolonged wetness fostering fungal growth). Drones are also invaluable for rapid and high-resolution post-event damage assessment, swiftly quantifying localized hail damage, wind damage, or the extent of flooding impacts immediately following an extreme weather event.

The chief advantages of UAVs are their extremely high spatial and temporal resolution, providing granular detail at the plant or sub-field level, coupled with flexible deployment frequency (e.g., daily or even multiple times a day if needed). This granularity translates into highly actionable insights for precision agriculture management. However, their coverage area is relatively limited per flight due to battery life and operational constraints, making them less suitable for very large regional assessments. Regulatory issues regarding airspace and drone operations can also pose limitations. Furthermore, the immense volume of high-resolution data they collect necessitates significant computational power and expertise for processing and analysis.

5.6.6 Ground-Based Proximal Sensing (e.g., PhenoCams)

These systems involve the deployment of automated digital cameras (often RGB or multispectral) at fixed positions within a field. These cameras capture time-lapse images continuously (e.g., every hour or every day) over extended periods.

Evaluation of Weather Effects: PhenoCams provide an exceptionally high temporal resolution visual record of plant development. They excel at monitoring subtle phenological changes (e.g., the exact timing of bud break, leaf emergence, first flower, or the onset of senescence) and changes in canopy greenness (using image-derived greenness indices). Their continuous data capture also allows for the observation of diurnal changes in plant appearance or stress levels in direct response to hourly weather fluctuations (e.g., wilting during the hottest part of the day, recovery overnight).

The key advantages of ground-based proximal sensing include its continuous, fine-scale detail, making it excellent for capturing rapid phenological transitions that might be missed by less frequent observations. They offer a direct visual record that is easily interpretable and verifiable. The primary limitation, however, is their very limited spatial coverage, as they are typically restricted to a small area immediately around the camera. While providing high temporal detail, these sensors often offer less spectral information compared to multispectral satellite or drone sensors.

5.7 Conclusion: A Synergistic Approach

In essence, evaluating the myriad effects of weather on crops is not an isolated task for a single scientific discipline or a solitary methodology. Instead, it demands a sophisticated, synergistic approach that strategically integrates insights from across these diverse techniques. The meticulous detail and control afforded by growth chambers complement the real-world complexities captured in field experiments. Statistical models reveal broad historical patterns, while mechanistic crop models provide the predictive power and what-if scenario testing capabilities essential for future foresight. Finally, remote and proximal sensing technologies offer the invaluable ability to monitor these responses across vast scales and with unprecedented resolution. By leveraging the unique strengths of each method in a complementary fashion, agricultural scientists and practitioners can achieve the most comprehensive and robust understanding of how crops interact with their dynamic weather environment, ultimately empowering them to diagnose past impacts, predict future responses, and develop increasingly resilient and productive cropping systems for a changing world.

5.7.1 Empirical Methods for Evaluating Weather Effects on Crops

In the broad spectrum of approaches used to understand how weather impacts crops, empirical methods stand out for their reliance on direct observation, historical data, and statistical analysis. Unlike mechanistic models that simulate underlying physiological processes, or controlled experiments that isolate specific factors, empirical methods seek to establish observable relationships between weather variables and crop performance outcomes (such as yield or phenological timing) without necessarily detailing the biological mechanisms. These methods are rooted in the principle that if a consistent pattern exists between weather and crop response over time or space, it can be statistically identified and used for evaluation or even prediction.

Empirical approaches are often the first line of investigation due to their relatively lower data demands (compared to mechanistic models) and their ability to quickly reveal broad trends and dominant weather influences. They are particularly valuable for large-scale assessments, historical analyses, and preliminary studies where detailed mechanistic data might be scarce.

5.7.2 Core Principles of Empirical Methods

The fundamental idea behind empirical methods is to find a mathematical or statistical relationship that describes the observed input-output behavior of the crop system. This relationship is derived directly from data, rather than being built from first principles of plant physiology.

Observation-Driven: They are entirely dependent on observed historical data (weather records, yield statistics, phenological observations).

Statistical Foundation: They utilize statistical techniques (e.g., regression, correlation, time-series analysis) to quantify relationships.

Black Box" Approach: Often treat the crop as a black box, focusing on inputs (weather) and outputs (yield) without necessarily explaining the internal biological workings.

Context-Specific: The derived relationships are valid primarily for the specific environmental conditions and agricultural practices from which the data were collected.

5.7.3 Key Empirical Methods Employed

- **Historical Weather-Yield Correlations and Regression Analysis**

Explanation: This is perhaps the most common empirical method. It involves compiling long-term historical records of crop yields (e.g., annual district or national average yields) and matching them with corresponding weather data for the same period and region. Weather variables typically include seasonal totals (e.g., total rainfall during the growing season), averages (e.g., mean growing season temperature), or accumulated indices (e.g., Growing Degree Days (GDD) for specific periods).

Process: Statistical techniques, most commonly regression analysis (linear or non-linear), are used to model the relationship between yield (dependent variable) and one or more weather variables (independent variables). For example, a simple linear regression might be used to assess how maize yield changes with increasing rainfall during the critical flowering period, or a multiple regression could include temperature and solar radiation as well.

Yield Deviation Analysis: A common refinement is to analyze yield *deviations* from a long-term trend (e.g., technology trend) rather than absolute yields. This helps to isolate the weather effect by removing the influence of gradual improvements in genetics and management over time.

Evaluation of Weather Effects

Identifies the magnitude and direction of the statistical relationship (e.g., a 10% increase in monsoon rainfall correlates with a 5% increase in rice yield).

Pinpoints critical periods during the growing season when certain weather factors have a disproportionate impact on final yield.

Quantifies the contribution of different weather variables to overall yield variability in a region over historical periods.

Advantages

Data Availability: Leverages existing, often extensive, historical datasets, making it relatively inexpensive.

Ease of Implementation: Statistical software makes these analyses straightforward.

Broad-Scale Insights: Can provide valuable insights into dominant climatic controls on agricultural productivity at regional or national scales.

Quick Assessment: Offers a rapid way to gain initial understanding of weather-yield sensitivities.

Limitations

Correlation Does Not Imply Causation: This is the most significant limitation. A strong correlation doesn't mean a direct biological mechanism; other unmeasured factors or indirect relationships might be at play.

Limited Extrapolation: The derived statistical relationships are only valid within the range of the historical data used. They perform poorly when extrapolated to novel climate conditions (e.g., future climate change scenarios) or different crop types/management practices for which the relationship was not established.

Lack of Mechanistic Detail: Provides no insight into the underlying physiological processes (e.g., how water stress reduces photosynthesis).

Confounding Factors: It's often challenging to fully disentangle the effects of weather from other evolving factors like technological advancements, changes in varieties, or socio-economic shifts.

- **Agro-climatological Indices and Zonation**

Explanation: This method involves developing indices that combine multiple weather parameters into a single value that represents the suitability of a region's climate for a particular crop, often over a specific growing period. These indices are based on empirically observed thresholds and optimal ranges.

Examples

Aridity Index: Ratio of precipitation to potential evapotranspiration, indicating dryness.

Climatic Water Balance: Net difference between precipitation and evapotranspiration.

Crop-specific thermal accumulation indices: Like GDD, but often with more complex weightings or thresholds reflecting stress (e.g., effective GDD, accounting for heat stress above optimal).

Zonation: These indices are then used to delineate agro-climatic zones where a particular crop is expected to perform well or poorly, based on historical weather patterns.

Evaluation of Weather Effects

Provides a **synthetic measure** of climatic suitability for a crop in a given area.

Helps in **identifying climatically analogous regions** for crop introduction or technology transfer.

Assesses the **risk of climatic limitations** (e.g., drought risk, frost risk) for a crop in a particular zone.

Advantages

Simplifies Complex Interactions: Reduces multiple weather variables into a single, interpretable index.

Spatial Analysis: Ideal for mapping and zonation, aiding in regional planning.

Relatively Simple to Calculate: Often uses readily available aggregated weather data.

Limitations

Oversimplification: Can mask important daily or sub-daily weather events and their critical timing.

Empirical Basis: The thresholds and weightings in the indices are often derived from historical observations and may not hold universally or under future climate extremes.

Limited Predictive Power: Primarily descriptive of suitability based on historical norms, less robust for predicting responses to unprecedented weather.

- **Phenological Event Date Analysis (from Historical Records)**

Explanation: This method focuses on analyzing the dates of key phenological events (e.g., emergence, flowering, harvest) for a crop over long historical periods, correlating these dates with prevailing weather conditions (especially temperature and photoperiod). This often involves historical farmer records, experimental station logs, or citizen science data.

Evaluation of Weather Effects

Quantifies the shift in crop calendars in response to inter-annual weather variability or long-term climatic trends. For example, documenting an earlier flowering date for wheat in response to warmer spring temperatures over the last few decades.

Identifies the **sensitivity of specific developmental stages** to particular thermal or photoperiodic cues.

Advantages

Direct Observation of Timing: Provides tangible evidence of how crop development responds to environmental cues.

Valuable for Climate Change Studies: Offers direct evidence of historical shifts in crop phenology attributable to climate change.

Relatively Accessible Data: Historical records of planting and flowering dates can sometimes be found.

Limitations

Data Availability and Quality: Long-term, consistent phenological observation records are rare and can be subjective.

Focus on Timing Only: Primarily evaluates developmental timing, not quantitative growth or yield directly.

Confounding Factors: Changes in cultivar, management, or planting date decisions over time can influence observed phenological shifts, making it difficult to attribute changes solely to weather.

5.7.4 Conclusion on Empirical Methods

Empirical methods provide a pragmatic and often efficient means to evaluate the historical effects of weather on crops. They are excellent for identifying broad correlations, critical periods, and climatic suitability zones, particularly useful for large-scale assessments and historical trend analysis. Their primary strength lies in their direct use of observed data and relative simplicity. However, their major weakness is their inherent inability to explain underlying mechanisms, their limited reliability for extrapolation beyond the conditions from which they were derived, and their susceptibility to confounding factors. For deeper mechanistic understanding, precise predictions under novel conditions, and detailed what-if scenario analysis, empirical methods are often complemented by, or serve as a foundational step for, more advanced process-based modeling approaches.

5.8 Statistical Methods for Evaluating Weather Effects on Crops

As a crucial subset of the broader empirical approaches, **statistical methods** form a foundational pillar for evaluating how weather influences crop performance. These methods systematically employ mathematical statistics to identify, quantify, and model the relationships between observed weather variables (inputs) and crop outcomes (outputs, such as yield, phenological timing, or biomass). Unlike mechanistic models, which delve into the biological why, statistical models focus on the what and how much, establishing patterns and predictions based purely on historical data without explicit representation of underlying physiological processes. They treat the complex crop-weather system as a black box, where the internal workings are less important than the observable correlations and regressions. Statistical methods are invaluable for:

Identifying Significant Relationships: Pinpointing which weather variables exert the most influence on crop outcomes.

Quantifying Impact: Determining the magnitude and direction of these influences.

Forecasting and Risk Assessment: Using established relationships to predict future outcomes or assess risks based on current or forecast weather.

Initial Screening: Serving as a rapid and cost-effective preliminary assessment tool before investing in more complex mechanistic modeling.

5.8.1 Key Statistical Techniques Employed in Crop-Weather Analysis

- **Correlation Analysis**

Explanation: Correlation analysis measures the strength and direction of a linear association between two (or more) variables. The result is a correlation coefficient (e.g., Pearson's r), typically ranging from -1 to +1. A value close to +1 indicates a strong positive linear relationship, -1 indicates a strong negative linear relationship, and 0 suggests no linear relationship.

Application: It is often the first step in exploring weather-crop relationships. For instance, a researcher might calculate the correlation between total growing season rainfall and maize yield, or between mean temperature during flowering and wheat yield. It can be applied to daily, weekly, or monthly weather data to identify critical periods where specific weather variables show the strongest association with the final yield.

Evaluation of Weather Effects: It helps in identifying potentially important weather variables or periods for further, more detailed analysis. For example, if rainfall in July shows a high positive correlation with soybean yield, it suggests July rainfall is a key driver.

Limitations: Correlation only indicates association, not causation. A strong correlation might be coincidental or driven by an unmeasured third variable. It only captures linear relationships.

- **Regression Analysis (Simple, Multiple, and Non-linear)**

Explanation: Regression analysis moves beyond simple association to build a mathematical equation that describes how a dependent variable (e.g., crop yield) changes as one or more independent variables (e.g., weather parameters) change.

Simple Linear Regression: Models the relationship between one dependent variable and one independent variable using a straight line: $Y=a+bX$. For example, $Yield=a+b\times Seasonal\ Rainfall$.

Multiple Linear Regression: Extends this to include several independent variables: Y=a+b1 X1 +b2 X2 +...+bn Xn . This allows for the simultaneous assessment of multiple weather factors (e.g., Yield=a+b1 ×Rainfall+b2 ×Mean Temp). The coefficients (bn) indicate the estimated change in yield for a one-unit change in the corresponding weather variable, holding others constant.

Non-linear Regression: When relationships are not linear (which is common in biology, where optimal points or saturation occur), non-linear functions (e.g., quadratic, exponential, sigmoid) are used. For example, yield might increase with temperature up to an optimum, then decrease (a quadratic relationship), or show a diminishing return to increasing rainfall.

Application

Quantifying Sensitivity: Directly quantifies how much yield is expected to change for a given change in a weather variable.

Identifying Critical Periods: By including weather variables from different growth stages as independent predictors, regression models can statistically pinpoint periods of high sensitivity. For example, separate variables for rainfall during vegetative stage and rainfall during reproductive stage can be included.

Yield Forecasting: Once calibrated with historical data, the regression equation can be used with current or forecast weather data to predict expected yields.

Evaluation of Weather Effects: Provides specific coefficients that denote the estimated impact of each weather factor. Model fit statistics (e.g., R^2) indicate how much of the yield variability can be explained by the included weather variables.

Limitations: Similar to correlation, regression does not prove causation. It assumes the historical relationships will hold true in the future. Non-linear relationships can be complex to model. Multicollinearity (when independent weather variables are highly correlated with each other) can complicate interpretation.

- **Time Series Analysis**

Explanation: This category of methods is specifically designed for analyzing data points collected sequentially over time. It aims to identify patterns, trends, seasonality, and cycles within the data.

Application: Can be applied to long-term series of crop yields or phenological dates (e.g., annual flowering date). By analyzing these series in conjunction with weather records, time series models can:

Identify Trends: Distinguish long-term trends in yield (e.g., due to technological advancements) from short-term fluctuations caused by weather. This is crucial for detrending yield data before attributing variability to weather.

Detect Seasonality: Confirm annual recurring patterns in weather and crop behavior.

Model Cycles: Identify multi-year cycles in weather patterns (e.g., El Niño/La Niña effects) and their corresponding impact on crop yields.

Forecasting: Use past patterns to forecast future yield or phenological events based on expected or observed weather anomalies. Methods include Moving Averages (MA), Autoregressive (AR) models, or combined ARIMA (Autoregressive Integrated Moving Average) models.

Evaluation of Weather Effects: Helps separate systematic non-weather influences from weather-induced variability. Allows for dynamic forecasting based on evolving weather conditions.

Limitations: Requires very long and consistent time-series data. Can be complex to model accurately, especially for non-stationary data.

- **Principal Component Analysis (PCA) and Factor Analysis**

Explanation: These are multivariate statistical techniques used to reduce the dimensionality of large datasets with many inter-correlated variables. They identify underlying principal components or factors that capture most of the variance in the original data.

Application: When dealing with numerous daily or weekly weather variables (e.g., maximum temperature, minimum temperature, rainfall, solar radiation, humidity, wind speed) across different periods of the growing season, PCA can synthesize these into fewer, uncorrelated principal components. These components often represent dominant weather patterns (e.g., a hot and dry component or a cool and wet component). Regression can then be performed using these components as predictors of crop yield.

Evaluation of Weather Effects: Helps to identify the most significant composite weather patterns that affect crop performance, rather than just individual variables in isolation. This can provide a more holistic understanding of climate influence.

Limitations: The principal components themselves may lack direct physical interpretability. Still an empirical approach, not providing mechanistic detail.

- **Cluster Analysis**

Explanation: Cluster analysis is a data mining technique used to group a set of objects (e.g., years, locations) in such a way that objects in the same group (cluster) are more similar to each other than to those in other groups.

Application:

Agro-Climatic Zonation: Can be used to group regions or years that experience similar weather patterns or similar crop responses. This allows for identifying distinct agro-climatic zones with shared characteristics and vulnerabilities.

Identifying Analog Years: Grouping historical years based on their weather patterns to find analog years to the current season, which can then inform expected crop performance.

Evaluation of Weather Effects: Helps categorize different weather scenarios and their typical associated crop outcomes, providing insights into discrete weather-response types.

Limitations: The choice of clustering algorithm and number of clusters can be subjective. Requires careful interpretation of what defines each cluster.

- **Machine Learning Approaches (Advanced Empirical)**

Explanation: Increasingly, more advanced machine learning algorithms are being employed. These include Decision Trees, Random Forests, Support Vector Machines (SVMs), and Artificial Neural Networks (ANNs). These methods are highly adept at identifying complex, non-linear relationships within large datasets.

Application: Can model highly non-linear weather-yield relationships that simpler regression models might miss. Often used for sophisticated yield forecasting or crop suitability mapping.

Evaluation of Weather Effects: Offer high predictive power based on patterns learned from historical data. Some (like Decision Trees/Random Forests) can provide measures of variable importance, indicating which weather inputs were most influential.

Limitations: Often operate as black boxes (especially Neural Networks), making it difficult to understand the exact nature of the relationship or the underlying physiological reasons. Require very large and high-quality datasets for training. Prone to overfitting if not properly validated.

- **Data Requirements for Statistical Methods:**

The success of any statistical method hinges on the quality and availability of data:

Long-term Historical Data: Years of consistent, high-quality records for both weather variables and crop outcomes (yields, phenology, disease incidence).

Spatial Consistency: Weather data should be representative of the areas from which crop data is sourced.

Detrended" Yields: For yield analysis, it is crucial to remove the long-term trend caused by technological advancements (e.g., new varieties, improved fertilizers) to isolate the weather effect. This often involves regressing yield against time and using the residuals.

- **Advantages of Statistical Methods**

Leverage Existing Data: Utilize readily available historical climate and agricultural statistics, often making them relatively quick and inexpensive to implement.

Identify Significant Relationships: Powerful for pinpointing key weather drivers and critical periods of influence.

Useful for Broad-Scale Assessment: Excellent for regional, national, or even global assessments of climate impact on agriculture.

Directly Quantify Impact: Provide quantifiable measures of how much yield (or other outcomes) is associated with changes in weather.

- **Limitations of Statistical Methods**

Correlation vs. Causation (Crucial Limitation): Cannot explain the biological mechanisms driving the relationships. They tell us *what* happened, not *why*.

Limited Extrapolation: The models are built on historical patterns and perform poorly when applied to future climate scenarios that lie outside the range of observed historical weather, or to new crop varieties or management practices.

Sensitivity to Data Quality: Gaps, errors, or inconsistencies in historical data can severely compromise the reliability of the results.

Cannot Handle Novelty: Struggle with unprecedented extreme weather events or novel interactions that have no historical precedent.

No Mechanistic Insight: Do not contribute to the fundamental understanding of how plants physiologically respond to weather.

In conclusion, statistical methods remain a cornerstone for evaluating weather effects on crops, particularly for historical analysis, broad-scale risk assessment, and preliminary investigations. They provide valuable empirical insights into observed relationships. However, their inherent limitations, especially the lack of mechanistic explanation and poor extrapolation capability, underscore the necessity of complementing them with or transitioning to process-based modeling for a deeper, more predictive understanding of crop responses to an ever-changing climate.

5.9 Analytical (Mechanistic) Methods for Evaluating Weather Effects on Crops

Following our discussion on empirical and statistical approaches, we now turn to analytical methods, which represent a fundamentally different and often more powerful paradigm for evaluating the effects of weather on crops. In the context of crop modeling, analytical methods are synonymous with process-based models or mechanistic models. These approaches do not merely seek statistical correlations between inputs and outputs; instead, they attempt to simulate the underlying biological, physical, and chemical processes that govern crop growth and development based on scientific understanding of how plants function.

The core idea is to break down the complex crop system into its constituent, definable processes (e.g., photosynthesis, respiration, water uptake, nutrient cycling) and represent these processes with mathematical equations derived from fundamental biophysical and physiological laws. These equations quantify how each process responds to environmental factors, including various weather parameters, thus providing a mechanistic explanation for observed crop responses.

5.9.1 Core Principle: Simulating Underlying Mechanisms

The defining characteristic of analytical methods is their commitment to simulating the why behind crop responses. Rather than simply observing that higher temperatures lead to faster development, a mechanistic model will include equations that describe how enzyme kinetics, cell division rates, and specific phenological triggers (like Growing Degree Days) are directly influenced by temperature. This allows the model to predict behavior even under conditions not encountered in historical records, as long as the underlying biological principles hold true.

These models typically operate on a daily or sub-daily time step, integrating the effects of dynamic weather inputs over the entire growing season. Each day, the model updates the plant‘s state variables (e.g., accumulated biomass, leaf area index, soil water content, developmental stage) based on the rates of various processes that occurred in response to the day‘s weather.

• Key Components and Processes Modeled in Detail

Light Interception and Photosynthesis (Carbon Assimilation): This is the engine of crop growth. Models calculate:

Incoming Solar Radiation (PAR): The model first converts total solar radiation into Photosynthetically Active Radiation (PAR), the portion of light plants use for photosynthesis.

Light Interception by Canopy: This is crucial. The model uses the Leaf Area Index (LAI) (total green leaf area per unit ground area) and canopy light extinction coefficients to determine how much PAR is intercepted by the leaves at different layers of the canopy. As the crop grows, LAI increases, leading to more light interception.

Gross Photosynthesis Rate: The rate at carbon is fixed per unit of intercepted light. This rate is influenced by:

Light Intensity: Follows a light response curve, typically saturating at high light.

Temperature: Each plant type (C3 vs. C4) has an optimal temperature range for photosynthesis, with rates declining rapidly outside this range.

Carbon Dioxide (CO_2) Concentration: As CO_2 is a substrate, higher atmospheric concentrations can enhance photosynthesis, particularly for C3 plants (the CO_2 fertilization effect).

Water Stress: Drought significantly reduces photosynthesis by inducing stomatal closure (limiting CO_2 entry) and directly impairing photosynthetic machinery.

Nutrient Stress: Deficiencies in essential nutrients (especially nitrogen) impair the photosynthetic capacity of leaves.

Respiration (Carbon Loss): Not all assimilated carbon contributes to growth; a portion is consumed for energy. Models distinguish:

Maintenance Respiration: Energy required to maintain existing plant tissues. This rate is highly dependent on existing biomass and increases exponentially with temperature.

Growth Respiration: Energy consumed to convert newly assimilated carbohydrates into new structural biomass. This is typically proportional to the rate of new growth.

Net Growth: The difference between gross photosynthesis and total respiration.

Biomass Partitioning (Allocation): The newly fixed carbon (net growth) is then distributed among different plant organs.

Dynamic Allocation: The fractions allocated to leaves, stems, roots, and reproductive organs (grains, fruits, tubers) change dynamically as the plant progresses through its developmental stages. For instance, more biomass goes to leaves early on, then to roots, and finally a large proportion is directed to the harvestable product during grain filling.

Environmental Influence: Stress factors (e.g., water deficit, nutrient deficiency) can alter these partitioning coefficients, sometimes favoring root growth under drought or reducing allocation to reproductive organs under severe stress.

Water Balance and Transpiration: These models meticulously track water movement within the soil-plant-atmosphere continuum.

Soil Water Dynamics: Simulates infiltration of rainfall/irrigation, run-off, drainage below the root zone, and evaporation from the soil surface. Soil characteristics (texture, bulk density, organic matter, water retention curves) are crucial inputs here.

Root Water Uptake: The model calculates how much water the roots extract from different soil layers, dependent on root density, soil water potential, and transpiration demand.

Transpiration: Water loss from the plant through stomata. This is driven by atmospheric demand (Vapor Pressure Deficit - VPD, wind speed) and modulated by stomatal conductance, which responds to light, CO_2 , and water stress.

Water Stress Calculation: Based on the balance between water supply (from soil) and demand (transpiration), the model calculates a water stress factor that reduces rates of photosynthesis, leaf expansion, and overall growth.

Nutrient Dynamics and Uptake: Models represent the cycling of key nutrients (e.g., Nitrogen, Phosphorus, Potassium) in the soil and their uptake by the crop.

Soil Nutrient Pools: Simulates nutrient mineralization from organic matter, immobilization, leaching, and denitrification.

Nutrient Uptake: Calculates how much nutrient the roots absorb, dependent on soil nutrient availability, root growth, and water uptake.

Nutrient Stress: If nutrient uptake is insufficient for the plant's demand, a nutrient stress factor is applied, reducing photosynthesis, leaf expansion, and biomass partitioning.

Phenological Development: As detailed previously, models simulate the progression through life stages based primarily on:

Thermal Time (GDD): Accumulation of heat units, driven by daily temperature.

Photoperiodism: Sensitivity to day length for flowering.

Vernalization: Cold requirement for specific crops.

Stress Interaction: Water or heat stress can accelerate development (e.g., premature maturity), which can limit the duration of growth processes like grain filling, thus reducing yield.

- **How Weather Effects are Integrated and Evaluated**

Analytical models are inherently designed to evaluate weather effects because every process within them is explicitly linked to weather parameters.

Solar Radiation (Light): Directly drives photosynthesis. Changes in daily radiation sums directly alter the potential carbon gain.

Temperature: Controls the rates of all metabolic processes (photosynthesis, respiration, development). Extreme temperatures (heat/cold stress) apply stress factors that reduce process rates.

Rainfall and Humidity (VPD): Influence the soil water balance, root water uptake, and transpiration. Deficits lead to water stress (reduced photosynthesis, expansion). High VPD increases water demand and stomatal regulation.

Carbon Dioxide (CO_2): A direct substrate for photosynthesis; its atmospheric concentration directly influences carbon fixation rates, especially in C3 crops.

Wind Speed: Impacts the boundary layer around leaves, influencing transpiration rates and CO_2 exchange. High winds can also cause physical damage.

- **Evaluation Through Simulation and Scenario Analysis**
  - **What-If' Scenarios:** By changing weather inputs, modelers can explore hypothetical scenarios (e.g., a drought year, a warmer growing season, increased CO_2). The model then simulates the crop's response, showing the resulting changes in phenology, biomass accumulation, water use, and final yield.
  - **Sensitivity Analysis:** Systematically perturbing individual weather parameters within the model (e.g., increasing mean temperature by 1^0C or decreasing rainfall by 20% across the season) allows for a quantitative assessment of the crop's sensitivity to each factor and the identification of critical thresholds.
  - **Climate Change Impact Assessment:** Running models with future climate projections from GCMs/RCMs allows for detailed predictions of how specific crops will respond to future temperature increases, altered rainfall patterns, and elevated CO_2, providing critical information for adaptation strategies.
  - **Identification of Limiting Factors:** By tracing the flow of resources and applying stress factors, models can identify *when* and *why* a particular weather factor became limiting (e.g., water stress during flowering, heat stress during grain filling).

- **Advantages of Analytical Methods**

Mechanistic Explanations: Provide deep insights into *why* a crop responds in a certain way to weather, based on the underlying physiological processes. This moves beyond mere correlation.

Predictive Power for Novel Conditions: Because they are based on fundamental principles, they can extrapolate and predict crop behavior under weather conditions or combinations of conditions that have not been historically observed. This is crucial for climate change impact assessments.

What-If' Scenario Analysis: Allow for virtually limitless testing of hypothetical scenarios (e.g., new varieties, altered management practices, different climate futures) without the cost, time, or risk of field trials.

Identification of Limiting Factors: Can pinpoint precisely when and where a specific resource (water, nutrient, light, optimal temperature) becomes limiting to growth and yield.

Foundation for Improvement: The modular nature allows for continuous refinement as scientific understanding of plant processes evolves.

- **Limitations of Analytical Methods**

Data Intensive: Require a vast amount of high-quality input data for both parameterization (defining the crop's genetic coefficients) and validation (checking model outputs against observed data).

Complexity and Computational Demand: These models can be highly complex to build, calibrate, and run, requiring significant computational resources and specialized expertise.

Uncertainty from Simplifications: All models are simplifications of reality. While mechanistic, they still contain assumptions and simplifications that can introduce uncertainty. Complete knowledge of all biological processes is still elusive.

Expertise Required: Building, calibrating, and interpreting these models requires a strong understanding of both plant science and modeling principles.

In conclusion, analytical or process-based methods form the cutting edge of crop-weather modeling. By simulating the fundamental biological and physical processes that link weather to crop growth and development, they provide an unparalleled level of mechanistic insight and predictive capability. While demanding in terms of data and expertise, their ability to conduct detailed what-if analyses and forecast responses to unprecedented climatic conditions makes them indispensable tools for advancing agricultural science and addressing global food security challenges.

5.10 Determination of Optimum Weather Conditions for Crops in a Given Region

The concept of optimum weather conditions for a crop is not a universal constant; rather, it is highly region-specific, crop-specific, and even cultivar-specific. It refers to the particular combination of meteorological factors—temperature, solar radiation, precipitation, humidity, and wind—that, over the course of a growing season, maximizes the growth, development, and ultimately, the yield of a chosen crop in a particular geographical area, given its unique soil conditions and typical management practices. Determining these optimum conditions is a crucial step in sustainable agricultural planning, allowing for informed decisions on crop selection, planting dates, and resource management.

This determination is a complex undertaking that integrates insights from various methodologies, moving beyond simple correlations to a more holistic understanding.

5.10.1 Why is Determining Optimum Conditions Challenging?

- **Interacting Factors:** Weather parameters don't act in isolation. The optimal temperature for photosynthesis, for instance, might shift depending on light intensity or water availability.
- **Temporal Dynamics:** A crop's sensitivity to weather changes throughout its life cycle. Optimal conditions during vegetative growth might differ drastically from those during flowering or grain filling.
- **Genotype-Environment Interaction:** Different crop varieties (cultivars) have varying responses to the same weather conditions. A heat-tolerant maize cultivar will have different optimums than a more sensitive one.
- **Trade-offs:** Optimizing one weather factor (e.g., maximum temperature for rapid growth) might come at the expense of another (e.g., increased water demand).
- **Stochastic Nature of Weather:** Weather is inherently variable and unpredictable, making the ideal rarely perfectly realized. The goal is often to find conditions that optimize *long-term average* performance and minimize risk.

5.10.2 Methodologies for Determining Optimum Weather Conditions

Determining optimum weather conditions requires a multi-pronged approach, drawing upon the methods discussed in previous sections and combining them for deeper insights:

- **Analysis of Long-Term Historical Data (Empirical & Statistical Basis):**

The most fundamental step involves a thorough analysis of decades of historical weather and crop performance data for the region of interest.

Yield-Weather Regression

Process: Collect high-quality, spatially representative historical yield data (e.g., for maize, wheat, rice) and corresponding daily or weekly weather data (temperature, rainfall, solar radiation, humidity) for the same period.

Analysis: Employ statistical regression models (simple, multiple, or non-linear) to identify which weather variables, and during which specific crop growth stages, explain the most variation in historical yields. This might involve:

Identifying the optimal temperature range for grain filling by finding the range that correlates with highest yields.

Determining critical rainfall amounts or drought periods during specific months (e.g., flowering, critical for grain number).

Assessing the impact of extreme events like heatwaves (e.g., number of days above 35□C) on yield reduction.

Output: Provides quantitative relationships (e.g., Yield increases by X kg/ha for every 10mm of rain in July, up to Y mm) and identifies periods of high sensitivity.

Growing Season Characterization

Process: Analyze long-term temperature data to determine the average length of the frost-free period and the accumulated Growing Degree Days (GDD) during this period.

Analysis: Compare these climatic characteristics with the known GDD requirements and frost sensitivities of different crops and cultivars.

Output: Helps in identifying if a region has sufficient heat units and a long enough growing season for a particular crop to mature reliably.

Rainfall Reliability Analysis

Process: Analyze historical daily or weekly rainfall data for onset, cessation, duration, and probability of dry/wet spells during the growing season.

Analysis: Determine the likelihood of receiving adequate rainfall during critical crop developmental stages, helping to assess rainfed suitability and irrigation needs.

Output: Maps of water availability reliability and risk of drought.

- **Controlled Environment and Field Experiments**

While expensive, these provide fundamental data for understanding specific crop responses to isolated weather factors, which then informs modeling and regional assessment.

Cardinal Temperature Determination

Process: Using growth chambers or controlled greenhouses, grow a specific crop cultivar under a range of constant or varying temperatures while keeping other factors optimal.

Analysis: Precisely measure rates of germination, leaf expansion, photosynthesis, and development to determine the Tbase , Topt , and Tmax for these processes.

Output: Provides physiological parameters essential for mechanistic models and for identifying the ideal temperature ranges for different crop types.

Water Stress Response Curves:

Process: Grow crops under varying levels of induced water stress (e.g., different soil moisture tensions) in controlled environments or using rain-out shelters in the field.

Analysis: Measure the impact on stomatal conductance, photosynthetic rate, biomass accumulation, and yield.

Output: Quantifies the yield reduction and physiological responses to different levels and timings of water deficit, helping to define optimal soil moisture conditions.

CO_2 Enrichment Studies (FACE experiments)

Process: Conduct large-scale FACE experiments in the field to expose crops to elevated CO_2 levels under natural weather conditions.

Analysis: Measure the enhancement of photosynthesis, water use efficiency, and yield.

Output: Provides empirical data on the CO_2 fertilization effect, which is crucial for determining optimal CO_2 benefits in a regional context.

5.10.3 Process-Based Crop Models (Analytical Methods)

Crop simulation models are perhaps the most powerful tools for defining optimum weather conditions for a given region due to their ability to integrate complex interactions and simulate hypothetical scenarios.

Calibration and Validation

Process: First, calibrate a generic crop model to accurately represent the specific crop cultivar‘s growth and development under local soil and management conditions, using observed field data from the region.

Validation: Test the model's accuracy against independent datasets from the same region to ensure it reliably predicts yield and phenology under historical weather variability.

Scenario Analysis and Optimization

Process: Once validated, run the model repeatedly with diverse combinations of weather parameters, representing a range of possible conditions (e.g., varying planting dates, temperature shifts, rainfall amounts/patterns). This creates a response surface showing yield outcomes across the weather spectrum.

Analysis: Identify the weather conditions (e.g., mean daily temperature, total radiation, rainfall distribution) that consistently result in maximum simulated yields for the given crop and soil. This can involve:

Optimal Planting Window Determination: Simulating yields for different planting dates over many years of historical weather to find the date that optimizes average yield and minimizes risk, considering frost, heat stress, and water availability.

Ideal Water Regimes: Simulating yields under different irrigation schedules or rainfall scenarios to determine optimal soil moisture levels throughout the season.

Heat/Cold Stress Thresholds: Identifying the frequency or intensity of high/low temperatures that significantly reduce simulated yield or accelerate maturity beyond optimal.

Output: Precise, dynamic definitions of optimal weather profiles (e.g., For this maize cultivar on this soil type, the optimum conditions are X GDD accumulated by Y date, with Z mm rainfall in the reproductive phase, and fewer than N days above 35^{0}C.).

5.10.4 Agro-Climatic Zonation

This is a macro-level application that builds on the understanding of optimal conditions.

Process: Divide a large geographical area into distinct zones based on similar climatic conditions (e.g., temperature ranges, rainfall patterns, length of growing season) that are relevant to agriculture. This involves overlaying climatic data with geographical information.

Analysis: For each zone, determine its suitability for various crops by comparing the zone's climatic characteristics with the known optimal requirements of different crops and their cultivars (derived from methods 1, 2, and 3).

Output: Maps delineating regions as highly suitable,moderately suitable, or marginally suitable for specific crops, aiding in regional agricultural planning and policy.

- **Practical Implications of Determining Optimum Conditions**

Crop Selection: Guiding farmers and policymakers on which crops (and cultivars) are best suited for a particular region's climate.

Optimized Planting Dates: Advising on the ideal time to plant to maximize the chances of favorable weather during critical growth stages.

Resource Allocation: Informing irrigation scheduling, fertilizer timing, and other management decisions to align with the plant's needs as dictated by the weather.

Risk Management: Identifying periods or conditions of high risk (e.g., heat stress during flowering) to implement mitigation strategies.

Climate Change Adaptation: Guiding the development of new crop varieties or altered farming systems that are better adapted to shifting optimal conditions in a changing climate.

In essence, determining optimal weather conditions for crops in a given region is a dynamic, iterative process that integrates historical observation, controlled experimentation, and advanced simulation modeling. It shifts agriculture from a reactive to a proactive paradigm, enabling more efficient resource use, higher yields, and greater resilience in the face of climatic variability.

5.11 Climate Variability and Change: Implications for Crop Evaluation

Understanding the dynamics of the Earth's atmosphere is paramount in agriculture, yet it's crucial to distinguish between two related but distinct concepts: climate variability and climate change. Both exert profound influences on crop growth, development, and yield, necessitating specific approaches to their evaluation and adaptation strategies in agricultural systems. While climate variability refers to short-term, natural fluctuations, climate change signifies long-term, directional shifts in climatic patterns. Both present unique challenges for crop production and require sophisticated evaluation methods.

5.11.1 Climate Variability

Definition: Climate variability refers to the natural, short-term (from inter-seasonal to inter-annual or multi-decadal) fluctuations in climatic parameters around their long-term average. It encompasses the normal, expected deviations from average conditions that characterize a region's climate. These variations are driven by natural processes within the climate system, such as oceanic oscillations (e.g., El Niño-Southern Oscillation (ENSO), Indian Ocean Dipole (IOD), North Atlantic Oscillation (NAO)).

- **Characteristics**

Fluctuations around a mean: The overall long-term average climate remains relatively stable, but individual years or seasons can be warmer or cooler, wetter or drier than average.

Predictable cycles (to some extent): Some modes of climate variability (like ENSO) have a semi-regular periodicity, allowing for seasonal forecasts that can inform agricultural planning.

Recurrent events: Includes phenomena like droughts, floods, heatwaves, or cold snaps that occur with varying frequency and intensity from year to year as part of natural climatic patterns.

- **Effects on Crops:** Climate variability is the primary source of inter-annual yield fluctuations in most agricultural regions. Crops are well-adapted to the *average* conditions of their native climate, but deviations can induce stress.

Yield Volatility: Years with unusually high or low rainfall, or extreme temperatures (hotter summers, colder winters) directly translate to variations in crop yields. For instance, a strong El Niño year might bring drought to some regions, severely impacting rainfed crops, while bringing beneficial rains to others.

Phenological Shifts (within a season): A warmer-than-average spring might accelerate crop development, leading to earlier flowering and maturity, potentially exposing the crop to late-season stresses. Conversely, a colder spring delays development.

Increased Vulnerability to Pests and Diseases: Unusual weather patterns can favor the proliferation or spread of certain pests and diseases, which can then negatively impact crop health and yield.

Water Availability Fluctuations: Variable rainfall directly affects soil moisture, influencing irrigation demand and the success of rainfed agriculture.

- **Evaluation of Effects:** Evaluating the impacts of climate variability heavily relies on historical data analysis and empirical methods (as discussed in previous section) as well as short-term forecasting.

Historical Yield-Weather Regressions: Statistical models identify correlations between past yield fluctuations and variations in temperature, rainfall, or other weather indices. This helps quantify the impact of typical climatic deviations.

Probabilistic Risk Assessment: Analyzing historical frequency and intensity of extreme events (droughts, floods, heatwaves) to quantify the probability of their occurrence and the associated yield losses for specific crops.

Seasonal Forecasting: Using climate models that predict large-scale climate patterns (e.g., ENSO states) to anticipate regional weather anomalies (e.g., warmer/drier, cooler/wetter seasons) and adjust planting decisions or management practices accordingly.

Field Data Analysis: Long-term field trials provide empirical evidence of how local crop varieties respond to the natural year-to-year variation in weather conditions.

5.11.2 Climate Change

- **Definition:** Climate change refers to a **long-term, fundamental, and persistent shift in the mean state or the variability of climate** over decades to centuries. While natural factors contribute, the current climate change is largely attributed to anthropogenic activities, primarily the emission of greenhouse gases (GHG) into the atmosphere, leading to global warming.

- **Characteristics**

Shift in the mean: The long-term average temperature is increasing, and average precipitation patterns are shifting globally (some regions becoming wetter, others drier).

Altered Extremes: Not just a shift in the mean, but also changes in the frequency, intensity, duration, and spatial extent of extreme weather events (e.g., more frequent and intense heatwaves, heavier downpours, prolonged droughts).

Irreversible trends (on human timescales): Many of the changes are projected to be long-lasting and, in some cases, effectively irreversible.

Global phenomenon: While impacts are regionally specific, the underlying cause is global.

- **Effects on Crops:** Climate change introduces a new baseline of environmental conditions to which current agricultural systems may not be fully adapted.

Overall Warming and Phenological Acceleration: Rising average temperatures often lead to faster accumulation of Growing Degree Days (GDD), resulting in accelerated crop development, shorter growing seasons, and earlier maturity. While seemingly beneficial in some colder regions (longer growing season), in already warm regions, this can mean crops mature before reaching their full yield potential due to reduced time for biomass accumulation (e.g., grain filling).

Altered Water Availability: Changes in global precipitation patterns mean some regions face increased drought frequency and severity, while others experience more intense rainfall and associated flooding. This directly impacts water-limited crop production.

Increased Heat Stress: More frequent and intense heatwaves, particularly during critical developmental stages (e.g., flowering, grain filling), can cause pollen sterility, reduced grain number/size, and significant yield losses, often exceeding the benefits of CO_2 fertilization.

CO_2 Fertilization Effect: Increased atmospheric CO_2 can enhance photosynthesis and water use efficiency, particularly for C3 crops (like wheat, rice, soybeans). However, this benefit can be offset by concurrent heat and water stress, and can also lead to reduced nutritional quality (e.g., lower protein and mineral content).

Shifting Agro-Climatic Zones: The geographical areas suitable for specific crops may shift, necessitating changes in regional cropping patterns.

Enhanced Pest and Disease Pressure: Warmer winters allow pests and diseases to overwinter more successfully or expand their geographical range, increasing their pressure on crops.

Mismatches in Biological Interactions: Changes in the timing of flowering due to warming might lead to phenological mismatches with pollinators or pests, disrupting ecosystem services or increasing damage.

- **Evaluation of Effects:** Evaluating climate change impacts requires **forward-looking, sophisticated modeling approaches**, as historical data alone may not capture the magnitude or novelty of future conditions.

Process-Based Crop Models (Analytical Methods): These are indispensable tools. They are driven by projected future weather data (downscaled from Global Climate Models (GCMs) under various GHG emission scenarios) to simulate crop growth and yield. They can capture the physiological responses to novel combinations of temperature, CO_2 , and water availability.

Scenario Analysis: Running crop models with different climate change scenarios allows for projecting future yields, identifying vulnerable regions, and testing adaptation strategies (e.g., shifting planting dates, introducing heat-tolerant cultivars, optimizing irrigation).

Impact Assessments: Combining crop model outputs with economic and social models to assess the broader implications of climate change for food security, farm income, and regional economies.

Adaptation Strategy Evaluation: Using models to evaluate the effectiveness of various adaptation measures (e.g., cultivar shifts, improved water management, changes in cropping systems) in mitigating negative climate change impacts or harnessing potential benefits.

5.11.3 Interplay Between Climate Variability and Climate Change

It's crucial to understand that climate change does not replace climate variability; rather, it alters the baseline around which variability occurs, and can also change the nature and frequency of extreme events. For example, a region that historically experienced droughts due to natural variability might now experience more frequent, intense, and prolonged droughts due to climate change. Evaluating these combined effects is paramount, as it's often the altered extremes, superimposed on a warming trend, that pose the greatest risk to agriculture.

In conclusion, both climate variability and climate change present significant challenges for crop production. Climate variability drives inter-annual fluctuations and requires robust risk management based on historical patterns and seasonal forecasts. Climate change, however, represents a fundamental and persistent shift, demanding long-term adaptation strategies guided by the predictive power of mechanistic crop models and forward-looking impact assessments. Effective crop evaluation in the face of these atmospheric dynamics is essential for building resilient and productive agricultural systems for the future.

5.12 Weather Response Traits in Crop Breeding Programs

The success of any crop cultivar hinges on its ability to perform optimally within its prevailing environment, a significant part of which is dictated by weather conditions. Consequently, modern crop breeding programs are increasingly prioritizing the development of cultivars endowed with superior **weather response traits**. These traits encompass a plant's inherent genetic capacity to respond favorably, or at least tolerate, various meteorological factors such as temperature, water availability, solar radiation, and even wind, throughout its life cycle. The ultimate goal is to breed crops that exhibit enhanced resilience, stability, and productivity under diverse and often challenging climatic conditions, a necessity amplified by the escalating realities of climate variability and change.

5.12.1 The Imperative of Weather Response Traits

The importance of focusing on weather response traits in crop breeding stems from several critical factors:

- **Maximizing Yield Potential:** Even with ideal soil and nutrient management, adverse weather can severely limit a crop‘s ability to express its full genetic yield potential. Breeding for favorable weather responses ensures that the plant can efficiently convert environmental resources into biomass and economic yield.
- **Ensuring Yield Stability:** Climate variability introduces significant year-to-year fluctuations in weather. Cultivars with robust weather response traits can buffer against these variations, providing more stable yields and reducing the risk of catastrophic crop failures for farmers.
- **Adapting to Climate Change:** Long-term shifts in mean temperatures, altered precipitation patterns, and increased frequency and intensity of extreme weather events (heatwaves, droughts, floods) necessitate crops that can specifically cope with these new climatic realities. Breeding for climate resilience is no longer an option but a strategic imperative for global food security.
- **Optimizing Resource Use:** Traits that enhance a crop‘s efficient use of water, nutrients, and solar radiation under varying weather conditions contribute to sustainable agriculture, reducing input requirements and environmental footprints.
- **Expanding Cultivation Zones:** By improving tolerance to environmental stresses, breeding programs can potentially expand the geographical areas where certain crops can be successfully grown, contributing to food production in marginal lands.

5.12.2 Key Weather Response Traits Targeted in Breeding Programs

Crop breeding programs focus on a diverse array of traits, each contributing to a cultivar‘s overall resilience and performance under specific weather conditions:

- **Temperature Response Traits**

Heat Tolerance: This encompasses a plant's ability to maintain physiological function (e.g., photosynthesis, pollen viability) and yield under high temperatures. Traits include efficient stomatal regulation for evaporative cooling, production of heat shock proteins to protect cellular components, and maintaining membrane integrity. Breeding programs aim for cultivars that can set seed and fill grains effectively even when temperatures exceed optimal thresholds during critical reproductive phases.

Cold Tolerance/Frost Resistance: The ability to withstand low temperatures, chilling injury (damage above freezing but below optimal), or freezing temperatures. This involves traits like membrane fluidity at low temperatures, accumulation of cryoprotectants (e.g., sugars, proteins), and rapid recovery from cold stress. This is crucial for expanding cultivation into cooler regions or mitigating risks from early/late frosts.

Optimal Thermal Time Requirements: Breeding for cultivars with phenological development (e.g., flowering, maturity) that aligns optimally with the thermal regime of a specific region, avoiding sensitive stages during likely periods of extreme heat or cold. This involves careful selection for specific Growing Degree Day (GDD) requirements for each developmental phase.

- **Water Response Traits (Drought and Waterlogging Tolerance)**

Drought Tolerance: This is a complex trait involving multiple mechanisms that allow plants to maintain growth and yield under water deficit. Traits include:

Water Use Efficiency (WUE): The amount of biomass produced per unit of water transpired.

Early Maturity (Drought Escape): Completing the life cycle before severe drought sets in.

Deep Rooting: Accessing deeper soil moisture.

Stomatal Control: Efficiently closing stomata to conserve water.

Osmotic Adjustment: Accumulating solutes to maintain cell turgor.

Maintaining Photosynthesis under Stress: Sustaining carbon assimilation even with reduced water availability.

Waterlogging Tolerance: The ability to withstand excessive soil moisture and anaerobic conditions in the root zone. Traits include aerenchyma formation (air channels in roots), adventitious root development, and efficient oxygen transport to roots. This is vital for flood-prone areas or regions with heavy, poorly drained soils.

- **Light Response Traits**

Photoperiod Sensitivity: Breeding cultivars with appropriate sensitivity to day length to ensure flowering and maturity occur at the optimal time for a given latitude and planting season. This is crucial for adapting crops like soybean and maize to different geographical zones.

Radiation Use Efficiency (RUE): Maximizing the efficiency with which intercepted solar radiation is converted into dry matter. Breeders select for canopy architecture (leaf angle, leaf size, leaf distribution) that optimizes light interception and for photosynthetic machinery that converts light energy into biomass efficiently.

Shade Tolerance: Important for intercropping systems or environments with frequently overcast conditions, enabling better performance under reduced light.

- **Wind Response Traits**

Lodging Resistance: Breeding for stronger stems and shorter plant stature to prevent crops from falling over (lodging) due to strong winds, especially in cereals with heavy grain heads.

Reduced Water Loss: Traits that minimize excessive transpiration under windy conditions.

5.12.3 Incorporating Weather Response Traits into Breeding Programs

The integration of these traits involves a multi-pronged strategy:

Phenotyping (Measuring Traits): This is the precise and often high-throughput measurement of a plant's observable characteristics (phenotype) in response to different weather conditions.

Field Screening: Exposing large populations of breeding lines to natural drought, heat stress, or cold events in target environments and selecting superior performers. This requires multi-location and multi-year trials.

Controlled Environment Studies: Using growth chambers, greenhouses, or rain-out shelters to precisely control and manipulate specific weather parameters, allowing for detailed physiological phenotyping (e.g., measuring stomatal conductance under drought, photosynthetic rates at high temperatures).

Remote and Proximal Sensing: Employing drones, satellites, and ground-based sensors (e.g., thermal cameras, multispectral cameras) to rapidly assess large numbers of plants for stress indicators (e.g., canopy temperature, vegetation indices) and developmental timing across vast breeding nurseries.

- **Genotyping (Identifying Genetic Markers):** Advanced molecular techniques are used to identify specific genes or DNA markers associated with desirable weather response traits.

Quantitative Trait Loci (QTL) Mapping: Locating regions on the genome that influence complex traits like drought tolerance or heat tolerance.

Genome-Wide Association Studies (GWAS): Identifying genetic variants (SNPs) across the entire genome that are statistically associated with trait variation in diverse germplasm.

Genomic Selection: Using genome-wide marker data to predict the breeding value of individuals more accurately and efficiently, accelerating the selection process, especially for complex traits influenced by many genes.

- **Selection and Hybridization**

Traditional Breeding: Cross-breeding parents with desirable weather response traits, followed by repeated selection over generations in target environments.

Marker-Assisted Selection (MAS): Using DNA markers identified through genotyping to select for desired traits early in the breeding process, even before phenotypic expression, significantly speeding up breeding cycles. For example, selecting for a cold-tolerance gene in seedlings rather than waiting for them to grow and be exposed to cold.

Speed Breeding: Accelerating plant development cycles (e.g., through extended photoperiods and controlled temperatures in greenhouses) to achieve multiple generations per year, shortening the time needed to develop new varieties.

- **Germplasm Evaluation and Utilization**

Screening vast collections of genetic resources (landraces, wild relatives, existing cultivars) for novel weather response traits. These sources often hold valuable genetic diversity for adaptation that can be introduced into elite breeding lines.

5.12.4 Conclusion: Breeding for a Resilient Future

The increasing volatility of weather patterns and the undeniable trajectory of climate change have transformed weather response traits from desirable characteristics into indispensable attributes for future crop varieties. Breeding programs worldwide are shifting their focus from merely maximizing yield under ideal conditions to enhancing the resilience and stability of crops across a broader range of environmental challenges. By integrating sophisticated phenotyping and genotyping technologies with traditional breeding strategies, plant breeders are working diligently to develop climate-smart crops—cultivars equipped with the genetic fortitude to thrive amidst heatwaves, resist drought, tolerate floods, and synchronize their life cycles with shifting seasons. This ongoing effort in developing robust weather response traits is a cornerstone of ensuring sustainable agricultural productivity and global food security in an era of unprecedented environmental change.

5.13 Weather Effects on Crop Yield Components

Crop yield, the ultimate output of an agricultural system, is not a single, monolithic entity but rather a complex product determined by several interacting yield components. These components are established at different, often sensitive, stages throughout the crop‘s life cycle. Consequently, weather conditions experienced during these specific developmental windows can have a profound and disproportionate impact on the final yield, even if overall growing season conditions appear favorable. Understanding how various weather parameters affect these individual components is crucial for diagnosing yield limitations, optimizing management practices, and developing resilient crop varieties.

For most grain crops, yield (e.g., in kg/ha) can be conceptualized as the product of:

Yield=Plant Population×Number of Reproductive Organs per Plant×Number of Grains/Seeds per Reproductive Organ×Individual Grain/Seed Weight

Let‘s break down how different weather parameters influence each of these critical components.

5.13.1 Plant Population (Stand Density)

This refers to the number of viable plants established per unit area (e.g., plants/m^2). It is primarily determined during the **germination and emergence phases**.

Rainfall/Soil Moisture

Deficit (Drought): Insufficient soil moisture at planting or during early seedling growth can lead to poor or uneven germination, reduced seedling emergence rates, and increased seedling mortality. Seeds may fail to imbibe enough water to germinate, or young seedlings may die from desiccation. This directly reduces the final plant population.

Excess (Waterlogging): Excessive rainfall or irrigation leading to saturated soil conditions (waterlogging) can create anaerobic environments (lack of oxygen) around the germinating seeds and young roots. This inhibits respiration, promotes the accumulation of toxic compounds, and can lead to seed rot or seedling death, again reducing stand density.

Temperature

Low Temperatures: Temperatures below the crop‘s base temperature for germination significantly delay or prevent emergence. Extreme cold or frost events after planting but before emergence can kill germinating seeds or newly emerged seedlings.

High Temperatures: Excessively high soil temperatures can inhibit germination, especially if coupled with low moisture, by damaging enzymes or accelerating seed dormancy.

5.13.2 Number of Reproductive Organs per Plant

This component refers to the count of yield-bearing structures on each plant (e.g., number of tillers or ears per plant in cereals, number of pods per plant in legumes, number of fruits per plant in fruit crops). This is primarily determined during the early vegetative to early reproductive transition phases.

Solar Radiation (Light)

Deficit: Low light intensity, particularly during the early vegetative phase (e.g., due to dense planting, cloudiness), can reduce tillering in cereals (fewer productive stems) or branching in other crops. Insufficient light limits carbon assimilation, providing less energy for the initiation and development of these structures.

Optimal: Adequate solar radiation ensures sufficient energy for vigorous vegetative growth and the successful initiation of a high number of reproductive sites.

Temperature

Low Temperatures: Can delay or reduce the initiation of tillers, branches, or flower primordia, thereby limiting the potential number of reproductive organs.

High Temperatures: Can accelerate development, shortening the duration of tillering or branching phases, leading to fewer initiated reproductive structures. Extreme heat can also cause abortion of early flower primordia.

Rainfall/Soil Moisture

Deficit (Drought): Water stress during tillering or branch development can significantly reduce the number of productive tillers/branches that form, as the plant prioritizes survival over vegetative expansion. It can also cause abortion of early flower buds.

Excess (Waterlogging): Can lead to stunted growth and reduced tiller/branch formation due to root damage and oxygen deprivation.

5.13.3 Number of Grains/Seeds per Reproductive Organ

This component refers to the quantity of individual grains, seeds, or fruits within each reproductive structure (e.g., kernels per ear of maize, grains per spikelet in wheat, seeds per pod in soybean). This component is critically determined during the flowering/anthesis and early grain set stages. This is often considered the most sensitive stage to environmental stresses.

Temperature

High Temperatures (Heat Stress): This is one of the most critical weather impacts. High temperatures, even for short durations (a few days), around flowering/anthesis can:

Reduce pollen viability (pollen becomes sterile).

Impair stigma receptivity (the female part cannot receive pollen).

Decrease pollen tube growth rate.

Lead to embryo abortion after fertilization.

All these factors result in reduced fertilization success and fewer grains/seeds setting, significantly reducing this yield component.

Low Temperatures (Cold Stress): Can also impair pollen development, fertilization, and initial seed set, especially in crops adapted to warmer climates.

Rainfall/Soil Moisture

Deficit (Drought): Water stress during flowering can cause severe yield losses due to:

Asynchronous pollen shed and silk emergence (e.g., in maize, silks emerge before pollen is viable).

Reduced pollen viability and stigma receptivity, similar to heat stress.

Abortion of developing florets or young seeds.

Excess (Waterlogging): Can also lead to poor flower development and seed set due to general plant stress and compromised root function.

Solar Radiation

Deficit: Low solar radiation during spikelet differentiation or flowering can limit the carbohydrates available for the development of potential grain sites, reducing the sink capacity.

5.13.4 Individual Grain/Seed Weight (Size/Quality)

This component refers to the average dry weight of a single grain or seed. It is primarily determined during the **grain (or fruit/tuber) filling period**, which follows successful fertilization and seed set.

Solar Radiation (Light)

Optimal: High solar radiation during the grain filling period is crucial for maximizing photosynthesis and, consequently, the supply of carbohydrates (assimilates) for deposition into the developing grains. This leads to heavier, well-filled grains.

Deficit: Prolonged cloudy weather or low light conditions during grain filling can significantly reduce photosynthesis, limiting assimilate supply and resulting in smaller, shriveled grains with reduced individual weight.

Temperature

High Temperatures: Heat stress during grain filling often shortens the duration of the filling period (forcing early maturity) and/or reduces the rate of assimilate accumulation, leading to lighter grains. The plant rushes to complete its life cycle, sacrificing grain quality and size. High night temperatures are particularly detrimental as they increase respiration, consuming carbohydrates that would otherwise go to grain filling.

Optimal/Cool Temperatures: Moderately cool temperatures (within the optimal range for grain filling) often allow for a longer duration of grain filling, contributing to heavier grains.

Rainfall/Soil Moisture

Deficit (Drought): Water stress during grain filling reduces photosynthesis, impairs the transport of carbohydrates to the grains, and can cause premature senescence of leaves (reducing the source of assimilates). This results in poorly filled, lighter grains.

Excess (Waterlogging): Can cause root damage and general plant stress, indirectly reducing assimilate supply to the grains.

Vapor Pressure Deficit (VPD): High VPD can increase transpirational demand to the point where the plant closes stomata to conserve water, even if soil water is available. This reduces photosynthesis, limiting assimilate supply to the grains and thus reducing grain weight.

5.13.5 The Interplay: Source and Sink Relationships

The impact of weather on yield components can be understood through the source-sink relationship:

Source: Primarily the leaves, which produce carbohydrates through photosynthesis.

Sink: The developing reproductive organs (grains, fruits, tubers) that accumulate these carbohydrates.

Yield is maximized when there is a strong and sustained supply (source) of assimilates from the leaves and a strong demand (sink) from a sufficient number of developing grains that can effectively import and store these assimilates. Weather conditions impact both aspects:

Early season weather often determines the sink size (plant population, number of reproductive organs, potential number of grains per organ).

Late season weather (especially during grain filling) primarily determines the filling of that sink (individual grain weight).

Adverse weather at any stage can disrupt this balance, leading to reductions in one or more components, and ultimately, a lower final yield. This complex interplay underscores the challenge and importance of understanding crop-weather interactions for effective agricultural management and breeding for future resilience.

5.14 Effect of Extreme Weather Events on Crops: A Comprehensive Analysis

Extreme weather events, defined as meteorological phenomena that deviate significantly from historical averages in terms of intensity, duration, or timing, are becoming increasingly common and severe. These events, which include prolonged droughts, intense heatwaves, devastating floods, damaging hailstorms, unseasonal frosts, and powerful winds, represent a profound and escalating threat to global agriculture. While crops possess a degree of inherent resilience to natural climatic variability, the heightened frequency and intensity of these extremes, often exacerbated by ongoing climate change, push physiological limits, leading to substantial yield reductions, compromised quality, and amplified economic risks for farming communities worldwide.

The precise impact of an extreme weather event on a crop is highly conditional, depending critically on the **nature and intensity of the event**, its **duration**, and crucially, the crop's specific developmental stage at the time of occurrence. Different phases of a crop's life cycle exhibit varying sensitivities to distinct stresses, leading to diverse and often devastating consequences for the final yield and its constituent components.

• Specific Impacts of Key Extreme Weather Events on Crops

5.14.1 Heatwaves (Extreme High Temperatures)

Heatwaves are characterized by periods where ambient temperatures surge significantly above a crop's optimal range for an extended duration.

Physiological Impacts: Heat stress drastically curtails photosynthesis as vital enzymes, like RuBisCO, become denatured, and chloroplasts (the sites of photosynthesis) are damaged. Simultaneously, plant respiration rates surge, consuming more carbohydrates for basic maintenance than are produced, leading to a net depletion of stored energy and even biomass loss. The high temperatures also accelerate transpirational water loss from leaves; if water uptake from the soil cannot keep pace, severe internal water stress ensues, often forcing stomatal closure and further inhibiting carbon dioxide uptake. Cellular membranes can also be damaged, impairing their function and leading to cell death.

Impacts on Yield Components

Reproductive Stage (Most Critical): This is the most sensitive period. Heatwaves around flowering (anthesis) can cause:

Pollen Sterility: Pollen grains become inviable or die.

Stigma Receptivity Impairment: The female floral parts lose their ability to receive pollen effectively.

Ovule Abortion: Developing ovules (potential seeds) abort.

These combined effects lead to a severe reduction in the number of grains/seeds per reproductive organ, manifesting as blasted maize ears or empty rice spikelets, significantly reducing final yield.

Grain/Fruit Filling Stage: Heat stress during this crucial phase accelerates the plant's physiological maturity, drastically shortening the duration available for grain filling. This results in reduced individual grain/seed weight, leading to shriveled, lighter grains as carbohydrate accumulation is curtailed. High night temperatures are particularly damaging, as they continue to drive high respiration rates without concurrent photosynthesis, further depleting carbohydrate reserves meant for the developing yield.

5.14.2 Droughts (Extreme Low Water Availability)

Droughts are characterized by prolonged periods of abnormally low precipitation, leading to severe and sustained deficits of soil moisture.

Physiological Impacts: Plants primarily respond to drought by closing their stomata to conserve water, which inevitably limits CO_2 uptake for photosynthesis. This immediate physiological adjustment drastically reduces carbon assimilation. Water deficit also inhibits cell expansion, leading to smaller leaves and reduced Leaf Area Index (LAI), thereby shrinking the plant's overall photosynthetic factory. Nutrient uptake is impaired as water is the medium for their transport in the soil and within the plant. Prolonged and severe drought stress can induce oxidative stress, leading to cellular damage.

Impacts on Yield Components

Germination & Emergence: Insufficient soil moisture at planting or during early seedling growth causes poor or uneven germination, reduced seedling emergence rates, and increased seedling mortality, directly impacting the plant population.

Vegetative Stage: Drought significantly reduces tillering in cereals, branching in other crops, and overall biomass accumulation, thereby limiting the potential number of reproductive organs per plant.

Flowering/Reproductive Stage (Highly Sensitive): Water stress during flowering often leads to:

Asynchronous Development: A mismatch between pollen shedding and stigma receptivity (e.g., maize silk delays).

Reduced Pollen Viability and Stigma Receptivity: Similar to heat stress.

Abortion of Developing Florets/Young Seeds: Leading to a significant reduction in the number of grains/seeds per reproductive organ.

Grain/Fruit Filling Stage: Water stress during this period curtails current photosynthesis, impairs the transport of carbohydrates to the developing grains, and can cause premature senescence of leaves, resulting in poorly filled, lighter grains with reduced **individual grain/seed weight**.

5.14.3 Floods and Waterlogging (Extreme High Water Availability)

Flooding occurs when agricultural land is submerged by water, while waterlogging refers to the saturation of soil pores, displacing air and creating anaerobic conditions.

Physiological Impacts: The primary physiological impact is oxygen deprivation (anaerobiosis/hypoxia) for plant roots, which require oxygen for respiration, crucial for water and nutrient uptake. Prolonged lack of oxygen leads to root cell damage and death, severely impairing the roots‘ ability to function and increasing susceptibility to root diseases. Under anaerobic conditions, soil microorganisms produce phytotoxic (plant-toxic) compounds that further harm roots and inhibit growth. Waterlogging also accelerates the loss of mobile nutrients like nitrogen through processes like denitrification and leaching.

Impacts on Yield Components

Germination & Emergence: Can cause seed rot and seedling death due to oxygen starvation, significantly reducing the plant population.

Vegetative Stage: Leads to stunted growth, widespread yellowing (chlorosis) from nutrient deficiencies (especially nitrogen), and reduced tillering/ branching, thereby limiting overall biomass and the potential number of reproductive organs.

Reproductive Stage: Overall plant stress from waterlogging can result in flower abortion and poor grain set, impacting the number of grains/seeds.

5.14.4 Cold Snaps and Frost

Cold snaps are sudden, sharp drops in temperature, while frost occurs when temperatures fall to or below the freezing point (0^0C).

Physiological Impacts

Chilling Injury (Above Freezing): For sensitive tropical/subtropical crops (e.g., maize, rice seedlings), temperatures above 0^0C but below their optimal range (e.g., 10–12^0C) can cause physiological damage, including disruption of cell membranes and inactivation of enzymes. This manifests as stunted growth, yellowing, and increased vulnerability to diseases.

Freezing Injury (Below Freezing): Ice crystals form within plant tissues, physically rupturing cell membranes and organelles, leading to irreversible tissue damage and, in severe cases, plant death.

Impacts on Yield Components

Germination & Emergence: Early season frosts or prolonged cold can kill newly planted seeds or emerged seedlings, directly reducing the plant population.

Vegetative Stage: Results in significantly slowed growth, stunted development, and reduced tillering/branching.

Reproductive Stage (Extremely Sensitive): Late spring frosts are particularly devastating to developing flower buds, open flowers, and newly formed fruits. This can cause the complete loss of **number of grains/seeds** or **number of fruits**, as reproductive tissues are often highly susceptible to freezing damage. For example, frost during wheat heading or maize silking can lead to significant sterility.

Grain/Fruit Filling Stage: Early autumn frosts can prematurely terminate the grain/fruit filling period, resulting in significantly reduced individual grain/ fruit weight and poorer quality, as the plant does not have enough time to fully mature and fill its yield.

5.14.5 Severe Storms (Including Strong Winds and Intense Rainfall)

Severe storms encompass a destructive combination of high wind speeds, torrential rainfall, and sometimes lightning.

Physical Impacts

Lodging: High winds, especially when accompanied by heavy rainfall, can cause cereal crops (e.g., wheat, maize, rice) to fall over or become flattened. Lodging makes mechanical harvesting extremely difficult or impossible, leading to significant pre-harvest yield losses.

Defoliation: Strong winds can strip leaves from plants, directly reducing the photosynthetic area, thereby limiting the plant's capacity for biomass accumulation and subsequent yield.

Stem Breakage: Brittle stems can snap, resulting in the partial or complete loss of individual plants.

Root Damage/Uprooting: Powerful winds can loosen soil around the roots or even uproot entire plants, particularly shallow-rooted crops or trees.

Pollination Disruption: Intense rainfall and strong winds during the flowering period can physically wash away pollen or severely hinder pollinator activity, leading to reduced fertilization success and ultimately fewer grains/seeds per reproductive organ.

Soil Erosion: Torrential rainfall associated with severe storms can cause significant soil erosion, washing away nutrient-rich topsoil, depleting soil fertility, and leading to long-term productivity decline in affected areas.

5.14.6 Hailstorms

Hailstorms involve precipitation in the form of solid ice pellets, which cause direct, blunt-force physical damage to crops.

Physical Impacts

Leaf Shredding/Tearing: Hailstones can extensively shred, bruise, and tear leaves, drastically reducing the photosynthetic surface area available for carbon assimilation.

Stem Bruising/Breaking: Stems can be bruised or broken, disrupting the vascular system and impairing the transport of water and nutrients throughout the plant.

Fruit/Vegetable Bruising/Pitting: Directly damage the harvestable product, reducing its quality, marketability, and storage life, even if the plant itself survives.

Increased Disease Susceptibility: The physical wounds inflicted by hail provide easy entry points for various fungal, bacterial, and viral pathogens, increasing the risk of secondary infections that further compromise plant health and yield.

Impacts on Yield Components: Hailstorms directly reduce total biomass and potential yield by destroying photosynthetic organs and physically damaging developing reproductive structures. They can cause direct loss of reproductive organs or reduce individual grain/fruit weight through physical impact and subsequent stress.

5.14.7 Compound Extreme Events: The Amplified Threat

A rapidly growing concern in agriculture is the increasing frequency and severity of compound extreme events. These are scenarios where multiple extreme weather conditions occur simultaneously, in close succession, or in combinations that amplify their individual destructive potential. The combined impact is often far more severe than the sum of their parts.

Example: Heatwave and Drought: A heatwave occurring concurrently with a drought significantly exacerbates water stress by increasing atmospheric evaporative demand to extreme levels. This leads to even more severe impacts on photosynthesis, plant development, and yield components than either heat or drought experienced in isolation.

Example: Flood followed by Heatwave: A crop that has survived waterlogging might then be exposed to high temperatures, making recovery difficult and stressing an already weakened root system.

Example: Strong Winds and Heavy Rainfall: The combination causes much more severe lodging and physical damage than either event alone.

These compound events present a formidable and complex challenge to agricultural resilience, as crops are forced to cope with multiple, interacting stresses that often push their physiological and adaptive capacities to their absolute limits. Understanding these intricate interactions is paramount for developing effective adaptation and mitigation strategies to safeguard future food production.

5.15 Conclusion

In conclusion, the intricate dance between crops and weather is a defining dynamic in agriculture, holding the key to both bountiful harvests and devastating losses. This chapter has underscored that the evaluation of crop response to weatheris not a simplistic task, but a multi-faceted scientific endeavor demanding diverse methodologies. From meticulously analyzing historical yield trends and applying advanced statistical models to unraveling complex physiological mechanisms through process-based simulations and leveraging cutting-edge remote sensing technologies, each approach offers invaluable insights into how temperature, radiation, water, and atmospheric conditions dictate plant life from germination to maturity.

Understanding these multifaceted interactions, including the precise impact on vital yield components and the heightened vulnerabilities during specific developmental stages, is paramount. This knowledge allows us to not only diagnose past limitations but, more critically, to anticipate future challenges posed by escalating climate variability and change, particularly the increased frequency and intensity of extreme weather events. Ultimately, the continuous and sophisticated evaluation of crop responses to weather provides the foundational intelligence necessary for developing resilient crop varieties through targeted breeding programs, optimizing management practices for a given region's optimum weather conditions, and forging adaptive strategies that are indispensable for ensuring sustainable agricultural productivity and global food security in an uncertain climatic future.

Multiple Choice Questions (Bold means correct answer)

1. Crop growth, representing biomass accumulation, is fundamentally the net result of:
   a) Transpiration minus Respiration
   b) Gross Photosynthesis minus Transpiration
   c) Gross Photosynthesis minus Total Respiration
   d) Light Interception minus Nutrient Uptake
2. Which part of the solar radiation spectrum is primarily utilized by plants for photosynthesis?
   a) Infrared radiation
   b) Ultraviolet radiation
   c) Photosynthetically Active Radiation (PAR)
   d) X-rays
3. High night temperatures are generally detrimental to crop growth because they:
   a) Promote excessive water uptake
   b) Increase maintenance respiration, depleting carbohydrates
   c) Lead to chilling injury
   d) Reduce light interception
4. How does severe water stress primarily reduce photosynthesis?
   a) By increasing stomatal opening and CO_2 uptake
   b) By directly denaturing photosynthetic enzymes
   c) By inducing stomatal closure, limiting CO_2 entry
   d) By increasing soil oxygen levels

5. The ‚CO_2 fertilization effect' is generally more pronounced in which type of plants?
   a) C4 plants
   b) CAM plants
   c) C3 plants
   d) Both C3 and C4 equally
6. Moderate wind speeds can positively influence crop growth by:
   a) Directly increasing leaf temperature
   b) Accelerating the onset of water stress
   c) Increasing the supply of carbon dioxide to the plants
   d) Causing physical damage to the plant canopy
7. Which evaluation method offers the highest precision and control over environmental parameters, making it ideal for studying fundamental physiological responses?
   a) Field Experiments
   b) Historical Weather-Yield Correlations
   c) Growth Chambers / Phytotrons
   d) Remote Sensing
8. The main limitation of historical weather-yield correlations is that they:
   a) Are too expensive to conduct
   b) Cannot be replicated in different locations
   c) Show correlation, not necessarily causation
   d) Require real-time data, which is often unavailable
9. Free Air CO2 Enrichment (FACE) experiments are primarily designed to:
   a) Simulate frost damage
   b) Measure the CO2 fertilization effect under realistic field conditions
   c) Analyze long-term phenological shifts
   d) Control soil moisture levels precisely
10. Which method provides large-scale, non-destructive monitoring of crop responses, often using Vegetation Indices (VIs)?
   a) Growth Chambers
   b) Field Experiments
   c) Process-Based Crop Models
   d) Satellite Imagery

11. What is a key advantage of process-based crop models?
    a) They are completely free of uncertainty.
    b) They only require minimal historical data.
    c) They provide mechanistic insights into *why* weather affects crop performance.
    d) They eliminate the need for any field validation.
12. Empirical methods primarily rely on:
    a) Simulations of physiological processes
    b) Direct observation and statistical analysis of historical data
    c) Precise control of all environmental variables
    d) Genetic engineering to test weather responses
13. In statistical weather-yield regression, if the coefficient for ‚July rainfall‘ is positive and significant, it implies:
    a) July rainfall directly causes increased yield.
    b) Higher July rainfall is statistically associated with higher yields.
    c) July rainfall is the only factor affecting yield.
    d) The relationship is always linear.
14. What is a significant drawback of purely statistical crop-weather models?
    a) They are too complex to interpret.
    b) They require extensive computational power.
    c) They cannot reliably extrapolate to novel climate conditions.
    d) They only work for a single crop type.
15. Which statistical technique is used to measure the strength and direction of a linear association between two variables?
    a) Regression Analysis
    b) Time Series Analysis
    c) Correlation Analysis
    d) Principal Component Analysis
16. Analyzing yield *deviations* from a long-term trend in historical data helps to:
    a) Account for changes in pest and disease incidence.
    b) Remove the influence of technological advancements and management improvements.
    c) Directly measure the impact of soil fertility.
    d) Predict future market prices.

17. A limitation of agro-climatological indices and zonation is that they can:
    a) Be too complex to calculate for large areas.
    b) Oversimplify complex interactions and mask daily weather effects.
    c) Only apply to perennial crops.
    d) Directly predict economic returns.
18. When dealing with numerous inter-correlated weather variables, which multivariate statistical technique can synthesize them into fewer, uncorrelated components?
    a) Simple Linear Regression
    b) Principal Component Analysis (PCA)
    c) Cluster Analysis
    d) Time Series Analysis
19. Which of the following is NOT a core principle of empirical methods?
    a) Observation-driven from historical data
    b) Statistical foundation for relationships
    c) Explicit representation of biological mechanisms
    d) Often treats the crop as a black box
20. What is a key requirement for reliable statistical analysis of weather effects on crops?
    a) Data from only one growing season
    b) Short-term forecasts rather than historical data
    c) Long-term, consistent, and high-quality records for both weather and crop outcomes
    d) Exclusion of all non-weather-related variables
21. Analytical methods for evaluating crop response primarily aim to:
    a) Find statistical correlations between yield and weather.
    b) Simulate the underlying biological and physical processes.
    c) Determine the economic impact of weather on farmers.
    d) Classify climate zones for crop suitability.

22 In a process-based crop model, which component is typically calculated by considering Leaf Area Index (LAI) and light extinction coefficients?
    a) Respiration Rate
    b) Nutrient Uptake

c) Light Interception by Canopy

d) Root Water Uptake

23. The carbon lost by a plant for maintaining existing tissues and synthesizing new ones is accounted for by:

a) Gross Photosynthesis

b) Transpiration

c) Respiration

d) Light Use Efficiency

24. What does a process-based crop model do with assimilated carbon (net growth)?

a) It assumes it's all converted to yield immediately.

b) It allocates it dynamically among different plant organs.

c) It releases it back into the atmosphere.

d) It only considers carbon gained by roots.

25. What critical process does a process-based model meticulously track between the soil, plant, and atmosphere?

a) Wind patterns

b) Hail formation

c) Water balance and Transpiration

d) Solar flare activity

26. A key advantage of analytical (mechanistic) crop models over statistical models is their ability to:

a) Be simpler to calibrate

b) Provide mechanistic explanations (the ‚why')

c) Require less input data

d) Be less computationally demanding

27. When using process-based models for what-if scenarios, what kind of inputs are typically varied?

a) Only historical yield data

b) Combinations of weather parameters, soil types, or management practices

c) Economic market prices only

d) Genetic sequences of pests

28. Which of the following is a major limitation of analytical crop models?
    a) They cannot simulate physiological processes.
    b) They are too simplistic for real-world applications.
    c) They require vast amounts of high-quality input data for parameterization and validation.
    d) They can only be used for short-term forecasting.
29. The concept of optimum weather conditions for a crop is:
    a) A universal constant for all crops globally.
    b) Independent of the crop cultivar or region.
    c) Highly region-specific, crop-specific, and even cultivar-specific.
    d) Determined solely by temperature.
30. Why is determining optimum weather conditions a complex task?
    a) Weather parameters never interact.
    b) Crops have uniform sensitivities throughout their life cycle.
    c) It involves interacting factors, temporal dynamics, and genotype-environment interactions.
    d) It only requires average annual data.
31. In analyzing long-term historical data, comparing GDD accumulation with known crop requirements helps to:
    a) Determine soil nutrient levels.
    b) Assess if a region has sufficient heat units for a crop to mature reliably.
    c) Predict pest outbreaks.
    d) Evaluate market demand for the crop.
32. What is a primary contribution of controlled environment studies (like growth chambers) to determining optimum weather conditions?
    a) They provide large-scale yield forecasts.
    b) They allow for precise determination of cardinal temperatures for specific processes.
    c) They assess economic viability directly.
    d) They predict global climate change impacts.
33. When using process-based crop models to determine optimal planting windows, what is the typical approach?
    a) Only using current year‘s weather forecast.

b) Simulating yields for different planting dates over many years of historical weather.

c) Observing phenological events visually.

d) Performing laboratory nutrient analysis.

34. Agro-climatic zonation, as a macro-level application, primarily helps in:

a) Predicting daily weather fluctuations.

b) Delineating regions based on climatic suitability for specific crops.

c) Measuring individual grain weight.

d) Controlling pest populations.

35. Which term refers to the natural, short-term fluctuations in climatic parameters around their long-term average?

a) Climate Change

b) Global Warming

c) Climate Variability

d) Anthropogenic Climate

36. An increase in the frequency and intensity of heatwaves over several decades is an example of:

a) Normal climate variability

b) Climate change

c) Seasonal weather anomaly

d) Microclimate effect

37. What is a key characteristic of the current climate change with respect to extreme events?

a) A decrease in their frequency.

b) A shift in their type but not intensity.

c) Changes in their frequency, intensity, duration, and spatial extent.

d) No impact on extreme events.

38. Which of the following is a potential effect of rising average temperatures due to climate change on crops?

a) Prolonged growing seasons and delayed maturity everywhere.

b) Accelerated crop development and shorter growing seasons.

c) Decreased pest pressure globally.

d) Guaranteed increase in nutritional quality.

39 The CO2 fertilization effect is a positive impact of climate change, but its benefits can be offset by:

a) Increased rainfall.

b) Concurrent heat and water stress.

c) Decreased solar radiation.

d) Reduced pest populations.

40. Evaluating climate change impacts on crops primarily relies on:

a) Long-term historical yield-weather correlations only.

b) Short-term weather forecasts.

c) Forward-looking, sophisticated process-based crop models driven by future climate projections.

d) Field observations of isolated weather events.

41. What is the fundamental difference in effect on crops between climate variability and climate change?

a) Variability causes irreversible shifts, while change causes annual fluctuations.

b) Variability causes only positive impacts, while change causes only negative impacts.

c) Variability refers to fluctuations around a mean, while change refers to a persistent shift in the mean itself and altered extremes.

d) Variability is human-induced, while change is purely natural.

42. The primary goal of incorporating ‚weather response traits' in crop breeding is to enhance:

a) Post-harvest processing efficiency

b) Resilience, stability, and productivity under diverse climatic conditions

c) Crop aesthetic appeal

d) Market access for farmers

43. Which of the following is a key weather response trait targeted for heat tolerance?

a) Deep rooting

b) Aerenchyma formation

c) Efficient stomatal regulation for evaporative cooling

d) Increased chilling hours

44. Breeding for ‚drought escape' involves selecting cultivars that:
    a) Can maintain high water use efficiency throughout the season.
    b) Complete their life cycle before severe drought sets in.
    c) Develop very deep root systems.
    d) Accumulate solutes for osmotic adjustment.
45. Why is ‚photoperiod sensitivity' an important trait for breeders to consider for light response?
    a) It directly impacts individual grain weight.
    b) It ensures flowering and maturity occur at the optimal time for a given latitude.
    c) It improves soil nutrient uptake.
    d) It reduces wind damage.
46. What does ‚lodging resistance' primarily address in cereals?
    a) Damage from hailstorms
    b) Resistance to pests
    c) Prevention of crops falling over due to strong winds
    d) Tolerance to waterlogging
47. Which advanced molecular technique helps identify specific genes associated with desirable weather response traits, accelerating the selection process?
    a) Traditional Cross-breeding
    b) Manual Phenotyping
    c) Marker-Assisted Selection (MAS)
    d) Direct Weather Forecasting
48 A severe reduction in plant population is most likely caused by adverse weather during which stage?
    a) Grain filling
    b) Flowering/Anthesis
    c) Germination and emergence
    d) Early senescence
49. High temperatures around flowering (anthesis) primarily reduce yield by impacting which component?
    a) Individual grain weight
    b) Plant population
    c) Number of grains/seeds per reproductive organ
    d) Number of tillers

50. Prolonged cloudy weather during the grain filling period is most likely to reduce:
    a) Plant population
    b) Number of tillers
    c) Individual grain/seed weight
    d) Number of reproductive organs
51. Waterlogging during the early vegetative stage most directly impacts which yield component?
    a) Individual grain weight
    b) Number of grains per reproductive organ
    c) Stunted growth and reduced tillering/branching, limiting potential reproductive organs
    d) Pollen viability
52. Frost occurring during the grain filling period is likely to cause:
    a) Increased individual grain weight
    b) Premature termination of grain filling, reducing individual grain weight
    c) Increased plant population
    d) Enhanced photosynthesis
53. If a crop experiences severe drought stress during flowering, which yield component is most likely to be significantly reduced?
    a) Plant Population
    b) Number of Grains/Seeds per Reproductive Organ
    c) Individual Grain/Seed Weight (with no impact on number)
    d) Root biomass
54. The most critical and often devastating impact of a heatwave on cereal crops occurs during:
    a) Early vegetative growth
    b) Tillering
    c) Flowering/Anthesis
    d) Post-harvest
55. How do strong winds primarily reduce crop yield?
    a) By increasing soil temperature.
    b) By causing lodging, defoliation, and stem breakage.

c) By enhancing nutrient uptake.

d) By reducing atmospheric CO2 concentration.

56. Flooding or waterlogging severely impacts plant roots primarily due to:

a) Excessive light exposure

b) Oxygen deprivation (anaerobiosis)

c) Increased soil pH

d) Accelerated nutrient absorption

57. A hailstorm causes direct physical damage to crops, which leads to:

a) Enhanced photosynthesis and growth

b) Shredded leaves, broken stems, and bruised produce

c) Increased soil organic matter

d) Improved nutrient translocation

58. What is a compound extreme event?

a) A single, very intense weather event.

b) Multiple extreme weather conditions occurring simultaneously or in close succession.

c) An extreme event that only affects one type of crop.

d) A long-term shift in average climate.

59. If a late spring frost occurs, which part of a fruit tree is most susceptible to damage leading to reduced fruit set?

a) Dormant woody stems

b) Mature leaves

c) Developing flower buds, open flowers, and newly formed fruits

d) Deep roots

60. Which type of extreme event can lead to an accumulation of toxic compounds in the root zone due to anaerobic conditions?

a) Heatwave

b) Drought

c) Waterlogging/Flood

d) Hailstorm

6

Elementary Crop Growth Models

6.1 Introduction

Building on our extensive exploration of how individual weather parameters intricately influence every facet of crop life and the diverse methodologies employed to evaluate these responses, we now embark on a journey into the mathematical heart of crop science. This chapter, Elementary Crop Growth Models," will introduce you to the foundational concepts and simpler mathematical frameworks used to quantify, simulate, and ultimately predict how crops grow under varying environmental conditions.

In the upcoming sections, we'll delve into the fundamental building blocks of crop modeling. You'll discover how these elementary models simplify the complex biological processes occurring within a plant to create manageable, yet insightful, representations. Unlike highly detailed mechanistic models, which require vast amounts of data and computational power, elementary models focus on capturing the most essential relationships between environmental drivers (like solar radiation, temperature, and water) and key aspects of crop growth, such as biomass accumulation or phenological progression.

These simpler models serve as a crucial stepping stone. They allow us to grasp core principles of crop-weather interaction in a quantifiable way, providing practical tools for initial assessments, quick predictions, and conceptual understanding, without getting lost in the overwhelming complexity of every single physiological pathway. By understanding these foundational models, you'll gain a clearer perspective on how agricultural scientists begin to predict the future performance of our food crops, even as the climate around them continues to evolve

6.2 Elementary Crop Growth Models

In the realm of crop science, elementary crop growth models represent the most fundamental and often simplified mathematical frameworks used to quantify and predict crop development and biomass accumulation in response to key environmental factors. Unlike the complex, highly detailed process-based models (which we will explore later), elementary models prioritize simplicity, ease of implementation, and a clear focus on the most dominant weather drivers. They aim to establish straightforward quantitative relationships, often empirical in nature, between environmental inputs and observed crop outputs, without necessarily delving into the intricate biological mechanisms at a cellular or subcellular level.

These models serve several crucial purposes:

First Approximation: They provide a quick, initial estimate of crop performance under various conditions.

Conceptual Understanding: They help illustrate fundamental principles of crop-weather interactions in a digestible manner.

Data Scarcity: They are particularly useful in situations where detailed physiological or soil data, required by more complex models, are limited or unavailable.

Resource Allocation: They can guide basic decisions regarding cultivar selection, planting dates, and overall suitability of a region for a particular crop.

While less mechanistic, elementary models form the conceptual bedrock upon which more sophisticated models are built, and they remain highly relevant for practical agricultural applications.

6.2.1 Characteristics of Elementary Crop Growth Models

- **Simplicity:** They use a limited number of input variables and relatively simple mathematical equations.
- **Empirical Basis:** Many are derived from observed correlations and relationships from field or controlled experiments, rather than solely from biophysical laws.
- **Focus on Key Drivers:** They typically concentrate on one or two dominant weather factors (e.g., temperature for development, radiation for biomass).
- **Less Mechanistic Detail:** They may not explicitly simulate complex physiological processes like stomatal conductance, nutrient uptake kinetics, or detailed carbon partitioning.
- **Output-Oriented:** They aim to predict specific crop outputs such as phenological dates, total biomass, or final yield.

6.2.2 Key Types of Elementary Crop Growth Models

• Thermal Time Models (Growing Degree Days - GDD or Heat Units)

This is perhaps the most widely used and simplest elementary model for predicting crop phenological development (Fig.6.1).

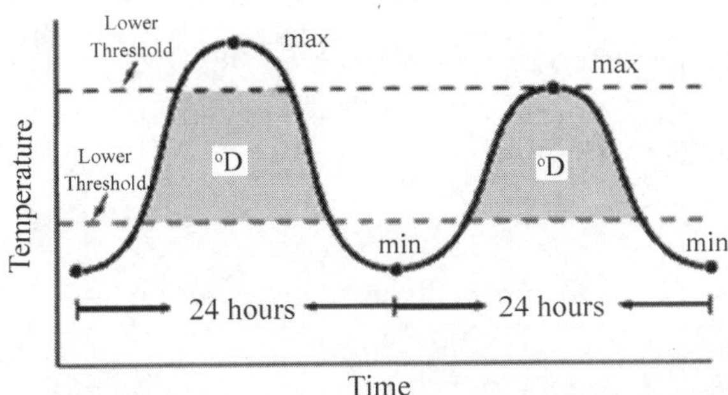

Fig. 6.1: Thresholds and accumulated degree-days

It is based on the premise that plant development is primarily driven by temperature and that crops require a certain accumulation of heat units (thermal time) to progress through specific developmental stages.

Concept: Plant metabolic rates, and thus developmental rates, accelerate with increasing temperature up to an optimum, and cease below a certain base temperature (Tbase). GDD models assume a linear relationship between temperature and development within a defined range.

Calculation: Daily GDD is typically calculated as: GDD=(Tmax +Tmin)/2 −Tbase Where:

Tmax = Daily maximum temperature

Tmin = Daily minimum temperature

Tbase = Base temperature (the minimum temperature below which development ceases, specific to each crop/cultivar).

Note: If the calculated mean temperature is below Tbase , GDD for that day is 0. Some models also include an upper optimum temperature (Topt) or a maximum temperature (Tmax) above which development either plateaus or declines.

Application: Accumulated GDD are used to predict the timing of key phenological stages like emergence, flowering, silking, physiological maturity, and harvest readiness. Each stage requires a specific, accumulated amount of

GDD from a reference point (e.g., planting).

Evaluation of Weather Effects

Temperature: Directly assesses the impact of daily temperatures on the speed of development. Warmer temperatures (above Tbase) lead to faster GDD accumulation and earlier phenological progression.

Frost Risk: Helps determine the length of the frost-free period and the likelihood of a crop maturing before the first killing frost or being exposed to early season cold stress.

Advantages

Simplicity: Easy to calculate and understand, requiring only daily temperature data.

Wide Applicability: Used across many crops and regions for scheduling and cultivar selection.

Forecasting: Can be used with historical average temperatures or climate forecasts to predict general timing.

Limitations

Ignores Other Factors: Does not account for the influence of solar radiation, water stress, nutrient availability, or photoperiod (though photoperiod can be added as a modifier).

Assumes Linearity: Assumes a linear response to temperature, which is not always physiologically accurate, especially at temperature extremes.

No Yield Prediction: Primarily predicts timing, not biomass or yield quantity directly.

6.2.3 Radiation Use Efficiency (RUE) Models

These models provide a simple framework for estimating total biomass accumulation based on intercepted solar radiation (Fig.6.2).

Concept: RUE models are based on the principle that total dry matter production (DM) of a crop is directly proportional to the amount of photosynthetically active radiation (PAR) intercepted by the canopy (IPAR), multiplied by a crop-specific efficiency factor (RUE, ϵ). $DM = \epsilon \times IPAR$ Where:

DM = Total dry matter accumulation (g/m^2)

ϵ = Radiation Use Efficiency (g DM / MJ IPAR) – a crop-specific constant.

IPAR = Total intercepted PAR over the growing period (MJ/m^2)

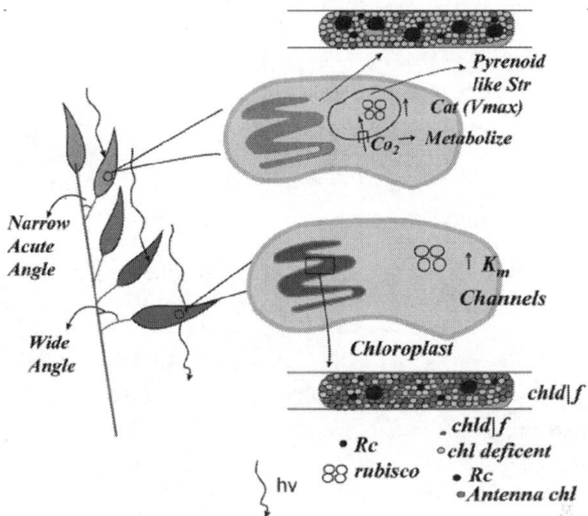

Fig. 6.2: Radiation use efficiency modeling for improving yield potential

Calculation of IPAR

Daily incident PAR is estimated (typically 45-50% of total solar radiation).

Intercepted PAR is then calculated based on the incident PAR and the Leaf Area Index (LAI) of the crop, using Beer's Law (e.g., IPAR = PARincident $\times(1-e^{-k*LAI})$, where k is a light extinction coefficient). The LAI itself needs to be estimated, sometimes using a simple growth curve driven by GDD.

Application: Used to estimate total biomass production over a season or specific growth phases, providing a simple estimate of crop productivity under non-stressed conditions.

Evaluation of Weather Effects

Solar Radiation: Directly assesses the impact of incident solar radiation and the crop's ability to intercept it (via LAI development) on biomass accumulation. More radiation and better canopy development lead to higher dry matter.

Advantages

Simple and Intuitive: Easy to understand the direct link between light and growth.

Good for Potential Yield: Provides a reasonable estimate of potential biomass accumulation under ideal, non-stressed conditions.

Low Data Requirement: Primarily needs solar radiation data and LAI development.

Limitations

Assumes Constant RUE: The major limitation is that RUE is assumed constant, but in reality, it varies with temperature, water stress, nutrient deficiency, and CO_2 concentration.

No Yield Components: Does not predict partitioning of biomass to harvestable yield directly, nor does it account for individual yield components.

Doesn't Handle Stress Well: Poorly accounts for reductions in growth due to temperature extremes or water/nutrient stress unless RUE is empirically adjusted for these factors.

6.2.4 Simple Water-Limited Yield Models

These models attempt to account for the impact of water availability on yield, often by applying a reduction factor to potential yield.

Concept: The actual yield (Yactual) is estimated by adjusting the potential yield (Ypotential) (which might be derived from an RUE model or empirical maximums) by a factor related to water stress. Yactual =Ypotential × Ky × (1−ETactual /ETpotential) (This is a simplified version of the FAO AquaCrop approach, or similar water balance models.) Where:

Ky = Yield response factor (crop-specific sensitivity to water deficit).

ETactual = Actual evapotranspiration (water consumed by the crop under current conditions).

ETpotential = Potential evapotranspiration (water consumed under non-limiting water).

Calculation: Requires calculating a daily or periodic water balance, tracking soil moisture based on rainfall, irrigation, and potential/actual evapotranspiration.

Application: Used to estimate yield reduction due to water deficit and to assess irrigation requirements.

Evaluation of Weather Effects

Rainfall/Soil Moisture: Directly assesses how rainfall patterns and soil water holding capacity translate into water stress and subsequent yield reductions.

Temperature/Humidity (VPD): Influence potential evapotranspiration, thereby affecting water demand and the severity of water stress.

Advantages:

Accounts for Water Stress: Provides a basic mechanism to incorporate the critical impact of water availability.

Improved Realism: Offers a more realistic yield prediction than models ignoring water.

Limitations

Still Simplified: May not capture complex interactions between water stress and other factors, or the timing of stress at critical developmental stages as finely as process-based models.

Requires ET Estimation: Accurate estimation of actual and potential evapotranspiration is crucial and can be complex.

6.2.5 Empirical Regression Models (as Elementary Models)

While discussed under statistical methods, simple regression models can also be considered elementary crop growth models when they are directly used for prediction rather than just correlation.

Concept: A mathematical equation derived from historical data that links crop yield directly to one or more aggregate weather variables over the growing season.

Example: Yield=a+b1×(Seasonal Rainfall)+b2×(Average Growing Season Temp)

Application: Used for quick yield forecasting, or to identify which aggregate weather variables have historically been most important for a region's yield.

Evaluation of Weather Effects: Provides coefficients that quantitatively describe the statistical relationship between weather inputs and yield outputs.

Advantages

Ease of Use: Simple to apply once the equation is established.

Low Data Demand: Only requires historical aggregate weather and yield data.

Limitations

No Causation: Cannot explain *why* the relationship exists.

Poor Extrapolation: Unreliable outside the range of historical data; cannot predict responses to unprecedented weather conditions or new cultivars.

Ignores Timing: Often uses seasonal averages, failing to capture the critical importance of weather timing during specific growth stages.

In essence, elementary crop growth models, despite their inherent simplifications, provide foundational tools for understanding and predicting crop responses to weather. They are valuable for initial assessments, conceptual understanding, and situations where data or computational resources are limited. They lay the groundwork for understanding the more detailed and mechanistic models that seek to emulate the full complexity of crop growth processes.

6.3 Applications and Limitations of Elementary Crop Growth Models

Having explored the fundamental principles and specific types of elementary crop growth models, it's essential to understand their practical utility and, crucially, their inherent limitations. While these models offer simplified representations of complex biological systems, their ease of use and modest data requirements make them valuable tools in various agricultural contexts. However, their very simplicity also dictates the boundaries of their applicability and the accuracy of their predictions, particularly when dealing with complex environmental interactions or unprecedented conditions.

6.3.1 Applications of Elementary Crop Growth Models

Elementary crop growth models, despite their simplicity, serve several important functions in agricultural research, planning, and management:

- **Phenological Forecasting and Scheduling**

GDD models are widely used to predict the timing of key developmental stages (e.g., emergence, flowering, maturity) for various crops in a given region. This information is invaluable for:

Optimal Planting Date Determination: Helping farmers choose the best planting window to avoid late frosts or ensure maturity before early frosts, or to synchronize with expected rainfall patterns.

Timing of Field Operations: Guiding the precise timing of critical management activities such as fertilizer application, pest and disease scouting, irrigation, and harvest.

Cultivar Selection: Matching cultivars with appropriate GDD requirements to the thermal regime of a specific growing season or region.

- **Potential Yield Estimation (under non-stressed conditions)**

RUE models provide a straightforward method to estimate the maximum possible biomass accumulation and, by extension, potential yield, given sufficient solar radiation and assuming no limitations from water, nutrients, pests, or extreme temperatures.

This potential yield serves as a benchmark against which actual yields can be compared, helping to identify the yield gap" attributed to various limiting factors in the field.

- **Water-Limited Yield Assessment and Irrigation Scheduling**

Simple water balance models offer a basic means to quantify the impact of water deficits on crop growth and yield. They are used to:

Estimate the expected yield reduction under rainfed conditions.

Determine the timing and amount of irrigation needed to prevent water stress and optimize water use efficiency, especially in regions with variable rainfall.

- **Early Warning Systems and Risk Assessment**

By continuously monitoring weather parameters and running simple models, agricultural agencies can identify periods when crops are experiencing significant temperature or water stress.

This allows for the issuance of early warnings for potential yield reductions, enabling timely interventions or triggering insurance mechanisms.

They can also be used for preliminary risk assessment of new crops or cultivars in a region by comparing their basic requirements to historical climate data.

- **Educational and Conceptual Tools**

Their simplicity makes elementary models excellent tools for teaching fundamental concepts of crop physiology and agroclimatology. They help students and practitioners grasp the direct relationships between environmental factors and plant responses before delving into more complex multi-factor interactions.

6.3.2 *Limitations of Elementary Crop Growth Models*

Despite their utility, the simplicity of elementary crop growth models comes with significant limitations that restrict their accuracy and applicability, particularly under complex or changing conditions:

- **Oversimplification of Biological Processes**

Lack of Mechanistic Detail: Elementary models do not explicitly represent the intricate biochemical and biophysical processes within the plant (e.g., enzyme kinetics, detailed nutrient cycling, complex hormonal signaling). They are often black box" models, describing what happens rather than *why* it happens.

Linearity Assumption: Many models (especially GDD) assume a linear response to environmental factors, which is often physiologically inaccurate, particularly at temperature extremes or under high light/water conditions where responses can plateau or decline.

Ignoring Interactions: They often treat weather factors in isolation or in very simple combinations, failing to capture the complex **synergistic or antagonistic interactions** between temperature, light, water, and CO_2 that profoundly affect real-world crop performance. For instance, high light combined with high temperature can exacerbate heat stress, which a simple GDD model would miss.

- **Poor Extrapolation Beyond Observed Data**

Because they are largely empirical, the relationships derived from historical data may not hold true under novel or unprecedented weather conditions (e.g., future climate change scenarios outside the historical range). They cannot reliably predict how a crop will respond to environments it has never experienced.

They are also less reliable when applied to new crop varieties with different genetic characteristics that might alter their responses to weather.

- **Inadequate Handling of Stress and Extremes**

Elementary models typically do not explicitly model the physiological impacts of stress factors (e.g., heat stress on pollen viability, severe drought on stomatal function, waterlogging on root respiration) that can dramatically reduce yield. They might use simple empirical reduction factors, but these often lack mechanistic basis and are less robust.

They are particularly weak at predicting the impact of extreme weather events (e.g., intense heatwaves, sudden frosts, prolonged floods) because these events often involve non-linear, threshold-based responses that are not well captured by simple linear or average-based models.

- **Limited Spatial and Temporal Resolution**

Many elementary models use aggregated weather data (daily averages, seasonal totals), potentially missing the critical importance of sub-daily fluctuations or the precise timing of a stress event (e.g., a few hours of very high temperature during anthesis can be devastating but might be averaged out in a daily mean).

- **No Feedback Mechanisms**

They typically lack feedback loops. For example, severe water stress might reduce leaf area, which in turn reduces light interception. Elementary models often don't dynamically adjust the plant's morphology or physiology in response to stress in such an integrated way.

In conclusion, elementary crop growth models provide accessible and useful tools for initial estimations and understanding fundamental crop-weather relationships. They are invaluable for educational purposes, basic regional planning, and situations with limited data. However, for a more comprehensive, accurate, and predictive understanding of crop responses, especially in the context of climate variability and change, their simplicity becomes a significant limitation. This highlights the necessity for more complex, process-based (mechanistic) models, which attempt to capture the intricate biological realities with greater fidelity.

Multiple Choice Questions (Bold means correct answer)

1. Elementary crop growth models are primarily characterized by their:
   a) Extreme complexity and detailed cellular representations.
   b) Simplicity, ease of use, and focus on the main weather factors.
   c) Exclusive reliance on genetic sequencing data.
   d) Ability to predict long-term global climate shifts.
2. What is a main function of elementary crop growth models?
   a) To perform highly precise mechanistic simulations of all plant processes.
   b) To provide an initial, quick estimate of crop performance.
   c) To manage international agricultural trade policies.
   d) To eliminate the need for any field observations.

3. Which statement accurately describes a characteristic of elementary crop growth models?
   a) They explicitly simulate nutrient uptake kinetics.
   b) They require vast computational power for daily runs.
   c) They typically focus on the influence of one or two dominant weather variables.
   d) They offer perfect predictions under all stress conditions.
4. Thermal Time Models, often represented by Growing Degree Days (GDD), are commonly used to predict:
   a) The total amount of biomass produced by a crop.
   b) Changes in soil water content over time.
   c) The timing of a crop's phenological developmental stages.
   d) The market price of agricultural commodities.
5. Thermal Time Models (Growing Degree Days - GDD) primarily use which weather input for their core calculation of developmental progress?
   a) Total daily solar radiation.
   b) Daily maximum and minimum temperatures.
   c) Wind speed and relative humidity.
   d) The total amount of rainfall.
6. In a Thermal Time Model, if the average daily temperature is below the crop's base temperature, how much developmental progress (GDD) is typically accumulated for that day?
   a) A negative amount.
   b) The full potential for that day.
   c) Zero.
   d) An amount equal to the maximum temperature.
7. A significant limitation of Thermal Time (GDD) models is that they:
   a) Are too difficult for farmers to understand.
   b) Do not typically account for the direct influence of factors like solar radiation or water stress on development rate.
   c) Can accurately predict final crop yield directly.
   d) Are only suitable for crops grown in tropical climates.

8. Radiation Use Efficiency (RUE) models estimate total dry matter production based on the concept that it's proportional to:
   a) The plant's genetic potential and soil type.
   b) The amount of photosynthetically active radiation (PAR) intercepted by the crop canopy.
   c) The daily average wind speed.
   d) The total amount of applied fertilizer.
9. What is a major drawback of Radiation Use Efficiency (RUE) models?
   a) They are overly complex to set up.
   b) They require too many specialized weather instruments.
   c) They assume a constant RUE value, ignoring how environmental stresses can reduce it.
   d) They can accurately predict individual grain weight.
10. Simple water-limited yield models are designed to assess the impact of:
   a) Soil temperature on root growth.
   b) The availability of water on overall crop yield.
   c) Air pollution on plant health.
   d) Light intensity on chlorophyll content.
11 In simple water-limited yield models, how is the actual crop yield typically determined?
   a) As a direct measure of seasonal rainfall totals.
   b) By adjusting a potential yield based on how well the crop's water demand is met.
   c) Exclusively through the plant's maximum daily temperature.
   d) Based solely on the crop's inherent genetic yield potential.
12. A key advantage of simple water-limited yield models is their ability to:
   a) Mechanistically simulate every water transport pathway in the plant.
   b) Provide a basic but important way to include the critical effect of water availability on yield.
   c) Precisely predict the economic returns from irrigation.
   d) Function without any weather data inputs.

13. When empirical regression models are categorized as elementary growth models, their primary role is to:
    a) Fully explain the biochemical reasons behind crop growth.
    b) Establish quantifiable relationships from historical data for predictive purposes.
    c) Conduct multi-century simulations of climate change impacts.
    d) Discover entirely new plant species.
14. A significant limitation when relying on empirical regression models for crop growth prediction is their:
    a) Ease of interpretation by non-experts.
    b) Difficulty in extrapolating reliably to novel or unprecedented weather conditions.
    c) Requirement for only a single year of historical data.
    d) Capacity to precisely model sub-daily weather fluctuations.
15. Beyond phenological forecasting, what is another practical application of GDD models in agriculture?
    a) Determining the precise amount of nutrients in the soil.
    b) Forecasting the optimal timing for critical field operations like harvest.
    c) Predicting the emergence of new insect pests.
    d) Calculating the efficiency of solar panels in the field.
16. Elementary models prove particularly useful in situations where:
    a) The highest level of scientific precision in predicting crop physiology is required.
    b) Extensive and detailed physiological or soil data are readily available.
    c) Detailed physiological or soil data are either limited or completely unavailable.
    d) The main objective is to design new genetic crop modifications.
17. Which of the following represents an inherent oversimplification within many elementary crop growth models?
    a) Their capability to dynamically adjust plant morphology.
    b) Their explicit modeling of complex carbon partitioning processes.
    c) Their assumption of linear responses to environmental factors within certain ranges.
    d) Their ability to accurately predict the impact of all extreme weather events.

18. Compared to more complex, process-based models, elementary models often lack:
    a) Any representation of temperature effects.
    b) The explicit representation of intricate physiological feedback mechanisms.
    c) Outputs such as phenological dates or total biomass.
    d) A connection to observed crop responses in the field.
19. When dealing with environmental stress, a major limitation of elementary models is that they:
    a) Are designed to perfectly predict threshold-based responses to extreme events.
    b) Often rely on simple, empirically derived reduction factors that may not capture complex physiological realities.
    c) Can precisely simulate changes in stomatal function under severe drought conditions.
    d) Require no adjustment for variability in soil properties.
20. Elementary crop growth models are considered valuable educational tools primarily because of their:
    a) Advanced mathematical formulations and computational requirements.
    b) Capacity to model global food production systems.
    c) Simplicity and clarity in illustrating fundamental crop-weather relationships.
    d) Unmatched accuracy in long-term climate predictions.

7

Calibration, Validation, Verification, and Sensitivity of Elementary Crop Models

7.1 Introduction

Having explored the fundamental principles and various types of elementary crop growth models in the previous chapter, we now arrive at a pivotal juncture in their journey from theoretical constructs to practical, dependable tools. It's one thing to conceptualize and build a mathematical representation of crop growth; it's quite another to ensure that this model accurately reflects real-world plant behavior and can be reliably used for predicting outcomes and informing critical agricultural decisions.

This chapter, Calibration, Validation, Verification, and Sensitivity of Elementary Crop Models," will guide you through the indispensable processes that establish a model's credibility and utility. We will delve into the systematic steps involved in calibration, where we meticulously fine-tune a model's parameters to accurately reflect a specific crop's response within a particular environment, effectively adapting the general model to local conditions. Following this, we will explore validation, a rigorous process of independently testing the model's predictive power against datasets *not* used during calibration, ensuring its robustness and generalizability. Beyond accuracy, we will also cover verification, which confirms the model's logical consistency and computational correctness, ensuring that its internal workings are sound. Finally, we will examine sensitivity analysis, a crucial technique for understanding how changes in various input factors—especially weather parameters—or internal model coefficients influence the model's outputs, thereby revealing the model's critical drivers and potential vulnerabilities.

These four processes are far more than mere technical formalities; they form the bedrock upon which the utility, accuracy, and trustworthiness of any crop model, no matter how elementary, are firmly built. By engaging with these essential steps, we can transform theoretical frameworks into powerful instruments capable of supporting sustainable agricultural planning, risk management, and adaptive strategies in a dynamic climate.

7.2 Calibration of Elementary Crop Models

Calibration is a foundational and indispensable step in making any crop growth model, including elementary ones, a reliable tool for specific applications. Essentially, calibration is the process of adjusting or tuning" certain model parameters (often called coefficients or constants) to ensure that the model's outputs closely match observed data for a particular crop under specific environmental conditions. It's about adapting a generalized model, which may be based on universal plant physiological principles, to the unique characteristics of a specific cultivar, soil type, and management practices in a given region (Fig.7.1).

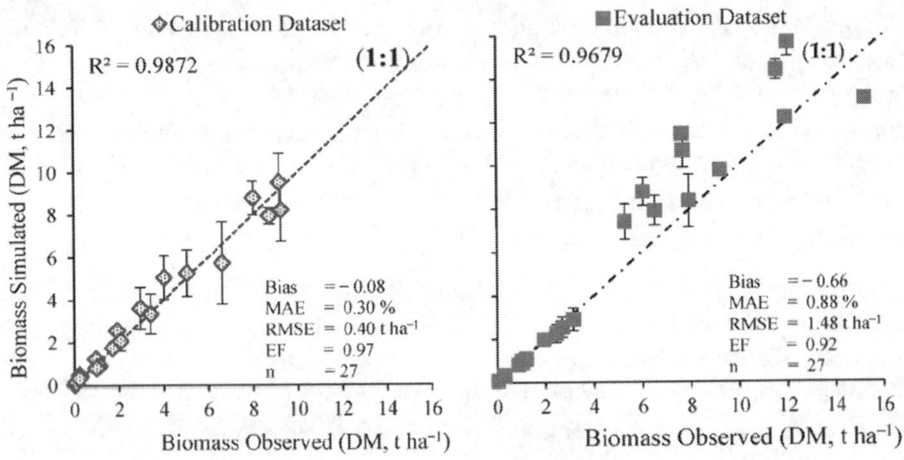

Fig. 7.1: Calibration and evaluation of the SIMPLE crop growth model

Think of it like tuning a musical instrument: an instrument (the model) has a basic structure, but it needs fine adjustments (calibration) to play the notes (predict crop behavior) accurately for a specific piece (cultivar in a region).

7.2.1 Why is Calibration Necessary?

- **Cultivar Specificity:** Even within the same crop species, different cultivars (varieties) can have distinct genetic characteristics that influence their growth rates, developmental timings, and responses to stress. A model initially developed for one maize hybrid, for instance, will likely need adjustment to accurately simulate another.
- **Environmental Variability (Local Adaptation):** While models incorporate weather, local soil properties (e.g., water holding capacity, nutrient levels), pests, diseases, and specific management practices (e.g., planting density, fertilization regimes) can all influence how a crop performs. Calibration helps account for these local nuances.

- **General vs. Specific:** Most crop models are built with general physiological principles. Calibration bridges the gap between these general principles and the unique realities of a particular agricultural system, making the model site-specific and useful for decision-making.
- **Improving Accuracy:** The ultimate goal of calibration is to minimize the difference between what the model predicts and what is actually observed in the field, thereby improving the model's accuracy and predictive power for the intended application.

7.2.2 Key Parameters to Calibrate in Elementary Models

In elementary models, the number of parameters to calibrate is typically small, aligning with their simplicity.

• Thermal Time Models (GDD)

Base Temperature (Tbase): The minimum temperature below which a crop's development ceases. This is crucial for accurately calculating GDD accumulation. For example, maize might have a Tbase of 10^0C, while winter wheat might be 0^0C.

Accumulated GDD for Phenological Stages: Each critical developmental stage (e.g., emergence to flowering, flowering to physiological maturity) requires a specific sum of GDD. These GDD sums are crucial calibration parameters. For example, a particular rice cultivar might need 800 GDD from planting to flowering, and another 1000 GDD from flowering to maturity.

• Radiation Use Efficiency (RUE) Models

Radiation Use Efficiency (ϵ): This is the core parameter: the efficiency with which intercepted PAR is converted into total dry matter (g DM / MJ PAR). While a general range exists for crop types (e.g., C3 vs. C4), specific cultivar performance and local environmental nuances (e.g., average CO_2 levels in greenhouse) can necessitate adjustment.

Light Extinction Coefficient (k): This parameter determines how much light penetrates the crop canopy as a function of LAI. It reflects the canopy's architecture (e.g., erect vs. prostrate leaves).

• Simple Water-Limited Yield Models

Yield Response Factor (Ky): This empirical coefficient quantifies a crop's sensitivity to water deficit. It defines how much yield is reduced for a given percentage reduction in actual evapotranspiration compared to potential evapotranspiration. This factor can vary with crop type and developmental stage.

Soil Water Holding Characteristics: While often considered input data, for a simplified model, some effective soil water holding capacity (e.g., field capacity, permanent wilting point) might be simplified and need adjustment based on local soil surveys.

7.2.3 The Calibration Process (Iterative Steps)

Calibration is typically an iterative process involving several steps:

- **Data Collection (Calibration Data Set)**

High-quality, observed field data from the target region and crop are essential. This includes:

Weather Data: Daily values of temperature (max/min), solar radiation, rainfall, humidity, wind speed for the growing season(s).

Crop Data: Observed dates of key phenological stages (e.g., emergence, flowering, maturity), measurements of total above-ground biomass at different times, and final grain yield.

Soil Data: Basic soil characteristics (e.g., texture, organic matter, bulk density, water holding capacity).

Management Data: Planting date, planting density, fertilizer application, irrigation events.

Crucially, the calibration data set should ideally represent the range of conditions and typical management practices for which the model is intended to be used.

- **Initial Parameter Estimation (Literature/Expert Knowledge)**

Start with parameter values from scientific literature, previous studies for similar crops/regions, or expert knowledge. These provide a sensible starting point.

- **Simulation and Comparison**

Run the model using the collected weather, soil, and management data and the initial parameter values.

Compare the model's predicted outputs (e.g., phenological dates, biomass, yield) with the actual observed data from the field.

- **Parameter Adjustment**

Based on the comparison, systematically adjust the model parameters. This is often done manually (trial and error) for elementary models, or using optimization algorithms for more complex ones.

Example (GDD): If the model consistently predicts flowering too late, you might reduce the required GDD sum for the flowering stage. If biomass is consistently underestimated, you might increase the RUE.

The goal is to minimize the difference between simulated and observed values using various statistical metrics (e.g., Root Mean Square Error (RMSE), Normalized Root Mean Square Error (nRMSE), R-squared (R^2), Mean Absolute Error (MAE), coefficient of determination).

- **Iteration and Refinement**

Repeat steps 3 and 4 until the model‘s predictions align acceptably with the observed data. This iterative process allows for continuous refinement of parameter values.

It‘s important not to over-calibrate– fitting the model too perfectly to a single dataset can make it perform poorly on new, independent data (lack of generalization).

7.2.4 Challenges and Considerations in Calibration

- **Data Quality:** Garbage in, garbage out. Poor quality or insufficient observed data will lead to poorly calibrated models.
- **Equifinality:** Different combinations of parameter values can sometimes lead to similar model outputs. This means multiple best parameter sets might exist, making the choice challenging.
- **Over-calibration/Overfitting:** Adjusting too many parameters or adjusting them too aggressively to fit a specific dataset can lead to a model that performs well only for that particular dataset but poorly for others. This compromises the model‘s generalizability.
- **Time and Resources:** Even for elementary models, rigorous calibration requires significant time, effort, and access to good quality data.
- **Uncertainty:** Calibration helps reduce uncertainty but does not eliminate it. There will always be some level of error or deviation due to inherent biological variability and unmodeled factors.

Calibration is not a one-time event; for continuous improvement, models might be re-calibrated or refined as more data become available or as agricultural practices/cultivars evolve. It is the critical process that makes a theoretical model a practical and reliable tool for specific agricultural contexts.

7.3 Validation of Elementary Crop Models

Once an elementary crop growth model has been calibrated, the next crucial step is validation. While calibration fine-tunes the model to a specific set of observed data, validation rigorously tests whether the calibrated model can accurately predict crop behavior for independent datasets—data that were *not* used during the calibration process. This step is paramount for establishing the model's credibility, reliability, and its ability to generalize to new, unseen conditions within its intended range of application.

Think of it as the ultimate test: after a student has studied thoroughly (calibration), validation is the exam where they apply what they've learned to new problems they haven't seen before.

7.3.1 Why is Validation Essential?

- **Assessing Generalizability:** The primary purpose of validation is to confirm that the model's calibrated parameters are robust enough to accurately simulate crop responses across different seasons, locations, or management practices, as long as they fall within the model's intended scope. It ensures the model isn't just memorizing the calibration data but truly understanding the underlying relationships.
- **Building Confidence:** Successful validation provides confidence to users (researchers, farmers, policymakers) that the model's predictions are reliable and can be trusted for decision-making.
- **Detecting Overfitting:** Validation is the key mechanism to detect overfitting– a situation where a model performs exceptionally well on the data it was trained on (calibrated with) but poorly on new data. Overfitting occurs when a model captures random noise or peculiarities in the calibration data rather than the true underlying patterns.
- **Identifying Model Weaknesses:** If a model performs poorly during validation, it signals potential weaknesses in its structure, underlying assumptions, or parameterization. This failure provides valuable feedback for further model improvement or refinement of calibration data.

7.3.2 The Validation Process

Validation requires an independent dataset, meaning data from different growing seasons, different fields (within the same agro-climatic zone), or even different cultivars (if the model is intended to be general enough for the species) that were **not** used during the calibration phase.

- **Independent Data Collection (Validation Data Set)**
  - Acquire a separate set of high-quality observed field data. This data should be collected with the same rigor as the calibration data (weather, crop phenology, biomass, yield, soil, management).
  - It is crucial that this dataset is truly independent of the calibration data. Using a split sample from the same experiment (e.g., 70% for calibration, 30% for validation) is a common practice, but ideally, data from entirely different years or sites provide stronger validation.
- **Model Simulation (without further adjustment)**
  - Run the *calibrated* model using the independent weather, soil, and management data from the validation dataset.
  - Crucially, no further adjustments to the model parameters are made during this stage. The model is run as is with its final calibrated parameter values.
- **Comparison of Predicted vs. Observed**
  - Compare the model's simulated outputs (e.g., predicted phenological dates, biomass accumulation curves, final yields) with the actual observed data from the validation dataset.
  - This comparison is quantified using various statistical metrics.

7.3.3 Statistical Metrics for Validation

Several statistical metrics are commonly used to assess the agreement between simulated and observed values. A combination of these metrics provides a comprehensive picture of model performance.

- **Graphical Comparisons**
  - **Scatter Plots (1:1 Plots):** Plot observed values on the X-axis against simulated values on the Y-axis. A perfect model would have all points falling on a 1:1 line (or 45-degree line). Deviations from this line indicate errors.
  - **Time Series Plots:** Plot observed and simulated values over time (e.g., daily biomass accumulation, LAI development). This helps visualize how well the model captures the dynamics of crop growth and development throughout the season.

- **Quantitative Metrics**
- **Root Mean Square Error (RMSE)**
  - Measures the average magnitude of the errors. It indicates the typical difference between predicted and observed values. Lower RMSE indicates better fit.
  - Units are the same as the variable being measured (e.g., kg/ha for yield, days for phenology).
- **Normalized Root Mean Square Error (nRMSE) or Relative RMSE**
  - RMSE divided by the mean of the observed values, expressed as a percentage. This provides a relative measure of error, useful for comparing model performance across different variables or studies with different scales.
  - nRMSE<10 is generally considered excellent, 10–20 good, 20–30 fair, and >30 poor.
- **Coefficient of Determination (R^2)**
  - Quantifies the proportion of the variance in the observed data that is explained by the model. A value closer to 1 (or 100%) indicates that the model explains a large proportion of the variability.
  - While important, a high R^2 alone does not guarantee good model performance if there‘s a systematic bias.
- **Mean Absolute Error (MAE)**
  - The average of the absolute differences between predicted and observed values. Less sensitive to outliers than RMSE.
- **Bias (or Mean Error)**
  - The average difference between predicted and observed values. A positive bias means the model generally overpredicts, a negative bias means it underpredicts. Indicates systematic errors.
- **d-index (Index of Agreement)**
  - Developed by Willmott, it measures the degree to which observed values are estimated by the simulated values. Ranges from 0 (no agreement) to 1 (perfect agreement). Less sensitive to extreme values than R^2.

7.3.4 Interpreting Validation Results

- **Good Agreement:** If the model performs well across multiple statistical metrics and graphical comparisons for the independent validation dataset, it suggests that the model is robust and can be reliably used for its intended purpose within the range of conditions represented by the validation data.
- **Poor Agreement:** If the model shows large errors, systematic biases, or low explanatory power during validation, it indicates that:
    - The model may have been overfitted during calibration.
    - The model structure or underlying assumptions may be inadequate for capturing the crop's behavior under the tested conditions.
    - The validation data itself might have quality issues.
    - This warrants further investigation, potentially leading back to recalibration, model refinement, or even a re-evaluation of the model's suitability for the specific application.

Validation is a continuous process. As new data become available (e.g., from new growing seasons or locations), a robust model should ideally be re-validated periodically to ensure its continued accuracy and relevance. It provides the empirical evidence that a calibrated model can indeed be trusted as a predictive tool.

7.4 Verification of Elementary Crop Models

While calibration (7.2) adjusts model parameters to match observed data and validation (7.3) assesses the model's predictive accuracy on independent data, verification is a distinct and crucial step that focuses on the internal logic and computational correctness of the model itself. Verification ensures that the model is performing as intended, free from errors in its coding, structure, and theoretical implementation (Fig.7.2).

It's about ensuring that the model's algorithms accurately reflect the conceptual design, independent of whether those algorithms perfectly mimic real-world processes. Think of it this way:

- **Calibration:** Is the instrument tuned to the specific concert pitch?
- **Validation:** Can the instrument play a new piece accurately after tuning?
- **Verification:** Is the instrument built correctly? Are all the strings, frets, and keys in the right place and functioning properly according to the blueprint?

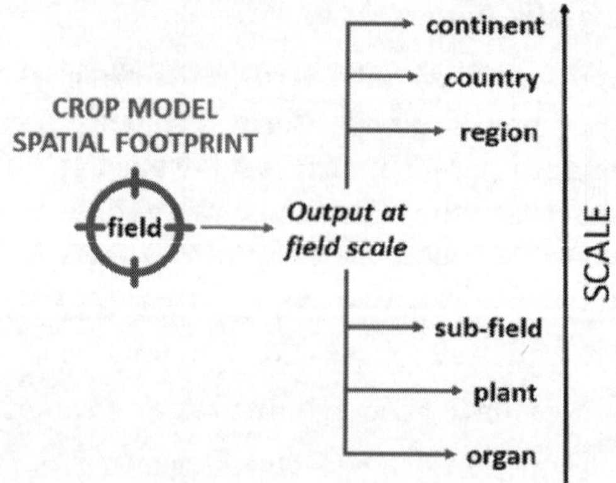

Fig. 7.2: Crop model designed at the field scale and beyond

Verification is often referred to as debugging or quality assurance of the model's code and logical flow. It answers the question: Is the model solving the equations correctly? not Are the equations the right ones for reality?

7.4.1 Why is Verification Necessary?

- **Ensuring Internal Consistency:** Verification confirms that the model's internal calculations are free of mathematical, logical, or programming errors. A model might seem to produce reasonable outputs during calibration, but subtle errors in its code could lead to incorrect results, especially under specific input conditions or at the extremes of its operational range.
- **Debugging and Error Detection:** It helps pinpoint specific bugs or logical flaws in the model's code or its underlying conceptual framework. Identifying and correcting these errors *before* calibration and validation saves significant time and effort.
- **Transparency and Trust:** A verified model is more transparent. If users understand that the model's internal mechanisms are sound, it builds trust in its predictions.
- **Foundation for Further Development:** A well-verified elementary model provides a stable and reliable foundation for future development, such as incorporating more complex processes or extending its applicability.

7.4.2 Key Aspects and Techniques of Verification

Verification involves a systematic review of the model's source code, equations, and internal logic. For elementary models, which are simpler, this process can be more straightforward but is still vital.

- **Code Review / Inspection**
  - **Process:** Carefully examining the model's source code (if available and accessible) line by line. This involves checking syntax, variable assignments, loops, conditional statements, and function calls to ensure they are logically correct and match the intended algorithms.
  - **Purpose:** To catch typos, logical inversions (e.g., > instead of <), incorrect loop conditions, or misapplied formulas. Even in elementary models, simple coding mistakes can lead to significant errors.
- **Dimensional Consistency Check**
  - **Process:** Ensuring that all equations in the model are dimensionally consistent. This means that the units on both sides of an equation match. For example, if an output is in kilograms, then all terms on the right-hand side, after all multiplications and divisions, should combine to produce kilograms.
  - **Purpose:** Catches errors in fundamental formulas or conversion factors (e.g., mixing metric and imperial units without proper conversion, or incorrect units for constants). This is particularly important for models dealing with physical processes like radiation, temperature, and water.
- **Analytical Solutions / Benchmarking with Simple Cases**
  - **Process:** For very simple scenarios or isolated components of the model, it might be possible to calculate the expected output manually or using a separate, trusted analytical solution. The model is then run with these specific inputs, and its output is compared to the manually calculated or benchmarked value.
  - **Purpose:** To confirm that individual equations or small sub-routines (e.g., GDD calculation for a constant temperature, or RUE for a fixed radiation input) are producing the correct numerical results. This is especially effective for elementary models due to their simplicity.
- **Boundary and Extreme Value Testing**
  - **Process:** Run the model with input values at the extreme ends of their expected range (e.g., minimum and maximum possible temperatures, zero rainfall, extremely high radiation, etc.).

- **Purpose:** To check if the model behaves logically and generates plausible outputs under these extreme conditions. Sometimes, errors or unexpected behaviors only manifest when inputs push the boundaries of the code's logic. This can reveal division-by-zero errors, overflow errors, or unexpected outputs from conditional statements.

- **Trace Analysis / Stepping Through Code**
  - **Process:** For more complex elementary models or specific sections, one might step through the code line by line (using a debugger or manual print statements) to observe the values of intermediate variables at each step.
  - **Purpose:** To verify that the calculations are progressing as expected and that variables are being updated correctly throughout the simulation. This helps identify where an error might first appear in a sequence of calculations.

- **Comparison with Published Algorithms**
  - **Process:** If the model implements algorithms or equations that are well-established and published in scientific literature (e.g., specific GDD formulas, simple ET equations), compare the model's implementation directly against these published methods.
  - **Purpose:** To ensure that the model is accurately translating the scientific understanding into executable code.

7.4.4 Verification vs. Validation - A Critical Distinction

It's crucial to understand that verification is distinct from validation:

- **Verification:**Are we building the model right? (Checking the internal consistency and correctness of the model's construction). A verified model might still be a poor representation of reality if its underlying theory or parameters are flawed.
- **Validation:**Are we building the right model? (Checking if the model's outputs accurately represent real-world phenomena). A valid model might have some minor internal inefficiencies but still produce good predictions.

7.4.5 Order of Operations

Verification should ideally precede or occur concurrently with early stages of calibration. It makes little sense to spend time calibrating parameters or validating a model's predictive power if its internal calculations are flawed.

A verified model provides a solid foundation for meaningful calibration and validation.

For elementary crop models, thorough verification ensures that their simple formulas and logic are correctly implemented, thus allowing calibration and validation to genuinely assess their ability to represent real-world crop behavior, rather than simply debugging computational mistakes.

7.5 Sensitivity of Elementary Crop Models

Once an elementary crop model has been verified (ensuring its internal correctness) and calibrated (tuned to observed data for a specific context), the next vital step is sensitivity analysis. Sensitivity analysis is the process of systematically assessing how the model‘s outputs change in response to variations in its input parameters or driving variables (like weather data). It‘s about understanding which inputs or internal coefficients have the most significant influence on the model‘s predictions.

Think of it as stress-testing the model: How much does the model‘s prediction of yield change if temperature increases by 1^{0}C or if rainfall decreases by 20%? Which input has the biggest ‚lever‘ on the output?

7.5.1 Why is Sensitivity Analysis Important?

- **Identifying Critical Inputs:** It helps pinpoint which weather variables, soil properties, or management practices are the most influential drivers of crop growth and yield within the model. This is invaluable for focusing research, data collection efforts, and management interventions.
- **Understanding Model Behavior:** It provides insights into how the model fundamentally works and its underlying assumptions. If a small change in one input causes a disproportionately large change in output, it highlights a highly sensitive (and potentially critical or vulnerable) aspect of the modeled system.
- **Prioritizing Data Collection:** Knowing which inputs are most sensitive allows researchers to prioritize collecting high-quality, accurate data for those specific variables. Less sensitive inputs can tolerate a wider margin of error without significantly impacting model predictions.
- **Assessing Uncertainty:** It helps quantify the uncertainty in model predictions stemming from uncertainty in input data. If an output is highly sensitive to an uncertain input, the prediction itself will carry a high degree of uncertainty.

- **Targeting Adaptation Strategies:** For climate change impact assessment, sensitivity analysis can identify how vulnerable a crop is to projected changes in specific weather parameters (e.g., increased temperature, reduced rainfall). This information is crucial for developing targeted adaptation strategies in breeding or management.
- **Parameter Estimation Guidance (Post-Calibration):** While calibration determines the best fit, sensitivity analysis can later highlight if small errors in a specific parameter‘s value during calibration could lead to large prediction errors.

7.5.2 Types of Sensitivity Analysis for Elementary Models

For elementary models, sensitivity analysis typically focuses on direct and relatively straightforward changes to inputs and parameters.

- **One-at-a-Time (OAT) or Local Sensitivity Analysis**
  - **Process:** This is the simplest and most common method for elementary models. It involves varying one input parameter or driving variable at a time over a specified range, while keeping all other inputs constant at their baseline (reference) values. The model is run for each varied value, and the change in output is recorded.
  - **Example (GDD Model):** Vary Tbase by ±1^{0}C or vary daily mean temperature by ±2^{0}C and observe how the predicted flowering date changes.
  - **Example (RUE Model):** Vary the RUE (ϵ) by ±10 or the total solar radiation by ±20 and observe the change in total biomass.
  - **Visualization:** Results are often presented in tables or simple plots showing output change versus input change.
  - **Limitations:** This method does not account for interactions between parameters. If the effect of parameter A depends on the value of parameter B, OAT analysis won‘t fully reveal that.
- **Global Sensitivity Analysis (for more complex elementary models or a preliminary step for advanced models)**
  - **Process:** More advanced techniques that explore the entire input space simultaneously, varying multiple parameters and inputs at once. This can involve statistical methods like Monte Carlo simulations (randomly sampling input values from their plausible ranges) or variance-based methods (e.g., Sobol‘ indices, which quantify how much of the output variance is due to each input and its interactions).

- **Purpose:** To capture interactions between inputs that OAT analysis misses.
- **Applicability to Elementary Models:** While often associated with complex models, simpler forms of global sensitivity can be applied even to elementary models, especially if they have several interacting parameters. For example, simultaneously varying rainfall and temperature in a simple water-limited yield model.
- **Limitations:** Can be more computationally intensive than OAT, even for elementary models.

- **Key Variables and Parameters for Sensitivity Testing in Elementary Models**
  - **Weather Inputs**
    - **Temperature:** Changes in Tmax , Tmin , mean temperature, and the frequency/intensity of extreme temperature events (e.g., heat stress days, frost days).
    - **Solar Radiation:** Changes in daily incoming solar radiation.
    - **Rainfall:** Changes in total amount, intensity, and distribution (e.g., frequency of dry spells).
    - **Humidity/VPD:** For water-related models.
  - **Crop-Specific Parameters**
    - Tbase **(for GDD models):** How sensitive is the predicted phenology to small changes in this value?
    - **Accumulated GDD for stages:** How sensitive is the timing of flowering or maturity to these calibrated sums?
    - **Radiation Use Efficiency (ϵ) (for RUE models):** How much does total biomass change if this efficiency varies?
    - **Yield Response Factor** (Ky) **(for water-limited models):** How sensitive is yield reduction to this factor?
    - **Phenological coefficients:** For models that explicitly use photoperiod or other non-temperature drivers for development.
  - **Interpreting Sensitivity Analysis Results**
  - **High Sensitivity:** If a small change in an input results in a large change in the output, the model is highly sensitive to that input. This input is critical and needs to be measured accurately. It also indicates a potential vulnerability of the real-world crop system to changes in that factor.

- **Low Sensitivity:** If even a large change in an input results in only a small change in the output, the model is less sensitive to that input. Less precision is needed in measuring this input, and the real-world crop system is more robust to variations in that factor.
- **Non-linear Responses:** Sensitivity analysis can reveal non-linear responses. For example, yield might be highly sensitive to rainfall below a certain threshold, but less sensitive above it (diminishing returns).
- **Interaction Effects:** Global sensitivity analysis can reveal that the effect of one input depends on the value of another (e.g., the sensitivity to temperature might change depending on water availability).

For elementary crop models, sensitivity analysis is a pragmatic way to understand their robustness and identify the most influential environmental drivers that underpin their simplified predictions. It helps users appreciate the model's strengths and weaknesses, guides data acquisition, and provides foundational insights for agricultural decision-making and adaptation strategies.

7.6 Advantages and Disadvantages of Elementary Crop Models

Elementary crop growth models, as discussed, represent simplified mathematical representations of crop growth and development. Their design inherently balances predictive power with ease of use and data requirements. Understanding their strengths and weaknesses is crucial for selecting the appropriate modeling approach for a given agricultural challenge.

7.6.1 Advantages of Elementary Crop Models

Elementary models offer several compelling benefits, particularly in situations where resources (data, computational power, expertise) are limited, or where a quick, broad understanding is sufficient.

- **Simplicity and Ease of Use**
    - **Low Complexity:** They involve fewer equations and parameters compared to complex mechanistic models. This makes them easier to understand, implement, and run, even for users with limited modeling experience.
    - **Accessibility:** Their straightforward nature makes them highly accessible to a wider audience, including extension agents, farmers (through simplified tools), and students.

  - **Faster Development and Computation:** Building, coding, and running these models is significantly faster and less resource-intensive. This allows for rapid prototyping and quick turnaround times for predictions.

- **Minimal Data Requirements**
  - **Less Input Data:** They typically require only basic, readily available weather data (e.g., daily maximum/minimum temperature, total rainfall, total solar radiation) and a few crop-specific constants. This is a major advantage in regions or situations where comprehensive, high-resolution data sets are scarce.
  - **Reduced Calibration Effort:** With fewer parameters, the calibration process is generally less demanding in terms of the amount and type of observed data needed, and the computational effort for parameter fitting.

- **Clear Attribution of Effects**
  - **Direct Relationships:** Because they often focus on one or two dominant weather drivers (e.g., GDD for temperature, RUE for radiation), it's easier to directly attribute changes in output to changes in specific inputs. This clarity aids in understanding fundamental crop-weather relationships.
  - **Conceptual Clarity:** They are excellent for illustrating basic ecological and physiological principles in a quantitative manner, serving as valuable educational and communication tools.

- **Cost-Effectiveness**
  - Lower demands for data collection, software, and computational infrastructure translate into lower overall costs for model development, implementation, and application. This makes them a feasible option for resource-constrained environments.

- **Suitable for Initial Assessments and Screening**
  - **First-Pass Predictions:** They can provide quick, indicative predictions of potential yield or phenological timing, useful for early-season forecasts or rough suitability assessments of new crops/cultivars for a region.
  - **Broad-Scale Planning:** Useful for macro-level agricultural planning, identifying general agro-climatic zones, or understanding broad regional patterns of crop performance.

7.6.2 Disadvantages of Elementary Crop Models

The very simplicity that confers advantages also introduces significant limitations, particularly when accuracy, mechanistic insight, and predictive power under complex or novel conditions are paramount.

- **Lack of Mechanistic Detail and Explanatory Power**
  - **Black Box" Nature:** They often describe *what* happens (e.g., yield decreases with drought) but not *why*(e.g., through stomatal closure, reduced photosynthesis, or impaired nutrient uptake). They don't explicitly simulate the underlying biological and physical processes.
  - **Limited Physiological Insight:** This lack of mechanism means they provide little understanding of the internal plant responses to environmental factors, hindering deeper scientific inquiry.
- **Poor Handling of Interactions and Stresses**
  - **Ignored Synergies/Antagonisms:** They typically struggle to capture the complex, non-linear interactions between multiple weather factors (e.g., the combined effect of heat and drought, or how high CO_2 modifies water use efficiency). The impact of one factor might depend on the level of another, which simple models often miss.
  - **Inadequate Stress Representation:** They use simplistic ways (if any) to represent stress effects (e.g., empirical reduction factors, often assumed constant). They cannot realistically simulate the physiological impacts of severe heat stress on pollen, or the effects of waterlogging on root oxygen supply and toxic compound accumulation.
  - **Weak for Extreme Events:** Because extreme weather events often induce threshold-based, non-linear responses and complex interactions, elementary models are generally poor at predicting their specific impacts accurately.
- **Limited Extrapolation and Predictive Power for Novel Conditions:**
  - **Empirical Constraint:** Since they are largely based on empirical relationships derived from historical data, they perform poorly when extrapolated to conditions outside the range of the data used for their development or calibration.
  - **Climate Change Challenge:** This is a major drawback for climate change impact assessments. They cannot reliably predict crop responses

to unprecedented future climates (e.g., higher temperatures, altered CO_2 levels, new combinations of stresses) for which no historical analog exists. They struggle with novel crop varieties or management systems.

- **Lower Accuracy and Precision**
  - **General vs. Specific:** Their simplicity sacrifices precision. While they might provide a reasonable first pass, they often cannot achieve the high level of accuracy needed for site-specific management decisions or detailed scientific investigations.
  - **Averages Mask Detail:** Relying on aggregate daily or seasonal weather data can mask the critical impact of hourly weather fluctuations or the precise timing of stress events within a day or a sensitive short period.
- **No Feedback Mechanisms**
  - They typically lack dynamic feedback loops. For example, if a plant experiences severe water stress, its leaf area might reduce, which in turn affects light interception and transpiration. Elementary models generally do not simulate such adaptive changes in real-time.

In essence, elementary crop growth models are valuable for their simplicity and accessibility, making them effective for initial assessments and educational purposes. However, their limitations in mechanistic detail, handling of complex interactions and stresses, and ability to extrapolate reliably mean that for comprehensive, accurate, and truly predictive applications, especially in the context of a rapidly changing climate, more sophisticated process-based models are often necessary. The choice of model depends heavily on the specific research question, data availability, and the desired level of accuracy and mechanistic insight.

Multiple Choice Questions (Bold means correct answer)

1. What is the primary purpose of **calibration** in elementary crop models?
   a) To prove the model's theoretical correctness.
   b) To adjust model parameters to match observed data for specific conditions.
   c) To assess the model's performance on entirely new datasets.
   d) To simplify the model's equations.
2. Why is calibration necessary for a crop model?
   a) To make it universally applicable without change.
   b) To account for cultivar specificity and local environmental nuances.

c) To reduce its computational requirements.

d) To eliminate all future sources of error.

3. In Thermal Time (GDD) models, which parameters are typically adjusted during calibration?

a) Daily rainfall totals.

b) The base temperature (Tbase) and accumulated GDD for phenological stages.

c) Solar radiation intensity.

d) Wind speed coefficients.

4. If a GDD model consistently predicts flowering too late compared to observations, which parameter might be adjusted during calibration?

a) Increase the daily maximum temperature.

b) Reduce the required GDD sum for the flowering stage.

c) Increase the total growing season length.

d) Change the soil type in the model input.

5. What is the core parameter calibrated in Radiation Use Efficiency (RUE) models?

a) Leaf Area Index (LAI)

b) Water Holding Capacity

c) Radiation Use Efficiency (ϵ)

d) Number of tillers per plant

6. The calibration process is typically:

a) A one-time, simple calculation.

b) An iterative process involving data collection, simulation, adjustment, and refinement.

c) Conducted only by programming experts.

d) Done without any need for observed field data.

7. What is the main purpose of **validation** in elementary crop models?

a) To fine-tune model parameters.

b) To rigorously test the model's predictive accuracy on independent datasets.

c) To debug coding errors.

d) To identify the most sensitive input variables.

8. Which type of data is *essential* for validation?

a) Economic market data.

b) Data used repeatedly during the calibration process.

c) Independent datasets not used during calibration.

d) Genetic sequences of crop pests.

9. If a scatter plot of observed vs. simulated values shows points consistently above the 1:1 line during validation, it suggests:

   a) Perfect model performance.

   b) The model has a systematic positive bias (overprediction).

   c) The model has a systematic negative bias (underprediction).

   d) The model is perfectly calibrated.

10. Which statistical metric measures the proportion of variance in observed data explained by the model?

    a) Root Mean Square Error (RMSE)

    b) Mean Absolute Error (MAE)

    c) Coefficient of Determination (R^2)

    d) Bias

11. A Normalized Root Mean Square Error (nRMSE) of less than 10% for a model's prediction is generally considered:

    a) Poor performance.

    b) Fair performance.

    c) Good performance.

    d) Excellent performance.

12. What does a high R^2 value in validation NOT guarantee by itself?

    a) That the model explains a large proportion of variability.

    b) That there is a systematic bias in predictions.

    c) That the model is perfectly accurate.

    d) That the observed data is reliable.

13. **Verification** of an elementary crop model focuses on:

    a) Adapting the model to new cultivars.

    b) Assessing the model's predictive accuracy against field data.

    c) The internal logic and computational correctness of the model itself.

    d) Determining economic returns of crop production.

14. Verification helps answer which question?

    a) Are we building the right model?

    b) Is the model performing as intended, free from coding errors?

    c) Is the model economically viable?

    d) Is the model applicable to all global regions?

15. What technique is used in verification to check if individual equations or sub-routines produce correct numerical results for simple cases?
    a) Statistical regression.
    b) Analytical solutions / Benchmarking.
    c) Sensitivity analysis.
    d) Validation with large datasets.
16. Running a model with input values at the extreme ends of their expected range (e.g., very high or very low temperatures) is a technique used in:
    a) Calibration.
    b) Routine prediction.
    c) Boundary and Extreme Value Testing (part of verification).
    d) Model simplification.
17. Which of the following is an ideal order for crop model development and assessment?
    a) Validation →Calibration →Verification
    b) Calibration →Validation →Verification
    c) Verification →Calibration →Validation
    d) Sensitivity →Verification →Calibration
18. **Sensitivity analysis** for elementary crop models systematically assesses:
    a) The overall accuracy of the model after calibration.
    b) How the model's outputs change in response to variations in its input parameters or driving variables.
    c) The internal logical consistency of the model.
    d) The ethical implications of using the model.
19. What is a primary benefit of sensitivity analysis?
    a) It eliminates all uncertainty in model predictions.
    b) It identifies which inputs or internal coefficients have the most significant influence on model predictions.
    c) It directly optimizes irrigation schedules.
    d) It guarantees successful crop breeding programs.
20. If a crop model's output (e.g., yield) changes disproportionately with a small change in one input parameter (e.g., temperature), the model is said to be:

a) Insensitive to that input. b) Overfitted.
c) Highly sensitive to that input. d) Perfectly calibrated.

21. Which type of sensitivity analysis involves varying one input parameter at a time while keeping all others constant?
    a) Global Sensitivity Analysis.
    b) One-at-a-Time (OAT) Sensitivity Analysis.
    c) Monte Carlo Simulation.
    d) Variance-Based Sensitivity Analysis.
22. Sensitivity analysis helps in prioritizing data collection efforts by:
    a) Showing which data are easiest to collect.
    b) Identifying inputs to which the model is highly sensitive, thus requiring accurate measurement.
    c) Demonstrating which data have no impact on the model.
    d) Reducing the overall amount of data needed.
23. A disadvantage of the One-at-a-Time (OAT) sensitivity analysis is that it:
    a) Is computationally very demanding.
    b) Does not account for interactions between parameters.
    c) Requires too much expert knowledge.
    d) Can only be applied to very complex models.
24. For a water-limited yield model, which parameter‘s sensitivity might be tested to understand a crop‘s response to water deficit?
    a) Photosynthetically Active Radiation (PAR)
    b) Base Temperature (Tbase)
    c) Yield Response Factor (Ky)
    d) Plant population density
25. If sensitivity analysis reveals that a model‘s predicted yield is highly sensitive to the exact value of daily maximum temperature during flowering, this suggests:
    a) That daily maximum temperature is irrelevant to yield.
    b) That accurate measurement of this temperature is crucial for model reliability, and the real crop is vulnerable to heat at this stage.
    c) That the model is flawed and should be discarded.
    d) That calibration was done incorrectly.

8

Potential and Achievable Crop Production

8.1 Introduction

The fundamental goal of agriculture is to harness the earth's resources to produce food, feed, fiber, and fuel to sustain a growing global population. At the heart of this endeavor lies the concept of crop production, a complex interplay of genetic potential, environmental factors, and human management. This chapter, Potential and Achievable Crop Production, delves into the crucial distinction between what a crop can produce under ideal circumstances and what it *actually* produces in the farmer's field. Understanding this dichotomy is paramount for addressing global food security challenges, optimizing resource utilization, and developing sustainable agricultural practices.

We will begin by exploring the theoretical maximum yield, often termed "potential production," which represents the inherent capacity of a crop variety to produce biomass and yield when all limiting factors are eliminated. This potential is primarily dictated by a crop's genetic makeup and the prevailing climatic conditions, such as solar radiation, temperature, and carbon dioxide concentration. Subsequently, we will shift our focus to achievable production, the real-world yield obtained by farmers. This actual yield invariably falls short of the potential due to a myriad of biotic (pests, diseases, weeds), abiotic (water scarcity, nutrient deficiencies, extreme temperatures, poor soil health), and socio-economic (lack of access to inputs, knowledge, or markets) constraints. The disparity between potential and achievable yield, known as the "yield gap," represents a significant opportunity for enhancing agricultural output without necessarily expanding cultivated land. By meticulously examining the factors that contribute to this gap, we can identify targeted interventions and innovative strategies to bridge it, thereby moving closer to realizing the full productive capacity of our agricultural systems. This chapter will therefore provide a comprehensive framework for understanding the science behind crop yields and the practical approaches required to optimize them for a more food-secure future.

8.2 Potential Production

Potential crop production represents the ultimate biological limit of a crop's yield under ideal growing conditions, free from any limitations or stresses. It is a theoretical maximum, serving as a benchmark against which actual farm yields are compared to identify yield gaps and opportunities for improvement.2

8.2.1 Meaning of Potential Production

Potential production is defined as the yield of a crop that can be achieved when all growth-limiting factors are at their optimum, and growth-reducing factors are absent. This includes:

- **Optimal genotype:** The use of the highest-yielding variety available, specifically bred for maximum biomass accumulation and partitioning to economic yield.
- **Optimal climatic conditions:** Sufficient solar radiation (light), ideal temperatures (day and night), and adequate atmospheric CO_2 concentration throughout the growing season.
- **Non-limiting water supply:** Crops receive all the water they need, precisely when they need it, with no water deficit or excess (waterlogging).
- **Non-limiting nutrient supply:** All essential macro and micronutrients are available in optimal concentrations, without deficiencies or toxicities.
- **Absence of pests, diseases, and weeds:** The crop is entirely free from any biological stresses that could reduce its growth or yield.
- **Ideal soil physical conditions:** The soil provides optimal aeration, structure, and root penetration, without compaction or other physical impediments.

In essence, potential production is what a crop could produce if it were grown in a perfect environment, allowing its full genetic potential to be expressed.

8.2.2 Significance of Potential Production

Understanding potential production holds immense significance for various aspects of agriculture and food security:

- **Benchmarking and Yield Gap Analysis:** Potential yield provides a crucial benchmark for evaluating the performance of actual farming systems. By comparing achievable yields to potential yields, scientists and policymakers can quantify the yield gap– the difference between what is produced and what could be produced. This gap highlights the inefficiencies in current agricultural practices and identifies areas where improvements are most needed.

- **Identifying Research Priorities:** A large yield gap often indicates that there is significant scope for increasing productivity through improved agronomic practices, better resource management, or the development of more resilient crop varieties.This understanding helps in prioritizing agricultural research efforts, focusing on factors that are currently limiting yields.
- **Food Security Planning:** For a rapidly growing global population, knowing the potential production capacity of different regions and crops is vital for long-term food security planning. It helps in assessing the maximum food output from existing arable land and understanding the potential for increasing food supply.
- **Resource Optimization:** By understanding the ideal conditions for a crop, we can identify how to optimally allocate resources like water, fertilizers, and land to maximize efficiency. This is particularly important in regions facing resource scarcity.
- **Environmental Sustainability:** Maximizing potential production often involves optimizing input use, which can lead to more efficient resource utilization and potentially reduce the environmental footprint per unit of output. For example, if a crop can yield more with the same amount of land, it reduces pressure to convert natural habitats into agricultural land.
- **Genetic Improvement Targets:** Plant breeders use potential yield as a target for developing new, higher-yielding crop varieties. They aim to enhance the genetic potential of crops to capture more solar energy, utilize nutrients more efficiently, and convert biomass more effectively into economic yield.
- **Policy Formulation:** Governments and international organizations use potential production estimates to formulate agricultural policies, investment strategies, and development programs aimed at improving farmer livelihoods and national food security.

8.2.3 Factors Determining Potential Production

The potential production of a crop is fundamentally determined by the interaction of genetic and climatic factors. When all other factors are non-limiting, these become the primary drivers:

- **Solar Radiation (Light)**
    - **Photosynthetically Active Radiation (PAR):** Plants convert light energy into chemical energy through photosynthesis. The amount

of PAR intercepted by the crop canopy directly influences the rate of photosynthesis and, consequently, biomass accumulation.

- **Light Interception:** The efficiency with which a crop canopy intercepts incoming solar radiation depends on its Leaf Area Index (LAI) and canopy architecture. A fully closed, healthy canopy that intercepts nearly all available PAR is crucial for maximizing potential production.
- **Duration of Sunshine:** Longer periods of bright sunshine generally lead to higher potential yields, assuming other factors are optimal.

- **Temperature**
    - **Optimal Temperature Range:** Each crop species and variety has an optimal temperature range for its various physiological processes, including photosynthesis, respiration, growth, and development (e.g., flowering, grain filling). Temperatures outside this range, whether too low (cold stress, frost) or too high (heat stress), can significantly reduce potential productivity.
    - **Day and Night Temperatures:** The difference between day and night temperatures (diurnal temperature range) can also affect growth and respiration rates, impacting net biomass accumulation.
    - **Accumulated Heat Units (Growing Degree Days):** The rate of plant development is often related to the accumulation of heat units above a base temperature. A sufficient accumulation of heat units during the growing season is necessary for the crop to complete its life cycle and reach its full yield potential.

- **Carbon Dioxide (CO_2) Concentration**
    - **Photosynthesis Substrate:** CO_2 is a fundamental substrate for photosynthesis. In the atmosphere, CO_2 concentration is around 430 ppm (as of mid 2025). While not a limiting factor in the sense of being absent, higher CO_2 concentrations (e.g., in controlled environments like greenhouses) can lead to increased photosynthetic rates, particularly for C3 plants, thus increasing potential biomass production.
    - **C3 vs. C4 Plants:** The response to CO_2 differs between C3 crops (e.g., wheat, rice, soybean) and C4 crops (e.g., maize, sorghum, sugarcane). C3 plants generally show a more significant increase in photosynthetic efficiency and yield under elevated CO_2, while C4 plants are less responsive as their photosynthetic pathway is already highly efficient at ambient CO_2 levels.

- **Crop Genetic Potential (Genotype)**
  - **Photosynthetic Efficiency:** Different crop varieties possess varying efficiencies in converting light energy into biomass.
  - **Biomass Partitioning:** The way a plant allocates its accumulated biomass to different organs (leaves, stems, roots, and reproductive parts) is genetically determined. Varieties bred for high yield typically have a high harvest index– the ratio of economic yield (e.g., grain) to total above-ground biomass.
  - **Growth Duration:** The length of the growing season required by a particular variety influences its potential to accumulate biomass.
  - **Resistance to Abiotic/Biotic Stresses:** While potential production assumes no stresses, a variety's genetic robustness against minor inherent variations in temperature or CO_2, or its ability to efficiently utilize light, contributes to its *inherent* potential.

8.2.4 Methods of Estimation of Potential Production

Estimating potential production is a complex task as it involves simulating ideal conditions that are rarely fully met in reality. However, several methods are employed, ranging from experimental approaches to sophisticated modeling:

- **Controlled Environment Experiments (Growth Chambers/ Greenhouses)**
  - **Method:** Crops are grown in highly controlled environments where light, temperature, humidity, CO_2 levels, water, and nutrients can be precisely regulated and maintained at optimal levels. Pests and diseases are meticulously excluded.
  - **Advantages:** Provides the most direct measurement of genetic potential under near-ideal conditions. Allows for isolation and study of individual environmental factors.
  - **Limitations:** High cost, limited scale, and difficulty in fully replicating complex field conditions (e.g., natural wind, soil microflora, spatial variability). Results may not be directly translatable to open field agriculture due to the artificial environment.
- **Field Experiments with Optimal Management (Maximum Yield Trials)**
  - **Method:** Conducted in research plots where every possible effort is made to provide optimal growing conditions. This includes:

- Selecting the best available crop variety for the region.
- Applying irrigation to ensure no water stress.
- Supplying nutrients in optimal amounts and timing based on soil tests and crop demand.
- Implementing comprehensive pest, disease, and weed control programs.
- Ensuring good soil structure and drainage.

- **Advantages:** Provides a more realistic estimate of potential production under field conditions, as it accounts for the natural variations in solar radiation and temperature for a given location.
- **Limitations:** Still challenging to eliminate *all* limiting factors perfectly. Some inherent soil issues or unpredictable weather events (e.g., hailstorms) can still affect yields. It's an *achievable maximum* for a specific location, not necessarily the absolute biological potential.

- **Crop Growth Simulation Models (Process-Based Models)**

Method: These are mathematical models that simulate the physiological processes of crop growth (photosynthesis, respiration, biomass partitioning, water uptake, nutrient uptake) in response to environmental factors. Inputs typically include daily weather data (solar radiation, temperature, rainfall), soil characteristics, crop genetic parameters, and management practices.

Examples: CERES-Maize, DSSAT (Decision Support System for Agrotechnology Transfer), WOFOST (World Food STudies), AquaCrop.

Procedure for Potential Yield Estimation: To estimate potential yield using these models, the input parameters for water and nutrients are set to non-limiting conditions, and the absence of pests/diseases is assumed. The model then predicts the maximum biomass accumulation and yield based solely on the prevailing climatic conditions (light, temperature, CO_2) and the crop's genetic parameters.

- **Advantages:** Can simulate potential production for different locations and climate scenarios without extensive physical experimentation. Useful for scenario analysis (e.g., impact of climate change). Can analyze the contribution of different factors.
- **Limitations:** Require extensive data for calibration and validation. Accuracy depends on the robustness of the underlying physiological equations and the quality of input data. May not fully capture all complex interactions in real-world systems.

- **Agro-Ecological Zone (AEZ) Methods**
  - **Method:** This approach divides land into zones based on their climate, soil, and landform characteristics. For each AEZ, the potential yield of various crops is estimated by considering the climatic resources (temperature, radiation) available during the growing period and making assumptions about optimal soil and management conditions (Fig.8.1).

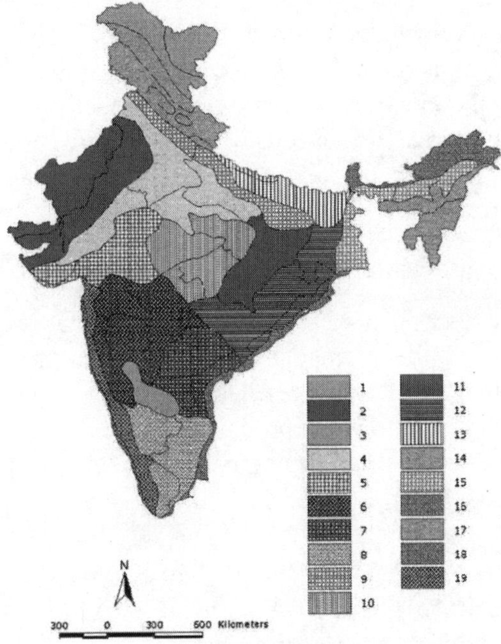

Fig. 8.1: Agro-ecological zones in India

  - **Advantages:** Useful for large-scale (regional to global) assessments of land productivity potential. Helps in identifying suitable areas for specific crops.
  - **Limitations:** Often relies on simplified assumptions and aggregated data, which may not capture local variability. Less precise than field-specific models or experiments.
- **Remote Sensing and Geospatial Analysis (for broad-scale assessment)**
  - **Method:** While not directly estimating potential in the same way as the above, remote sensing data (satellite imagery, drone imagery) can be used in conjunction with crop models or AEZ frameworks to assess and map areas with high potential. Vegetation indices (like NDVI) can reflect canopy health and biomass, which are indicators of how close a crop is to its potential.

- **Advantages:** Provides broad spatial coverage and can monitor crop performance over large areas. Useful for identifying areas performing well and those underperforming, indirectly indicating where potential is being realized or missed.
- **Limitations:** Requires ground truthing for validation. Cannot directly measure potential but rather current performance relative to a benchmark or model.

In conclusion, potential production is a theoretical yet powerful concept that helps agricultural scientists and practitioners understand the upper limits of crop productivity. By meticulously studying the factors that determine this potential and employing various estimation methods, we gain invaluable insights into how to push the boundaries of agricultural output, ensuring sustainable food production for future generations.

8.3 Achievable Production

Achievable production, also known as actual production or farm yield, represents the crop yield obtained by farmers under their prevailing field conditions and management practices. Unlike potential production, which is a theoretical maximum, achievable production is a practical reality, reflecting the multitude of environmental, biological, and socio-economic factors that interact to influence crop growth and yield in real-world agricultural systems.

8.3.1 Meaning of Achievable Production

Achievable production is the average yield of a crop that farmers regularly attain in a specific region or on a particular farm over a given period (e.g., a growing season or an average of several years). It is significantly influenced by:

- **Sub-optimal environmental conditions:** Unlike the ideal conditions assumed for potential yield, actual fields often face varying degrees of water stress (drought or waterlogging), nutrient deficiencies, and unfavorable temperatures.
- **Presence of biotic stresses:** Pests, diseases, and weeds are almost always present in agricultural fields and compete with or damage the crop, leading to yield losses.
- **Farmer management decisions:** The choices farmers make regarding variety selection, planting density, fertilization, irrigation, pest and weed control, and other cultural practices directly impact the achievable yield. These decisions are often constrained by economic resources, knowledge, labor availability, and market access.

- **Soil variability:** Even within a single field, soil properties can vary significantly, leading to differences in water holding capacity, nutrient availability, and root penetration, which in turn affect local yields.

In essence, achievable production is the output that results from the dynamic interaction between the crop's genetic potential, the actual environmental conditions, and the applied management practices, often under real-world constraints.

8.3.2 Significance of Achievable Production

Understanding and improving achievable production is critical for several practical and strategic reasons:

- **Direct Measure of Farmer Livelihood:** Achievable production directly translates to farmers' income and profitability. Higher achievable yields, especially when economically viable, lead to better livelihoods for agricultural communities.
- **Foundation for Food Security:** National and global food security relies on the collective achievable production of millions of farmers. Monitoring trends in achievable yields is essential for assessing food supply and identifying regions vulnerable to food shortages.
- **Yield Gap Quantification:** Comparing achievable production with potential production allows for the quantification of the yield gap. This gap highlights the unrealized potential and provides a clear target for research, extension, and development efforts aimed at increasing productivity.
- **Policy Formulation and Investment:** Data on achievable production informs agricultural policies, subsidies, and investments. Governments use this information to design programs that support farmers, improve infrastructure, and promote practices that can sustainably boost yields.
- **Benchmarking and Performance Evaluation:** Achievable yields serve as a benchmark for evaluating the effectiveness of new technologies, management practices, and extension programs introduced in a region.
- **Resource Allocation:** Understanding which factors are limiting achievable yields helps in prioritizing and allocating scarce resources (e.g., water, fertilizers, improved seeds) to achieve the greatest impact on productivity.
- **Sustainability Assessment:** While aiming for higher yields, it's crucial to ensure that the methods used to increase achievable production are environmentally sustainable, preventing soil degradation, water pollution, and biodiversity loss.

8.3.3 Factors Determining Achievable Production

Achievable production is a result of a complex interplay of various limiting and reducing factors. These can be broadly categorized as:

- **Abiotic Limiting Factors (Resource Scarcity)**
    - **Water Availability:** Insufficient rainfall (drought), erratic rainfall patterns, or inadequate irrigation infrastructure are major limitations in many rainfed systems. Conversely, excessive rainfall and poor drainage can lead to waterlogging, depriving roots of oxygen.
    - **Nutrient Availability:** Soil nutrient deficiencies (e.g., nitrogen, phosphorus, potassium, micronutrients) are widespread. Inadequate or imbalanced fertilizer application, low soil organic matter, and poor nutrient cycling can severely limit yield.
    - **Solar Radiation:** While generally not scarce during the day, cloudy weather, short growing seasons in certain climates, or high plant densities leading to mutual shading can reduce the amount of light intercepted by individual plants.
    - **Temperature Extremes:** Temperatures outside the optimal range for prolonged periods (e.g., heatwaves, cold snaps, frost) can cause significant physiological stress and damage, reducing growth and yield.
- **Soil Physical and Chemical Properties**
    - **Compaction:** Impedes root growth and water infiltration.
    - **Poor Drainage:** Leads to waterlogging and anaerobic conditions.
    - **Low Organic Matter:** Reduces nutrient retention, water holding capacity, and soil structure.
    - **pH Extremes:** Acidic or alkaline soils can render essential nutrients unavailable or toxic.
    - **Salinity/Sodicity:** High salt concentrations inhibit water uptake by plants.
- **Biotic Reducing Factors (Stress and Competition)**
    - **Pests:** Insect pests (e.g., stem borers, aphids, defoliators) cause direct damage to plant parts, reducing biomass and yield.
    - **Diseases:** Pathogens (fungi, bacteria, viruses) cause infections that impair physiological functions, leading to reduced growth, quality, and yield.

  - **Weeds:** Compete with crops for light, water, nutrients, and space, significantly reducing crop yield if not effectively managed.
  - **Other Organisms:** Nematodes, rodents, and birds can also cause significant yield losses.

- **Management Factors (Human Interventions)**
  - **Crop Variety Selection:** Use of varieties with lower genetic potential, poor adaptation to local conditions (e.g., drought susceptibility), or low disease resistance.
  - **Sowing Date and Method:** Incorrect planting time or method (e.g., too deep/shallow, improper spacing) can lead to poor germination, sub-optimal plant stand, and reduced growth duration.
  - **Plant Density:** Sub-optimal plant population (too sparse or too dense) affects light interception and resource competition.
  - **Nutrient Management:** Inaccurate fertilizer rates, incorrect timing, or improper application methods leading to nutrient loss or inefficiency.
  - **Water Management:** Inefficient irrigation systems, over- or under-irrigation, or lack of proper drainage.
  - **Pest, Disease, and Weed Control:** Ineffective or untimely application of control measures, leading to significant yield losses.
  - **Harvesting and Post-Harvest Losses:** Poor harvesting techniques, inadequate storage, and processing can lead to significant losses between field and consumption.
  - **Crop Rotation and Tillage Practices:** Monoculture or inappropriate tillage can degrade soil health and increase pest/disease pressure.

- **Socio-Economic Factors**
  - **Access to Inputs:** Limited access to quality seeds, fertilizers, pesticides, machinery, or irrigation facilities due to cost or availability.
  - **Knowledge and Information:** Lack of farmer awareness about best practices, new technologies, or timely weather advisories.
  - **Financial Constraints:** Limited access to credit or capital to invest in optimal inputs and management.
  - **Labor Availability:** Shortage or high cost of labor for timely field operations.
  - **Market Access and Prices:** Unfavorable market prices or lack of access to markets can disincentivize farmers from investing in higher production.

- **Policy Environment:** Inadequate government support, research, extension services, or unstable agricultural policies.

8.3.4 Methods of Estimating Achievable Production

Estimating achievable production is crucial for agricultural planning, policy making, and assessing the impact of interventions. Various methods are employed, ranging from traditional surveys to advanced technologies:

- **Farmer Surveys/Interviews (Recall or Prediction)**
  - **Method:** Directly asking farmers about their yields for a specific season or recalling past yields. Can be done before harvest (prediction) or after harvest (recall).
  - **Advantages:** Relatively low cost and quick for large areas. Captures local knowledge and practices.
  - **Limitations:** Subject to farmer memory bias, overestimation or underestimation, and difficulty in accounting for specific plot-level variations.May not be highly accurate.
- **Crop Cutting Experiments (CCEs)**
  - **Method:** This is a widely used and considered one of the most accurate ground-based methods. Small, representative plots (e.g., 5m x 5m or 10m x 10m) are randomly selected within a farmer's field (Fig.8.2). The crop within these plots is harvested manually, threshed, cleaned, and weighed. The yield is then extrapolated to a per-hectare basis.

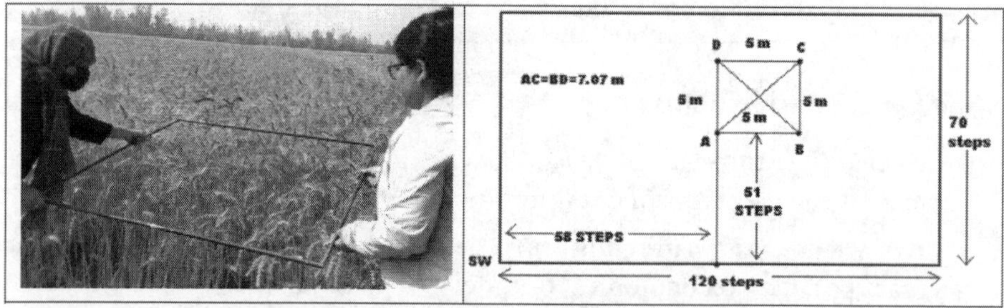

Fig. 8.2: Crop cutting experiment

  - **Advantages:** Provides direct and objective measurement of actual yield from specific field conditions. Widely accepted for official yield statistics.
  - **Limitations:** Labor-intensive, time-consuming, and can be expensive for large-scale surveys. Requires careful selection of plots to avoid bias.

- **Whole Plot/Field Harvest (for research/demonstration)**
    - **Method:** In specific research trials or demonstration plots, the entire plot or a significant portion of a small field is harvested and weighed to get a highly accurate measure of achievable yield under the applied treatments.
    - **Advantages:** Very accurate for the specific plot/field.
    - **Limitations:** Not practical for large-scale estimation across many farms.
- **Remote Sensing and Satellite Imagery**
    - **Method:** Using satellite data (e.g., Sentinel, Landsat) or drone imagery to assess crop health, biomass, and ultimately yield. Vegetation indices (like NDVI - Normalized Difference Vegetation Index) derived from spectral reflectance are correlated with crop vigor and yield. Machine learning and AI algorithms process these images to make predictions.
    - **Advantages:** Large spatial coverage, cost-effective for vast areas, non-invasive, provides real-time or near real-time data. Useful for early season yield forecasting.
    - **Limitations:** Requires ground-truthing for calibration and validation. Can be affected by cloud cover. Accuracy depends on the resolution of the imagery and the sophistication of the algorithms. May struggle to distinguish between biomass and actual economic yield (e.g., grain).
- **Ground Sensor-Based Estimation**
    - **Method:** Deploying sensors in the field (e.g., soil moisture sensors, temperature sensors, plant canopy sensors) to collect real-time data on environmental conditions and plant physiological responses. This data can be integrated with crop models to predict yield.
    - **Advantages:** Provides highly localized and continuous data. Enables precision agriculture.
    - **Limitations:** High initial investment in sensors. Maintenance and data management can be complex.
- **Crop Simulation Models (Calibrated for Actual Conditions)**
    - **Method:** Similar to potential yield estimation, but these models are run with actual weather data, specific soil properties of the farm, and *actual* farmer management practices (e.g., actual fertilizer application, irrigation schedules). The models can also incorporate parameters for pest and disease impact if quantified.

- **Advantages:** Can simulate yield under various real-world scenarios and help identify limiting factors. Useful for forecasting.
- **Limitations:** Requires detailed input data on local conditions and management. Model accuracy is highly dependent on calibration and validation with real farm data.

- **Expert Assessment**
  - **Method:** Experienced agronomists or local agricultural extension agents provide an estimate of expected or achieved yield based on their observations of crop health, growth stage, and knowledge of local conditions.
  - **Advantages:** Quick and low cost. Utilizes valuable practical experience.
  - **Limitations:** Subjectivity and potential for bias. Less precise than quantitative methods.

In conclusion, achievable production is the practical realization of agricultural output under real-world constraints. Its accurate estimation and continuous improvement are vital for ensuring food security, promoting sustainable agricultural development, and enhancing the livelihoods of farmers globally.

8.4 Yield Gap Analysis: Identification of Problems

Yield gap analysis is a powerful diagnostic tool in agricultural science. It systematically quantifies the difference between potential (or attainable) crop yields and the actual yields achieved by farmers in their fields. The primary purpose of this analysis is not just to measure the gap, but critically, to identify the underlying problems or constraints that prevent farmers from reaching higher productivity levels. By pinpointing these limitations, researchers, extension workers, policymakers, and farmers themselves can develop targeted interventions to bridge the gap and enhance agricultural output.

8.4.1 What is Yield Gap Analysis?

As discussed in previous sections, the yield gap (YG) is fundamentally defined as:

$$YG = YP - YA$$

Where:

- YP (Potential Yield): The maximum achievable yield under optimal conditions, determined by the crop's genetics and the climatic environment (solar radiation, temperature, CO_2).

- YA (Achievable/Actual Yield): The average yield obtained by farmers in their fields.

In some more refined analyses, especially for rainfed systems, a water-limited potential yield (YW) is used as the benchmark, representing the maximum yield achievable given the available water resources but with all other factors optimized. In such cases, the yield gap is YW −YA .

The concept can also be further broken down into two components:

- **Yield Gap I (YP −YEP or YW −YEP):** The difference between the potential yield (or water-limited potential yield) and the yield achieved in best-managed experimental stations or progressive farmer fields (sometimes termed exploitable potential yield, YEP). This gap often reflects fundamental biophysical and technological limitations, such as the inherent genetic potential of varieties or the limits of current optimal agronomic practices under specific climatic conditions.
- **Yield Gap II (YEP −YA):** The difference between the yield of best-managed fields (experimental or progressive farmers) and the average actual farm yield. This is often referred to as the extension gap or management gap, as it primarily reflects sub-optimal management practices and socio-economic constraints at the farmer level. This is where most actionable problems are identified.

8.4.2 Importance of Identifying Problems through Yield Gap Analysis

The real value of yield gap analysis lies in its ability to diagnose the specific constraints that limit crop productivity. This diagnostic capability is crucial for:

- **Prioritizing Research:** It helps pinpoint which areas of research (e.g., developing drought-tolerant varieties, improving nutrient use efficiency, pest-resistant crops) will have the greatest impact on increasing yields.
- **Tailoring Extension Services:** Identifies the specific knowledge and technology gaps among farmers, allowing extension workers to design relevant training programs and demonstrations.
- **Informing Policy Decisions:** Provides evidence for policymakers to allocate resources effectively, design input subsidies, improve infrastructure (e.g., irrigation), or address market failures.
- **Optimizing Resource Use:** By understanding why yields are low, strategies can be developed to use inputs (fertilizers, water, pesticides) more efficiently, reducing waste and environmental impact.

- **Targeting Investments:** Helps investors and development agencies direct funds to regions and interventions where the potential for yield improvement is highest.

8.4.3 Methods for Identifying Problems in Yield Gap Analysis

Identifying the specific problems or causes of the yield gap requires a systematic approach, often combining quantitative analysis with qualitative assessments. Here are common methods:

- **Agronomic Diagnosis (On-Farm Trials and Surveys)**
  - **Method:** This involves conducting on-farm trials where different management practices are tested against local farmer practices. Researchers implement best management practices (BMPs) side-by-side with common farmer practices. Detailed surveys are also conducted to gather information on farmer practices, resource availability, and perceptions of constraints.

Identification of Problems

- **Comparison of BMP vs. Farmer Practice:** If BMP yields are significantly higher than farmer yields, it indicates a *management gap*. Further analysis of which specific BMPs (e.g., improved variety, optimal fertilization, timely weeding) contribute most to the yield difference helps pinpoint the key agronomic limitations.
- **Nutrient Omission Trials:** Varying nutrient applications (e.g., all nutrients minus N, all nutrients minus P) can directly identify which nutrients are most limiting.
- **Pest/Disease/Weed Exclusion Trials:** Using effective control measures (e.g., pesticides, herbicides) in specific plots compared to untreated plots helps quantify yield losses due to biotic stresses.
- **Farmer Surveys:** Direct questions about challenges faced (e.g., high cost of inputs, lack of water, pest outbreaks, lack of knowledge) provide qualitative insights.
- **Statistical and Econometric Analysis**
  - **Method:** Using statistical techniques (e.g., multiple regression analysis, frontier production functions) to analyze relationships between yield and various input factors (fertilizer use, irrigation, labor, education, access to credit) and environmental variables (soil type, rainfall patterns).

- **Identification of Problems**
  - **Marginal Productivity Analysis:** Quantifies how much yield increases with an additional unit of input. Low marginal productivity of a key input might suggest inefficient use or other limiting factors.
  - **Constraint Prioritization:** Regression models can identify which factors have the most significant statistical impact on reducing yields. For example, a strong negative correlation between yield and weed density indicates weed infestation as a major problem.
  - **Efficiency Analysis (Stochastic Frontier Analysis):** This method estimates the technical efficiency of farms, showing how far individual farmers are from the best practice frontier, thereby indicating the scope for improvement through better management.

- **Crop Simulation Models (Scenario Analysis)**
  - **Method:** Process-based crop models (like DSSAT, APSIM, WOFOST) are powerful tools. After calibrating the model for a specific region and crop, various what-if scenarios can be run.
  - **Identification of Problems**
    - **Comparing Simulated Potential vs. Achievable:** Running the model with optimal water and nutrient inputs (potential yield) and then with actual farmer inputs (achievable yield) allows for quantifying the yield gap and attributing portions of it to specific factors (e.g., X% due to water stress, Y% due to nutrient deficiency).
    - **Virtual Omission Trials:** The model can be used to simulate the impact of removing a specific constraint (e.g., what if nutrient supply were optimal? or what if there were no water stress?). This helps quantify the yield loss attributable to each factor.
    - **Climate Change Impact:** Models can project how future climate scenarios might affect potential and achievable yields, highlighting future challenges.

- **Remote Sensing and Geospatial Analysis**
  - **Method:** Satellite and drone imagery, combined with GIS (Geographic Information Systems), can map spatial variability in yields and identify areas of low productivity. Data layers on soil type, topography, land use, and weather can be overlaid to correlate low yields with specific environmental or management indicators.

- **Identification of Problems**
  - **Hotspot Identification:** Visualizing yield maps can reveal hotspots of low yield, prompting further ground-based investigation in those specific areas.
  - **Correlation with Spatial Data:** For example, consistent low yields in saline soil areas point to salinity as a major constraint. Areas with low vegetation indices during dry spells indicate water stress.

- **Farmer Participatory Approaches / Participatory Rural Appraisal (PRA)**
  - **Method:** Engaging farmers directly in the analysis process through focus group discussions, community mapping, and direct observation. This provides valuable local knowledge and perspectives.
  - **Identification of Problems:** Farmers often have a deep understanding of their local challenges (e.g., lack of access to quality seeds,high price of pesticides,unreliable electricity for irrigation,lack of labor,wild animal damage). This method captures socio-economic and institutional constraints that quantitative models might miss.

8.4.4 Common Problems Identified by Yield Gap Analysis

Yield gap analysis frequently uncovers a range of limiting and reducing factors, which can vary significantly by region, crop, and farming system. However, some common problems identified include:

- **Agronomic/Biophysical Constraints**
  - **Nutrient Deficiencies:** Inadequate or imbalanced application of essential fertilizers (N, P, K, micronutrients), leading to widespread nutrient stress.
  - **Water Stress:** Insufficient or erratic rainfall, lack of irrigation facilities, inefficient irrigation methods, or poor soil water retention.
  - **Pest and Disease Infestations:** Uncontrolled outbreaks leading to significant yield losses due to direct damage, reduced photosynthetic efficiency, or plant mortality.
  - **Weed Competition:** High weed pressure due to inadequate or untimely weed control, leading to competition for resources.
  - **Sub-optimal Variety Selection:** Farmers using older, lower-yielding varieties or varieties not well-suited to local agro-climatic conditions or pest/disease pressures.

    - **Poor Soil Health:** Soil degradation issues like compaction, low organic matter content, soil erosion, acidity/alkalinity, and salinity reducing fertility and water infiltration.
    - **Sub-optimal Planting Practices:** Incorrect planting dates, spacing, or population density leading to inefficient resource capture.
    - **Poor Crop Residue Management:** Burning residues instead of incorporating them, leading to loss of organic matter and nutrients.

- **Socio-Economic and Institutional Constraints:**
    - **Lack of Access to Inputs:** Scarcity or high cost of improved seeds, fertilizers, pesticides, or agricultural machinery.
    - **Limited Access to Credit:** Farmers unable to invest in necessary inputs or technologies due to financial limitations.
    - **Inadequate Knowledge and Extension Services:** Farmers lacking awareness of best management practices, new technologies, or timely information.
    - **Labor Shortages/High Labor Costs:** Affecting timely operations like weeding, harvesting, or pest management.
    - **Poor Market Access and Price Volatility:** Lack of remunerative prices or stable markets can disincentivize farmers from investing in higher production.
    - **Insecure Land Tenure:** Disincentivizes long-term investments in soil health or infrastructure.
    - **Inadequate Infrastructure:** Poor irrigation systems, storage facilities, or transportation networks.
    - **Policy Gaps:** Ineffective or absent government policies to support agricultural development, research, and extension.

By systematically applying these methods, yield gap analysis moves beyond simply stating that a yield gap exists. It provides the crucial information needed to understand *why* it exists, thereby enabling the development of targeted, effective, and sustainable solutions to improve agricultural productivity.

8.5 Conclusion

This chapter has explored the critical concepts of potential and achievable crop production, highlighting the significant difference that often exists between them. We‘ve seen that potential yield represents the theoretical maximum a crop can produce under ideal conditions, dictated primarily by its genetic makeup and the prevailing climate. This serves as a vital benchmark, a vision of what's biologically possible. Conversely, achievable yield reflects the practical reality on farms, constrained by a myriad of environmental, biotic, management, and socio-economic factors. The disparity between these two—the yield gap—is not merely a statistic; it's a powerful indicator of unrealized agricultural potential and a direct challenge to global food security.

Understanding and quantifying this yield gap is the first crucial step towards sustainable agricultural intensification. By employing various analytical techniques, from on-farm trials to sophisticated crop simulation models, we can precisely identify the specific problems that prevent crops from reaching their full potential. Whether it's nutrient deficiencies, water stress, pest outbreaks, or limitations in farmer knowledge and access to resources, pinpointing these constraints allows us to move beyond general observations to targeted, effective interventions. As we look to feed a growing world population while contending with climate change and limited resources, bridging these yield gaps will be paramount.

Multiple Choice Questions (Bold means correct answer)

1. What is the primary definition of potential crop production?
   a) The average yield obtained by farmers in a region.
   b) The maximum yield achievable under ideal conditions with no limiting factors.
   c) The yield obtained in research plots under typical farmer management.
   d) The yield limited only by water availability.
2. Which of the following factors is least likely to be a determinant of potential crop production?
   a) Solar Radiation b) Optimal Temperature
   c) Carbon Dioxide Concentration **d) Pest and Disease Incidence**
3. The inherent capacity of a specific crop variety to produce biomass and yield under ideal conditions refers to its:
   a) Achievable yield b) Yield gap
   c) Genetic potential d) Management efficiency
4. What does achievable crop production represent?
   a) The theoretical maximum yield.
   b) The yield limited only by genetic factors.
   c) The actual yield obtained by farmers under real-world conditions.
   d) The yield that results from perfect management.

5. The difference between potential yield and achievable yield is known as the:
   a) Production surplus
   b) Harvest index
   c) Yield gap
   d) Environmental limitation
6. Which of the following is a biotic factor that commonly limits achievable crop production?
   a) Soil salinity
   b) Temperature extremes
   c) Waterlogging
   d) Weed competition
7. Which of the following is an abiotic factor that commonly limits achievable crop production?
   a) Insect pests
   b) Nutrient deficiency
   c) Fungal diseases
   d) Rodent damage
8. What is the primary significance of understanding potential production?
   a) It directly determines farmer income.
   b) It provides a benchmark for yield gap analysis.
   c) It indicates the success of current government policies.
   d) It predicts global food supply accurately without further analysis.
9. Which factor is considered optimal and non-limiting when defining potential crop production?
   a) Water supply
   b) Pest and disease control (applied by farmer)
   c) Sub-optimal soil pH
   d) Limited financial resources for inputs
10. The concept of Harvest Index is most directly related to:
    a) The efficiency of pesticide application.
    b) The ratio of economic yield to total above-ground biomass.
    c) The duration of the crop's vegetative phase.
    d) The total solar radiation intercepted by the canopy.
11. Crop growth simulation models estimate potential yield by assuming:
    a) Average farmer management practices.
    b) Significant yield losses due to pests and diseases.
    c) Non-limiting water and nutrient supply.
    d) Varying solar radiation based on cloud cover only.

12. Which of the following is NOT a common method for estimating potential production?
    a) Controlled environment experiments.
    b) Farmer surveys on past yields.
    c) Field experiments with optimal management.
    d) Crop growth simulation models.
13. In the context of crop modeling, what does RUE stand for?
    a) Root Uptake Efficiency
    b) Radiation Use Efficiency
    c) Rainfall Utilization Effectiveness
    d) Resource Underestimation Error
14. The extension gap or management gap primarily refers to:
    a) Yield Gap I (YP −YEP)
    b) Yield Gap II (YEP −YA)
    c) The difference between current year‘s yield and previous year‘s yield.
    d) The gap between research station yield and potential yield.
15. Nutrient omission trials are primarily used in yield gap analysis to identify:
    a) The most cost-effective fertilizer.
    b) Specific nutrient deficiencies limiting achievable yield.
    c) The genetic potential of a crop variety.
    d) The impact of pests on nutrient uptake.
16. Which of the following socio-economic factors can limit achievable production?
    a) Optimal temperature
    b) High soil organic matter
    c) Limited access to credit for farmers
    d) Efficient solar radiation interception
17. Crop cutting experiments (CCEs) are primarily used to estimate:
    a) Potential yield in controlled environments.
    b) Achievable yield in farmers‘ fields.
    c) The ideal harvest index of a crop.
    d) The maximum possible photosynthetic rate.

18. What is the main advantage of using remote sensing for estimating achievable production over traditional methods?
    a) It provides highly precise local nutrient deficiency data.
    b) It offers large spatial coverage and near real-time data.
    c) It directly measures harvest index.
    d) It eliminates the need for any ground-truthing.
19. Which concept is central to process-based crop models and assumes biomass is proportional to intercepted light?
    a) Liebig‘s Law of the Minimum
    b) Harvest Index
    c) Radiation Use Efficiency
    d) Transpiration Ratio
20. A significant yield gap often indicates a need for:
    a) Decreasing agricultural research funding.
    b) Importing more food.
    c) Prioritizing agricultural research and extension efforts.
    d) Reducing the cultivated area.
21. If a crop model is run in potential mode, what assumption is made regarding rainfall?
    a) Rainfall is minimal and supplementary irrigation is needed.
    b) Rainfall is exactly average for the region.
    c) Water supply is continuously optimal and non-limiting.
    d) Only natural rainfall is considered, no irrigation.
22. Which of the following is considered a management factor influencing achievable production?
    a) Soil pH
    b) Pest infestation
    c) Sub-optimal planting density
    d) Solar radiation
23. Which crop is generally known for having a higher potential for increased yield under elevated CO2 concentrations?
    a) Maize (C4 plant)
    b) Sorghum (C4 plant)
    c) Wheat (C3 plant)
    d) Sugarcane (C4 plant)

24. What information do yield maps generated from remote sensing and GIS primarily help identify?
    a) The genetic lineage of crop varieties.
    b) Hotspots of low productivity within a field or region.
    c) The exact historical yield of a farm from decades ago.
    d) The market price of a crop.
25. The method that involves direct interaction with farmers to understand their challenges and constraints is called:
    a) Crop cutting experiment
    b) Remote sensing analysis
    c) Farmer participatory approach / PRA
    d) Controlled environment experiment
26. Which of these is a typical output from a crop growth simulation model when estimating potential yield?
    a) Economic viability analysis
    b) Daily biomass accumulation
    c) Farmer decision-making patterns
    d) Market demand forecasts
27. The primary purpose of Yield Gap Analysis is to:
    a) Calculate average farm yields globally.
    b) Determine the exact amount of fertilizer applied by farmers.
    c) Identify the underlying problems preventing higher crop productivity.
    d) Predict future climate change impacts on agriculture without intervention.
28. When comparing achieved yields to a water-limited potential yield (YW), what does the resulting yield gap primarily indicate?
    a) The impact of optimal nutrient management.
    b) The yield loss due to factors other than water scarcity.
    c) The maximum yield that can be achieved without any irrigation.
    d) The difference between rainfed and irrigated potential.
29. A large yield gap between research station yields and average farmer yields (Yield Gap II) most often suggests issues related to:
    a) Genetic potential of the crop.
    b) Fundamental climatic limitations.
    c) Management practices and access to technology/resources.
    d) Global trade policies.

30. Which factor is assumed to be non-existent when calculating potential production using crop models?
    a) Solar radiation variability.
    b) Temperature fluctuations within optimal range.
    c) Presence of pests and diseases.
    d) Atmospheric CO_2 concentration.

9

Crop Production Under Moisture and Nutrient Limited Conditions

9.1 Introduction

The preceding chapter elucidated the concepts of potential and achievable crop production, highlighting the significant gap that often exists between a crop's theoretical maximum yield and what is realized in the field. While genetic potential and optimal climatic conditions define the upper limits, the vast majority of agricultural systems globally operate under less-than-ideal circumstances. Among the most pervasive and impactful constraints on crop productivity are moisture (water) and nutrient limitations. These two factors, either individually or in complex interaction, are frequently the primary drivers of the yield gap, preventing crops from expressing their full genetic potential and significantly impacting farm profitability and regional food security.

Water, as the universal solvent and a direct participant in photosynthesis and countless metabolic processes, is indispensable for plant life. However, its availability in agricultural systems is highly variable and often insufficient. From the rainfed fields dependent on increasingly erratic rainfall patterns to irrigated lands facing dwindling water resources, moisture stress – encompassing both drought and, paradoxically, waterlogging – poses an immense challenge. Drought alone accounts for a substantial portion of global crop losses, directly affecting plant physiological functions, growth, and ultimately, yield. Conversely, excessive moisture can lead to anaerobic soil conditions, nutrient leaching, and root damage, equally debilitating for crop performance.

Parallel to moisture, the availability and judicious management of essential plant nutrients are foundational to productive agriculture. Plants require a suite of macro- and micronutrients for healthy growth, enzymatic activity, and the synthesis of biomass. Soils, over time and with continuous cropping, often become depleted of these vital elements, necessitating external application through fertilizers or the enhancement of natural nutrient cycling processes. However, nutrient management in the field is rarely perfect. Deficiencies of primary nutrients like Nitrogen (N), Phosphorus (P), and Potassium (K) are widespread, but imbalances, toxicities, and inefficient nutrient use can be equally detrimental, reducing both yield and the economic returns on fertilizer investments.

This chapter delves into the intricate mechanisms by which crops respond to and are limited by sub-optimal moisture and nutrient availability. We will explore the physiological adaptations and challenges plants face under conditions of water scarcity and excess, examining how these translate into reduced growth and yield. Concurrently, we will investigate the critical roles of various macro and micronutrients, the symptoms of their deficiencies, and their impact on crop productivity. Crucially, we will move beyond simply identifying these limitations to discussing practical, science-based strategies for mitigating their adverse effects. This includes exploring efficient water management techniques, integrated nutrient management approaches, and the role of resilient crop varieties, all aimed at enhancing the efficiency of resource use and ensuring stable, productive agriculture in environments where moisture and nutrients are not always optimally supplied. Understanding these limitations is not just an academic exercise; it is an imperative for building more resilient, sustainable, and food-secure agricultural systems worldwide.

9.2 Indicators of Moisture and Nutrient Limitations in Crops

Identifying moisture and nutrient limited conditions in crop plants is crucial for timely intervention and improving crop yield. These limitations manifest through various indicators, ranging from visual symptoms to physiological changes and measurable parameters.

Here‘s a detailed breakdown of these indicators:

9.2.1 Indicators of Moisture Limited Conditions (Water Stress)

Moisture stress occurs when the plant‘s water demand exceeds the available water supply in the soil, or when roots cannot absorb water due to factors like high salinity or waterlogging.

- **Visual/Morphological Indicators (Often Late Stage)**
    - **Wilting:** This is the most common and easily observable symptom. Leaves and stems lose turgor pressure, causing them to droop and lose their rigidity. Early wilting might be transient (occurring during the hottest part of the day and recovering overnight), but persistent wilting indicates severe stress (Fig.9.1).

Fig. 9.1: Wilting as visual/morphological indicator

Leaf Rolling or Curling: Many crops, particularly cereals like maize and rice, roll or curl their leaves inward along the midrib. This reduces the exposed leaf surface area to direct sunlight, minimizing water loss through transpiration.

- **Changes in Leaf Color:** Leaves may appear duller, develop a bluish-green or grayish tint, or even become brittle. In severe, prolonged drought, older leaves may turn yellow or brown and eventually senesce (die prematurely).
- **Stunted Growth:** Reduced overall plant height, smaller leaves, and thinner stems are signs of chronic water stress, as cell expansion and division are inhibited.
- **Reduced Leaf Area:** Plants may produce fewer and smaller leaves to conserve water, leading to a less developed canopy.
- **Premature Senescence/Maturity:** Under severe water stress, plants may accelerate their development and mature earlier, often leading to significantly reduced yield and quality.
- **Shedding of Flowers/Fruits:** Plants may abort flowers or developing fruits to conserve water and energy for survival.

- **Physiological Indicators (More Sensitive & Earlier Detection)**
  - **Stomatal Closure/Reduced Stomatal Conductance:** Stomata are pores on the leaf surface that regulate gas exchange and transpiration. Under water stress, stomata close to reduce water loss, but this also limits CO_2 uptake, reducing photosynthesis. This can be measured with a porometer.

- **Reduced Transpiration Rate:** As stomata close, the rate of water vapor release from the leaves decreases.
- **Decreased Photosynthesis (CO_2 Assimilation Rate):** With reduced CO_2 uptake due to stomatal closure, and other metabolic disruptions, the rate of photosynthesis declines, impacting biomass accumulation.
- **Lower Leaf Water Potential/Turgor Pressure:** This is a direct measure of the water status within the plant. As water stress increases, leaf water potential becomes more negative. Measured using a pressure chamber (Scholander bomb).
- **Increased Canopy Temperature:** When plants are well-watered, evaporative cooling from transpiration keeps their leaf temperatures lower than the surrounding air. Under water stress, reduced transpiration leads to higher leaf and canopy temperatures, detectable with infrared thermometers or thermal cameras.
- **Accumulation of Stress Metabolites:** Plants produce specific compounds (e.g., abscisic acid - ABA, proline, sugars, betaines) as a response to stress. Elevated levels of these metabolites indicate water stress.

9.2.2 Soil-Based Indicators

- **Soil Dryness (Visual & Feel)**

Dry surface: Cracks appearing on the soil surface are a clear visual cue.

Hand-feel method: Taking a handful of soil from the root zone and squeezing it. Dry soil will crumble, while adequately moist soil will form a ball. Experienced farmers can estimate moisture content by feel.

- **Soil Moisture Sensors:** Devices like tensiometers, capacitance sensors, or Time Domain Reflectometry (TDR) measure soil water content directly and provide objective, continuous data. These are invaluable for precise irrigation scheduling.
- **Soil Water Potential:** Tensiometers directly measure soil water tension (suction), indicating how tightly water is held by the soil and how much energy roots need to extract it.

9.2.3 Advanced & Remote Sensing Indicators

- **Vegetation Indices (e.g., NDVI, NDMI):** Satellite or drone imagery can provide vegetation indices that reflect canopy vigor, greenness, and water content. Changes or low values compared to historical norms can indicate stress. NDMI (Normalized Difference Moisture Index) is specifically sensitive to plant water content.

- **Thermal Imaging:** Infrared cameras on drones or satellites can detect differences in canopy temperature, indicating areas experiencing water stress due to reduced transpiration (Fig.9.2).

Fig. 9.2: Moisture stress detection in plant using thermal imaging

Evapotranspiration (ET) Monitoring: Satellite-derived ET maps can show areas where actual ET is significantly lower than potential ET, indicating water limitation.

9.3 Indicators of Nutrient Limited Conditions (Nutrient Deficiencies)

Nutrient deficiencies occur when an essential nutrient is present in insufficient quantities in the soil or when plants cannot adequately absorb it. Symptoms often appear on specific parts of the plant and are related to the nutrient's mobility within the plant.

9.3.1 Visual/Morphological Indicators (Specific Symptoms)

General symptoms often include stunted growth, reduced vigor, abnormal leaf color (chlorosis, necrosis, purpling), distorted growth, and poor flowering/ fruiting. The specific location and pattern of these symptoms are crucial for diagnosis:

- **Nitrogen (N)**
  - **Symptoms:** General pale green to yellowing (chlorosis) of older leaves first, starting from the leaf tip and progressing along the midrib. Stunted growth, thin stems, reduced tillering/branching.
  - **Reason:** N is mobile in the plant, so it's retranslocated from older leaves to support new growth.
- **Phosphorus (P)**
  - **Symptoms:** Dark green or bluish-green leaves, often developing purple or reddish tints (especially on older leaves, sometimes stems). Stunted growth, poor root development, delayed maturity.
  - **Reason:** P is mobile. Purpling is due to accumulation of anthocyanins under P stress.
- **Potassium (K)**
  - **Symptoms:** Yellowing (chlorosis) and browning/scorching along the margins (edges) of older leaves. Leaves may appear crinkled or curled. Weak stalks, poor fruit/seed development, increased susceptibility to lodging and diseases.
  - **Reason:** K is mobile, translocated from older tissues.
- **Magnesium (Mg)**
  - **Symptoms:** Interveinal chlorosis (yellowing between the veins, while veins remain green) on older leaves. In severe cases, older leaves may turn reddish-purple and die.
  - **Reason:** Mg is mobile, moves to younger, actively growing parts.
- **Calcium (Ca)**
  - Symptoms: Symptoms appear on *younger leaves* and growing points first. Growing points (meristems) may die back, leading to stunted or distorted new growth. Leaf tips may stick together, or edges may appear ragged. Blossom-end rot in fruits (e.g., tomatoes, peppers).
  - **Reason:** Ca is largely immobile in the phloem; it cannot be retranslocated from older tissues.
- **Sulfur (S)**
  - **Symptoms:** General yellowing (chlorosis) of *younger leaves* first, sometimes the entire plant. Symptoms often resemble N deficiency but start on younger leaves. Stunted growth.

- **Reason:** S has limited mobility; new growth needs a continuous supply.

- **Iron (Fe)**
  - **Symptoms:** Pronounced interveinal chlorosis on *younger leaves*, with veins remaining distinctly green. In severe cases, the entire young leaf may turn pale yellow or almost white.
  - **Reason:** Fe is largely immobile in the phloem.
- **Manganese (Mn)**
  - **Symptoms:** Interveinal chlorosis on *younger leaves*, similar to Fe but often with finer mottling or speckling and less distinct green veins. In some crops, tiny necrotic spots may appear.
  - **Reason:** Immobile or very limited mobility.
- **Zinc (Zn)**
  - **Symptoms:** Stunted growth, little leaf (abnormally small new leaves). Interveinal chlorosis on younger leaves, often with a rosette appearance (shortened internodes). White banding on corn leaves along the midrib.
  - **Reason:** Immobile or very limited mobility.
- **Boron (B)**
  - **Symptoms:** Death of growing points (terminal buds), distorted or malformed new leaves, cracking of stems, poor flowering and fruit set. Hollow heart in some root crops.
  - **Reason:** Immobile.

9.3.2 Analytical/Scientific Indicators

- **Soil Testing**
  - **Method:** Collecting representative soil samples from the field and sending them to a laboratory for chemical analysis.
  - **Identification:** Lab results provide concentrations of available nutrients, pH, organic matter, and Cation Exchange Capacity (CEC). Comparing these values to established critical levels for the specific crop helps identify deficiencies or excesses.
  - **Significance:** Provides a baseline of nutrient availability *in the soil* before planting and helps in making fertilizer recommendations.

- **Plant Tissue Analysis (Leaf Analysis)**
  - **Method:** Collecting specific plant parts (e.g., youngest fully expanded leaves) at a defined growth stage and analyzing their nutrient content in a laboratory.
  - **Identification:** The concentration of nutrients in the plant tissue is compared to optimal or sufficiency ranges for that crop and growth stage.
  - **Significance:** Provides a direct measure of what nutrients the *plant has actually taken up* and is utilizing, even if soil levels appear adequate (e.g., due to pH issues affecting availability). Can diagnose hidden hunger where visual symptoms are not yet apparent but yield is already being limited.
- **Nutrient Omission Trials (On-Farm Experiments)**
  - **Method:** Setting up small experimental plots where all nutrients are supplied except for one (e.g., -N plot, -P plot). Comparing the yield and growth of these plots to a complete plot (all nutrients supplied optimally) reveals which specific nutrient is limiting growth.
  - **Significance:** Provides direct evidence of a limiting nutrient in a specific field context.
- **Remote Sensing (Less Direct for Specific Nutrients)**
  - **Method:** Satellite or drone imagery can detect broad patterns of stress and reduced vigor (e.g., low NDVI). While not directly identifying specific nutrient deficiencies, these stress maps can guide further ground-based investigation (soil and tissue testing) in problematic areas. Different spectral bands can sometimes indicate general chlorosis patterns.
  - **Significance:** Useful for large-scale monitoring and identifying areas where nutrient management might be sub-optimal.

By combining visual observations with analytical methods, farmers and agronomists can accurately diagnose moisture and nutrient limited conditions, enabling precise and effective management interventions to maximize crop productivity.

9.4 Effect of Moisture Stress on Crops during Different Stages of Growth

Moisture stress, particularly drought, has varying impacts on crop plants depending on the severity, duration, and crucially, the specific growth stage at which it occurs. Different physiological processes and developmental events have different sensitivities to water deficit, leading to varying levels of yield reduction.

Here's a breakdown of the effect of moisture stress at different stages of crop growth:

9.4.1 Seed Germination and Seedling Establishment

- **Impact:** This is a highly sensitive stage. Moisture stress (lack of sufficient water for imbibition) can severely inhibit or prevent seed germination altogether. If germination does occur, seedling establishment can be poor due to insufficient water for root penetration and shoot emergence.

- **Consequences**
    - **Reduced Plant Stand:** Fewer seeds germinate and fewer seedlings survive, leading to a thin, uneven crop stand.
    - **Delayed Emergence:** Germination and emergence may be delayed, pushing critical subsequent stages into less favorable environmental conditions.
    - **Weak Seedlings:** Even if they emerge, seedlings may be weak and more susceptible to other stresses (e.g., pests, diseases).
- **Reason for Sensitivity:** Germination requires adequate water uptake (imbibition) to activate metabolic processes. Seedlings have small, shallow root systems and limited water reserves, making them highly vulnerable to drying topsoil.

9.4.2 Vegetative Growth Stage

- **Impact:** Moisture stress during vegetative growth primarily affects biomass accumulation, leaf area development, and overall plant size.

- **Consequences**
    - **Stunted Growth:** Reduced cell expansion and division lead to shorter plants, fewer and smaller leaves, and less overall biomass.
    - **Reduced Leaf Area Index (LAI):** Decreased leaf production and expansion directly reduce the plant's photosynthetic factory, impacting its ability to capture sunlight and produce carbohydrates. Leaf rolling and wilting also reduce effective light interception.
    - **Reduced Tillering/Branching:** In tillering crops like cereals (wheat, rice) or branching crops (pulses, cotton), water stress can suppress the formation of new tillers or branches, directly impacting the number of potential reproductive units.

- **Delayed Flowering (sometimes):** Chronic stress might delay the onset of flowering as the plant prioritizes survival over reproduction, though severe stress can sometimes accelerate it.
- **Root System Changes:** Plants may reallocate more resources to root growth, developing deeper and more extensive root systems to explore for water. This is an adaptive response, but can come at the cost of shoot growth.

- **Sensitivity:** While detrimental, the vegetative stage is generally considered less critical than the reproductive stage for final economic yield, *provided* the stress is relieved before flowering. Plants often have some capacity for recovery (compensatory growth) if water becomes available again. However, severe early stress can irreversibly set back development.

9.4.3 Reproductive Growth Stage (Most Critical)

This stage, encompassing flowering, pollination, fertilization, and early grain/fruit development, is typically the most sensitive and critical to moisture stress for most crops. Even mild stress can cause significant yield losses.

- **Impact:** Moisture stress during this phase directly affects the formation and success of yield components.

- **Consequences**
  - **Flower Abortion/Shedding:** Plants may abort (drop) a large percentage of flowers and young fruits/pods to conserve resources, leading to a drastic reduction in the number of potential grains or fruits. This is a common response in many crops like cotton, pulses, and some vegetables.
  - **Pollination Failure:** Water stress can impair pollen viability, stigmatic receptivity, and anther dehiscence, leading to poor fertilization and thus fewer grains or seeds. In maize, for example, stress during silking can lead to barren cobs.
  - **Reduced Seed/Grain Set:** Even if pollination occurs, insufficient water can lead to fewer successful ovules developing into seeds.
  - **Shortened Reproductive Period:** Severe stress can prematurely terminate the reproductive phase, reducing the time available for grain/fruit filling.
  - **Reduced Number of Pods/Grains per Plant/Ear:** This is a direct consequence of flower/ovule abortion and poor fertilization.

- **Reason for Sensitivity:** Reproductive processes are highly metabolically demanding and require precise physiological timing and turgor. Any disruption to water status during this period can have irreversible impacts on yield formation.

9.4.4 Grain Filling/Fruit Development Stage

- **Impact:** Moisture stress during this stage primarily affects the size and quality of the economic yield.

- **Consequences**
  - **Reduced Grain/Fruit Size and Weight:** Water stress limits the supply of assimilates (sugars) to the developing grains/fruits, leading to smaller, shriveled, or lighter seeds/fruits. This reduces individual grain weight and test weight (in cereals).
  - **Reduced Yield Quantity:** Directly impacts the total weight of the harvested product.
  - **Reduced Quality:** Can affect nutrient content, oil content (in oilseeds), protein content (in pulses/cereals), and overall market quality.
  - **Premature Maturity/Forced Ripening:** In an attempt to complete its life cycle, the plant may prematurely senesce, leading to early, incomplete ripening.
- **Sensitivity:** While not as critical for determining the *number* of yield components as the flowering stage, this stage is crucial for *filling* those components. Significant yield reductions can still occur.

9.4.5 Maturity/Ripening Stage

- **Impact:** Mild moisture stress can sometimes be beneficial at this stage for crops like cereals, aiding in drying down and facilitating harvest. However, severe, prolonged stress can still have negative impacts.

- **Consequences**
  - **Forced Maturity:** Extremely dry conditions can accelerate maturity, which might reduce the time for final assimilate accumulation.
  - **Reduced Test Weight/Quality:** In some cases, very rapid drying can still lead to reduced grain quality parameters.
    - **Sensitivity:** Generally the *least sensitive* stage to moisture stress compared to other stages for yield formation, as most of the yield has already been determined and filled.

9.4.6 Summary of Critical Periods

The exact "critical period" of highest sensitivity to moisture stress varies among crops. However, for most annual field crops, the reproductive phase (flowering, pollination, and early grain/fruit development) is universally recognized as the most sensitive to water deficit, with drought at this stage leading to the most significant and often irreversible yield losses.

Vegetative growth and grain filling are also important but generally less sensitive than the reproductive stages.

Understanding these stage-specific impacts is crucial for effective irrigation scheduling, selecting drought-tolerant varieties, and developing overall water management strategies to minimize yield losses in rainfed and irrigated agriculture.

The impact of nutrient deficiencies varies significantly depending on the specific nutrient and the crop‘s developmental stage. Plants have different nutrient demands at different times, and some stages are more sensitive to particular deficiencies than others. Understanding these stage-specific effects is crucial for effective diagnosis and timely intervention.

Here‘s how nutrient deficiencies can affect crops at different growth stages:

- **Early Growth Stages (Germination to Vegetative Establishment):**

This period is critical for establishing a strong foundation for the entire plant. Root development, leaf initiation, and early biomass accumulation are key.

- **Nitrogen (N)**
    - **Effect:** Severe stunting, pale green/yellowing of seedlings. Reduced leaf area and tiller/branching initiation. Slow, weak establishment.
    - **Reason:** N is vital for rapid cell division and expansion. Early N deficiency permanently limits the plant‘s size and photosynthetic capacity.
- **Phosphorus (P)**
    - **Effect:** Very critical. Stunted root development, delayed emergence, slow initial growth. Leaves may turn dark green or purplish.Without adequate P, seedlings often fail to thrive even if other nutrients are abundant.

- **Reason:** P is crucial for energy transfer (ATP), nucleic acids (DNA, RNA), and early root system establishment.

- **Potassium (K)**
  - **Effect:** Less critical than N or P in the very early seedling stage, but deficiency can still lead to reduced vigor, smaller leaves, and potentially mild chlorosis on older leaves.
  - **Reason:** K is mobile; the seed provides initial supply.Demand increases as the plant grows and develops its internal transport systems.

- **Calcium (Ca)**
  - **Effect:** Highly critical. Deficiency results in deformed or dying growing points (root tips, shoot tips) of young seedlings. New leaves may be misshapen or fail to unfurl properly.
  - **Reason:** Ca is immobile and essential for cell wall formation and membrane integrity in rapidly dividing and expanding cells at the growing tips.

- **Zinc (Zn)**
  - **Effect:** Significant stunting ("rosetting), abnormally small new leaves ("little leaf), and interveinal chlorosis on young leaves.
  - **Reason:** Zn is required for auxin (growth hormone) synthesis and enzyme activity critical for early growth and internode elongation.

- **Boron (B)**
  - **Effect:** Death of terminal buds in severe cases, leading to a bushy or stunted appearance. Deformed young leaves.
  - **Reason:** B is crucial for cell wall development and cell division at growing points.

- **Vegetative Growth Stage (Tillering/Branching to Pre-Flowering)**

During this phase, the plant builds its photosynthetic machinery and prepares for reproduction.

- **Nitrogen (N)**
  - **Effect:** Continued general yellowing of older leaves, reduced leaf area, and suppressed tillering/branching. Reduced biomass accumulation and overall plant size. Directly impacts the number of potential heads/pods.
  - **Reason:** High demand for protein and chlorophyll synthesis for rapid growth.

- **Phosphorus (P)**
    - **Effect:** Stunted growth, dark green/purplish coloration. Reduced photosynthetic efficiency. If severe early P deficiency occurred, recovery is limited even if P is supplied later.
    - **Reason:** Still crucial for energy metabolism and overall vigorous growth.
- **Potassium (K)**
    - **Effect:** Symptoms (marginal chlorosis/necrosis on older leaves) become more apparent as K demand increases. Weak stems, making plants prone to lodging.Reduced disease resistance.
    - **Reason:** K is vital for water regulation (stomatal function) and many enzyme activities, supporting robust growth.
- **Magnesium (Mg)**
    - **Effect:** Interveinal chlorosis on older leaves. Reduced photosynthetic rates due to lack of chlorophyll. Can lead to accumulation of carbohydrates in leaves.
    - **Reason:** Central to chlorophyll molecule; continuously needed as older leaves produce energy for new growth.
- **Sulfur (S)**
    - **Effect:** General yellowing of younger leaves and overall stunted growth. Similar to N but on new leaves.
    - **Reason:** Crucial for protein synthesis, and S is not highly mobile, so new growth suffers first.
- **Iron (Fe), Manganese (Mn), Zinc (Zn), Copper (Cu)**
    - **Effect:** Continue to show characteristic symptoms on (interveinal chlorosis, little leaf, pale tips), leading to reduced photosynthetic capacity and stunted growth. These deficiencies can set the plant back severely before reproductive stages.
    - **Reason:** All are relatively immobile and involved in photosynthesis, enzyme function, or hormone synthesis for new growth.
- **Reproductive Growth Stage (Flowering, Pollination, Fruit/Seed Set):**

This is often the most critical period, where nutrient deficiencies can lead to severe yield losses due to direct impacts on flowers, pollen, and fruit development.

- **Nitrogen (N)**
  - **Effect:** Reduced flower initiation, fewer viable pollen grains, poor grain/ seed set. Premature senescence, cutting short the grain filling period.
  - **Reason:** High demand for protein synthesis during pollen and ovule development, and for maintaining photosynthetic activity during grain fill.
- **Phosphorus (P)**
  - **Effect:** Crucial impact. Poor flowering, reduced fruit/seed set, impaired pollen viability, and reduced seed size. Delayed maturity can also occur.
  - **Reason:** P is essential for energy (ATP) required for flower and seed development, and for genetic material formation.
- **Potassium (K)**
  - **Effect:** Poor fruit/seed filling, leading to smaller, shriveled grains/fruits. Reduced sugar translocation to reproductive organs. Increased lodging can occur just before harvest.
  - **Reason:** K is vital for carbohydrate transport and storage in economic yield, and for maintaining turgor in reproductive structures.
- **Calcium (Ca)**
  - **Effect:Highly critical for fruit crops.** Blossom-end rot in tomatoes, peppers, melons. Stunted and distorted fruit development. Reduced seed viability.
  - **Reason:** Ca is required for cell wall integrity and membrane function in rapidly expanding fruit tissues, and its immobility means fruits need a continuous supply.
- **Boron (B)**
  - **Effect:Highly critical.** Impaired pollen germination and pollen tube growth, leading to poor fertilization and drastically reduced fruit/seed set or complete sterility. Hollow stems, corking in fruits.
  - **Reason:** B is crucial for pollen viability and successful fertilization.
- **Zinc (Zn)**
  - **Effect:** Reduced flowering, poor fruit set. Abnormally small fruits/seeds.
  - **Reason:** Zn is involved in hormone synthesis important for reproductive development.

- **Molybdenum (Mo)**
  - **Effect:** In legumes, severe reduction in nitrogen fixation due to impaired nodule function, leading to N deficiency symptoms during reproductive growth. Reduced grain fill in non-legumes due to impaired nitrate assimilation.
  - **Reason:** Essential for nitrogenase enzyme (legumes) and nitrate reductase.

- **Grain Filling/Fruit Maturation Stage**

This is the final stage where assimilates are translocated to fill the economic yield.

- **Nitrogen (N)**
  - **Effect:** Premature leaf senescence, shortening the grain filling period. Reduced protein content in grains. Smaller, lighter grains/fruits.
  - **Reason:** N is needed to maintain active photosynthesis and synthesize proteins in the developing seeds.
- **Phosphorus (P)**
  - **Effect:** Poorly filled grains/fruits, leading to low test weight and reduced overall yield. Delayed maturity.
  - **Reason:** Continues to be important for energy required for filling and for the synthesis of complex organic compounds in seeds.
- **Potassium (K)**
  - **Effect:** Severely impacts grain/fruit size and weight. Impaired sugar translocation from leaves to reproductive organs. Lower quality produce. Increased lodging just prior to harvest.
  - **Reason:** Essential for the transport of carbohydrates (sugars) and their storage in the harvested product.

In conclusion, while general deficiency symptoms can guide initial assessment, understanding the specific nutrient requirements and their roles at each growth stage allows for a more precise diagnosis and targeted nutrient management strategies. Timely and appropriate nutrient application, often guided by soil and plant tissue analysis, is paramount to mitigate the negative effects of deficiencies and ensure optimal crop productivity throughout the growing season.

9.5 Methods Used to Quantify the Effect of Moisture Stress on Crop Yields

Several methods are used to quantify the impact of water deficit on crop yield. These methods range from direct field measurements to remote sensing techniques and modeling approaches. Here's a summary:

- **Crop Water Stress Index (CWSI):** The Crop Water Stress Index (CWSI) is a widely used indicator that quantifies the level of water stress in plants.It's based on the relationship between canopy temperature, air temperature, and vapor pressure deficit (VPD).
- **How it works:** CWSI measures the relative transpiration rate of a plant. When a plant is water-stressed, its stomata close to conserve water, reducing transpiration and increasing canopy temperature. CWSI uses this principle to assess stress levels.
- **Measurement:** CWSI can be calculated using infrared thermometers to measure canopy temperature. More advanced methods use data from drones or satellites
- **Use:** CWSI is valuable for irrigation scheduling and water management, helping farmers to apply water only when and where it's needed.
- **Evapotranspiration (ET) Measurement:** Evapotranspiration (ET) is the total water lost from the soil and plants through evaporation and transpiration. Measuring or estimating ET helps in understanding the water balance of a crop.
- **How it works:** ET is influenced by weather parameters (radiation, temperature, humidity, wind), soil type, and crop characteristics.
- **Measurement**
    - **Direct methods:** Lysimeters, water balance studies, and Bowen ratio energy balance methods.
    - **Indirect methods:** Using weather data and crop coefficients (Kc) in models like the FAO Penman-Monteith equation. Remote sensing techniques also estimate ET.
- **Use:** ET measurements help in irrigation scheduling, assessing water use efficiency, and understanding crop water requirements.
- **Yield Response Factor (Ky):** The Yield Response Factor (Ky) relates relative yield decrease to relative evapotranspiration deficit.
- **How it works:** It quantifies the sensitivity of crop yield to water deficits. A Ky > 1 indicates that the crop is very sensitive to water stress. Ky < 1 indicates more tolerance.

- **Calculation:** Ky is used in a formula that relates actual yield (Ya), maximum yield (Ym), actual evapotranspiration (ETa), and maximum evapotranspiration (ETm).
- **Use:** Ky helps in predicting yield reductions under different water deficit scenarios and in optimizing irrigation strategies.
- **Soil Moisture Monitoring:** Monitoring soil moisture levels provides direct information about water availability to the crop.
- **How it works:** Various methods are used to measure soil water content, including:
    - **Tensiometers:** Measure soil water tension.
    - **Capacitance sensors:** Measure the dielectric constant of the soil, which is related to water content.
    - **Time Domain Reflectometry (TDR):** Measures the travel time of an electromagnetic pulse through the soil.
- **Use:** Soil moisture data is crucial for precise irrigation scheduling and preventing water stress.
- **Remote Sensing:** Remote sensing techniques, using satellite or drone imagery, provide a large-scale assessment of crop water status.
- **How it works:** Vegetation indices (e.g., NDVI, NDMI) and thermal imaging can detect water stress. NDMI is specifically sensitive to plant water content. Thermal cameras measure canopy temperature, which increases under water stress.
- **Use:** Remote sensing is useful for monitoring large areas, identifying water-stressed regions, and guiding ground-based investigations.
- **Crop Models:** Crop models simulate crop growth and yield under different water stress scenarios.
- **How it works:** These models use data on weather, soil, and crop characteristics to predict yield. CROPWAT, developed by the FAO, is a widely used model for irrigation scheduling and yield prediction under water deficit conditions.
- **Use:** Crop models help in planning irrigation strategies and estimating yield reductions.
- **Water Productivity:** Water productivity measures the amount of crop yield produced per unit of water used.
- **How it works:** It can be expressed as yield per unit of evapotranspiration or yield per unit of total water input (irrigation + rainfall).

- **Use:** Water productivity is a key indicator for evaluating the efficiency of water use in agriculture.

These methods, often used in combination, help in quantifying water deficit and its impact on crop yield, enabling informed decisions for irrigation management and crop production.

9.6 Methods Used to Quantify the Effect of Nutrient Deficiencies on Crop Yields

Quantifying the effect of nutrient deficiencies on crop yields is crucial for understanding yield gaps, optimizing fertilizer use, and improving crop productivity sustainably. Various methods, ranging from direct measurements to analytical and modeling approaches, are employed:

9.6.1 Field Experiments (Yield Response Trials)

These are the most direct and reliable methods for quantifying yield loss due to specific nutrient deficiencies.

- **Nutrient Omission Trials**
    - **How it works:** This involves setting up experimental plots where all essential nutrients are supplied in optimal amounts (control plot), except for the one being tested (e.g., a minus N plot, a minus P plot, etc.).
    - **Measurement:** Crop yield (grain, biomass, fruit, etc.) is measured at harvest from each plot.
    - **Quantification:** The difference in yield between the control plot (with all nutrients) and the omission plot (deficient in one nutrient) directly quantifies the yield reduction caused by that specific nutrient deficiency under the given environmental conditions.
    - **Advantages:** Provides direct, site-specific yield loss data.
    - **Limitations:** Time-consuming, labor-intensive, and results are specific to the tested soil and environmental conditions.
- **Fertilizer Rate Trials**
    - **How it works:** Different rates of a specific nutrient are applied to plots (e.g., 0, 50, 100, 150 kg N/ha), and the yield response is measured.
    - **Measurement:** Yield (and often biomass, nutrient uptake, etc.) is measured at each fertilizer rate.
    - **Quantification:** A yield response curve is generated, which shows how yield increases with increasing nutrient application until a plateau is

reached. This helps determine the economically optimum fertilizer rate and the yield potential at different nutrient levels.

- **Advantages:** Helps determine economic fertilizer rates and identifies the magnitude of yield increase possible with adequate nutrient supply.

9.6.2 Plant Tissue Analysis (Leaf Analysis)

This method directly assesses the nutritional status of the plant itself, which is a better indicator of actual uptake than soil tests alone.

- **How it works:** Representative plant tissue samples (e.g., youngest fully expanded leaves, whole plant at seedling stage) are collected at specific growth stages and sent to a lab for chemical analysis of their nutrient content.
- **Measurement:** Concentrations of various macro- and micronutrients (e.g., N%, P%, K%, Fe ppm, Zn ppm) in the plant tissue are determined.
- **Quantification:** These measured values are compared to established critical levels or sufficiency ranges for that particular crop and growth stage.
  - If the nutrient concentration falls below the critical level, a deficiency is indicated.
  - The degree of deviation from the optimal range can be correlated with potential yield loss, based on established relationships for specific crops.
  - **Hidden Hunger:** Tissue analysis can detect hidden hunger, where nutrient levels are suboptimal, leading to yield loss without visible deficiency symptoms.
- **Advantages:** Provides a direct assessment of what the plant has absorbed, useful for mid-season adjustments.
- **Limitations:** Sampling time and plant part are crucial. Results are interpreted based on empirical critical levels, which can vary. Does not directly measure *yield loss* but indicates *likelihood of loss*.

9.6.3 Soil Testing

Soil tests provide information on the available nutrient status in the soil.

- **How it works:** Soil samples are collected from the root zone of a field and analyzed for plant-available forms of nutrients, pH, organic matter, and other parameters.
- **Measurement:** Lab results provide concentrations (e.g., kg/ha or ppm) of available N, P, K, micronutrients, etc., and soil pH.

- **Quantification:** These values are compared to critical levels or sufficiency ranges established for specific crops and soil types.
    - A nutrient level below the critical threshold indicates a potential deficiency.
    - While soil tests don't directly quantify yield loss, they inform fertilizer recommendations that aim to prevent yield loss by ensuring adequate nutrient supply. The expected yield increase from fertilizer application is an indirect quantification of the deficiency's impact.
- **Advantages:** Helps predict nutrient needs before planting, guides fertilizer applications.
- **Limitations:** Does not account for nutrient uptake efficiency by the plant or interactions with other factors. Results are highly dependent on proper sampling.

9.6.4 Visual Symptom Scoring / Deficiency Maps

While not quantitative in terms of exact yield loss, visual assessments are often the first step in identifying potential nutrient deficiencies.

- **How it works:** Trained observers or automated systems (e.g., image analysis) score the severity of visual deficiency symptoms (chlorosis, necrosis, stunted growth, specific discoloration patterns).
- **Measurement:** A qualitative score (e.g., 0-5 scale for severity) or a percentage of affected leaf area.
- **Quantification (Indirect):** These scores can be correlated with yield reductions through previous calibration studies (e.g., a score of 3 for N deficiency in maize at V6 corresponds to X% yield loss). Remote sensing can generate stress maps based on color or spectral reflectance changes, indirectly indicating areas of nutrient deficiency and thus potential yield loss.
- **Advantages:** Quick, non-destructive, good for early detection and large-area screening.
- **Limitations:** Symptoms can be ambiguous, mimic other stresses, and often appear only after significant yield loss has already occurred (hidden hunger).

9.6.5 Crop Growth Simulation Models

Advanced mechanistic models can simulate crop growth and yield under various nutrient limitations.

- **How it works:** These models incorporate sub-routines that simulate nutrient uptake, partitioning, and their effect on physiological processes (photosynthesis, respiration, biomass allocation). They require inputs on soil nutrient availability, plant genetic parameters (nutrient uptake efficiency, critical nutrient concentrations), and weather data.
- **Quantification:** By running the model with and without nutrient limitations, the predicted yield difference quantifies the simulated yield loss due to deficiency. These models can also simulate the impact of different fertilizer application strategies.
- **Advantages:** Can simulate complex interactions, predict yield under various scenarios, and help in strategic planning (e.g., climate change impacts on nutrient demand).
- **Limitations:** Require extensive data for calibration and validation. Model accuracy depends on the quality of underlying physiological relationships and parameterization.

9.6.6 Yield Component Analysis

Examining the individual components that contribute to final yield can help pinpoint where the nutrient deficiency had its greatest impact.

- **How it works:** At harvest (or specific growth stages), components like:
  - Number of plants per unit area
  - Number of tillers/branches per plant
  - Number of pods/ears/fruits per plant
  - Number of grains/seeds per pod/ear/fruit
  - Individual grain/seed weight (e.g., 1000-grain weight)
  - Biomass (total dry matter) are measured.
- **Quantification:** By comparing these components in nutrient-deficient plants versus adequately fertilized plants, researchers can quantify how the deficiency affected specific yield-determining processes. For example, N deficiency often reduces tillering and grain number, while K deficiency might reduce grain fill.
- **Advantages:** Provides insight into the physiological basis of yield loss.
- **Limitations:** Requires careful experimental design and labor-intensive measurements.

By integrating several of these methods, a more comprehensive and accurate quantification of the effect of nutrient deficiencies on crop yields can be achieved, leading to better nutrient management practices and improved agricultural sustainability.

9.7 Methods used to quantify combined effect of moisture stress and nutrient deficiency on crop yields

Quantifying the combined effects of moisture stress and nutrient deficiencies on crop yields is complete , as these factors often interact synergistically or antagonistically. This means the total impact isn‘t simply the sum of individual stresses. Several advanced and integrated methods are used to understand and quantify these interactions:

9.7.1 Factorial Field Experiments

This is the gold standard for directly quantifying interaction effects.

- **How it works:** Experiments are designed with multiple levels of both water availability (e.g., full irrigation, deficit irrigation, rainfed) and nutrient supply (e.g., optimal nutrient levels, different levels of individual nutrient deficiencies, nutrient omission plots). All possible combinations of these factors are tested in replicated plots.
- **Measurement:** Crop yield, biomass, water use (e.g., through evapotranspiration measurements), nutrient uptake, and various physiological parameters (e.g., photosynthesis, stomatal conductance, leaf water potential, tissue nutrient concentrations) are measured for each treatment combination.
- **Quantification:**
    - **Statistical Analysis:** Advanced statistical methods (e.g., ANOVA with interaction terms) are used to determine if the effect of one factor (e.g., water stress) depends on the level of another factor (e.g., nutrient supply). A significant interaction term indicates that the combined effect is not additive.
    - **Yield Response Surfaces:** Data can be used to develop multi-dimensional yield response surfaces that illustrate how yield changes across a range of water and nutrient combinations.
- **Advantages:** Provides direct, site-specific data on the interactive effects and quantifies the magnitude of yield loss.
- **Limitations:** Highly labor-intensive, time-consuming, and expensive, especially with many factors and levels. Results are specific to the tested conditions.

9.7.2. Crop Simulation Models

Mechanistic crop models are powerful tools for simulating the complex interactions between water and nutrient dynamics and their impact on yield.

- **How it works:** Models like AquaCrop, DSSAT, WOFOST, and APSIM integrate various sub-models for soil water balance, nutrient cycling, plant growth, photosynthesis, and yield formation. They take inputs on weather, soil properties, crop characteristics, irrigation schedules, and fertilizer applications.
- **Quantification:**
    - **Scenario Analysis:** Models can simulate different scenarios of combined water stress and nutrient deficiencies (e.g., varying irrigation levels with different N or P application rates).
    - **Yield Prediction:** By running simulations under various combined stress conditions, the models predict yield outcomes, allowing researchers to quantify the magnitude of yield reduction due to the combined stresses.
    - **Water and Nutrient Use Efficiency:** Models can also help in quantifying water use efficiency (WUE) and nutrient use efficiency (NUE) under combined stress, providing insights into optimal management strategies.
- **Advantages:** Can simulate long-term effects, explore a wide range of scenarios (including future climate conditions), and provide insights into underlying physiological mechanisms. Less costly and time-consuming than extensive field trials.
- **Limitations:** Require extensive calibration and validation with local data to ensure accuracy. Model complexity and data requirements can be high.

9.7.3. Remote Sensing and Proximal Sensing

These technologies allow for large-scale and non-destructive assessment of crop health and stress indicators, which can then be correlated with yield.

- **How it works**
    - **Vegetation Indices (NDVI, NDMI, etc.):** Satellites or drones collect multispectral and hyperspectral imagery. Indices derived from these images (e.g., Normalized Difference Vegetation Index for greenness and vigor, Normalized Difference Moisture Index for water content) can reflect the combined impact of stresses on plant health.

- **Thermal Imaging:** Infrared cameras measure canopy temperature, which reflects transpiration rates and can indicate water stress. Nutrient deficiencies can also indirectly affect canopy temperature.
- **Fluorescence Sensing:** Measures chlorophyll fluorescence, which is directly related to photosynthetic efficiency and can be affected by both water and nutrient stress.

- **Quantification**
  - **Correlation with Yield:** Remote sensing data (e.g., time series of NDVI values) are correlated with ground-truth yield data from field experiments. Regression models can be developed to predict yield loss under combined stress.
  - **Stress Mapping:** Generating maps that identify areas experiencing combined stress, allowing for targeted intervention.
- **Advantages:** Non-destructive, provides spatial variability information, efficient for large areas.
- **Limitations:** Indirect measurement, requires ground truthing and calibration, spectral signatures can be ambiguous for specific stresses without other data. Cloud cover can be an issue for satellite imagery.

9.7.4. Physiological and Biochemical Assays

Detailed measurements of plant physiological and biochemical responses provide insights into how combined stresses affect plant function.

- **How it works**
  - **Gas Exchange Measurements:** Quantifying photosynthesis, transpiration, and stomatal conductance under different water and nutrient regimes.
  - **Leaf Water Potential:** Measuring the water status of leaves using pressure chambers.
  - **Tissue Nutrient Analysis:** Assessing the actual nutrient concentrations in plant tissues under varying water availability, as water stress can inhibit nutrient uptake and transport.
  - **Biochemical Markers:** Analyzing stress-related compounds like proline, abscisic acid (ABA), or antioxidant enzyme activity, which accumulate under stress conditions.
  - **Root System Architecture:** Studying root growth and distribution, as both water and nutrient availability profoundly influence root development.

- **Quantification (Indirect):** These parameters are often measured in factorial experiments and correlated with final yield. For instance, a significant drop in photosynthetic rate due to combined stress can be directly linked to reduced biomass and yield.
- **Advantages:** Provides mechanistic understanding of stress impacts at the plant level.
- **Limitations:** Labor-intensive, often destructive, and typically conducted on a smaller scale.

9.7.5 Water and Nutrient Use Efficiency Assessments

Quantifying how efficiently a crop uses water and nutrients under combined stress.

- **How it works**
  - **Water Use Efficiency (WUE):** Yield (or biomass) per unit of water consumed (e.g., kg grain/mm ET).
  - **Nutrient Use Efficiency (NUE):** Yield per unit of nutrient applied or taken up (e.g., kg grain/kg N applied).
- **Quantification:** These efficiencies are calculated from data obtained in field experiments. Comparing WUE and NUE under single stress vs. combined stress conditions helps quantify the synergistic or antagonistic effects. For example, water stress can reduce NUE by limiting nutrient uptake, and nutrient deficiency can reduce WUE by limiting photosynthetic capacity.
- **Advantages:** Directly links stress to resource efficiency, which is critical for sustainable agriculture.

By combining these methods, researchers and practitioners can gain a comprehensive understanding of the complex interplay between moisture stress and nutrient deficiencies and their cumulative impact on crop yields, leading to more effective integrated management strategies.

9.8 Components of Soil Water Balance

The soil water balance is a fundamental concept in hydrology and agriculture, essentially an accounting of all water inputs and outputs within a defined volume of soil over a specific period. Understanding these components is crucial for managing irrigation, predicting drought, assessing flood risk, and optimizing crop production.

The general soil water balance equation can be expressed as:

$\Delta S = P + I + CR - ET - R - D$

Where:

- **ΔS (Change in Soil Moisture Storage):** This represents the change in the amount of water stored within the soil profile (usually within the root zone) over the given time period.
  - **Increase (+ΔS):** Occurs when inputs exceed outputs.
  - **Decrease (−ΔS):** Occurs when outputs exceed inputs, leading to drier soil.

9.8.1 Inputs (Water Entering the Soil Profile)

- **Precipitation (P)**
  - **Definition:** Water falling from the atmosphere to the Earth's surface in any form (rain, snow, sleet, hail, dew, fog drip).
  - **Significance:** The primary natural input of water to most agricultural systems. Its amount, intensity, and distribution are critical.
- **Irrigation (I)**
  - **Definition:** Water artificially applied to the soil to supplement natural precipitation and meet crop water requirements.
  - **Significance:** A controlled input in many agricultural areas, essential for high-yield production, especially in arid and semi-arid regions or during dry spells.
- **Capillary Rise (CR)**
  - **Definition:** The upward movement of water from a deeper water table (or saturated zone) into the shallower soil profile (root zone) due to capillary forces (adhesion and cohesion between water molecules and soil particles).
  - **Significance:** Can be an important water source in areas with shallow water tables, especially during dry periods, effectively acting as natural sub-surface irrigation.

9.8.2 Outputs (Water Leaving the Soil Profile)

- **Evapotranspiration (ET)**
  - **Definition:** The combined process of water loss from the soil and plant surfaces to the atmosphere.

    - **Evaporation (E):** Direct loss of water vapor from the soil surface and any standing water.
    - **Transpiration (T):** Loss of water vapor from plant leaves through stomata.
  - **Significance:** Often the largest component of water loss from a cropped field. It represents the consumptive use of water by the crop. Potential evapotranspiration (PE) is the maximum water demand, while actual evapotranspiration (AE) is the water actually lost, limited by water availability.

- **Surface Runoff (R)**
  - **Definition:** Water that flows over the land surface when the rate of precipitation or irrigation exceeds the soil's infiltration capacity, or when the soil is saturated.
  - **Significance:** Represents a loss of water that could have been available to the crop. It can also lead to soil erosion and carry away nutrients and pesticides.

- **Deep Percolation / Drainage (D)**
  - **Definition:** The downward movement of water through the soil profile, beyond the effective root zone, typically driven by gravity. This water eventually recharges groundwater or moves into deeper subsurface flow paths.
  - **Significance:** Represents a loss of water from the root zone that is no longer available to the current crop. Excessive deep percolation can lead to nutrient leaching and may contribute to rising water tables and salinity issues in some regions.

- **Conceptual Considerations**
  - **System Boundaries:** Defining the specific volume of soil (e.g., the root zone) and the time period (daily, weekly, monthly, seasonal, annual) for which the balance is calculated is crucial.
  - **Soil Properties:** Soil texture, structure, organic matter content, and depth significantly influence how much water the soil can hold (field capacity, permanent wilting point) and how quickly water infiltrates and drains.7
  - **Plant Factors:** Rooting depth, crop type, and growth stage influence the plant's ability to extract water from the soil and its transpiration rate.
  - **Spatial and Temporal Variability:** All components can vary

significantly across a field and over time, making precise quantification challenging but essential for localized management.

By continuously monitoring and estimating these components, agronomists and water managers can develop more efficient irrigation schedules, select appropriate crops, and implement practices that conserve water and optimize yield.

9.9 Components of Nutrients Balance

The nutrient balance in a soil-plant system is an accounting of all nutrient inputs entering and all nutrient outputs leaving a defined system (e.g., a field, farm, or region) over a specific period. Maintaining a balanced nutrient budget is critical for sustaining soil fertility, optimizing crop productivity, and minimizing environmental pollution.

The general nutrient balance equation can be expressed as:

$$\Delta \text{Nutrientstorage} = \text{Nutrientinputs} - \text{Nutrientoutputs}$$

Where:

- **Δnutrientstorage (Change in Nutrient Storage):** Represents the net change in the amount of a specific nutrient stored within the soil and plant biomass within the defined system over the given period.
  - **Positive balance:** Inputs exceed outputs, leading to nutrient accumulation in the soil (potential for environmental loss).
  - **Negative balance:** Outputs exceed inputs, leading to nutrient depletion from the soil (potential for declining soil fertility and yield).
  - **Neutral/Balanced:** Inputs approximate outputs, indicating sustainable management.

9.9.1 Components of Nutrient Inputs (Nutrients Entering the System)

- **Mineral Fertilizers**
  - **Definition:** Commercially produced inorganic compounds (e.g., urea, DAP, MOP, SSP) providing specific nutrients (N, P, K, S, etc.).
  - **Significance:** The most common and often largest controlled input of nutrients in intensive agricultural systems. Applied to supplement soil nutrient pools and meet crop demand.
- **Organic Manures and Amendments**
  - **Definition:** Materials of biological origin applied to soil, including farmyard manure, compost, green manures, crop residues (if imported

from outside the system or not accounted for as internal cycling), and sewage sludge.

- **Significance:** Provide a slow-release source of various macro- and micronutrients, improve soil organic matter, and enhance soil physical and biological properties.

- **Biological Nitrogen Fixation (BNF)**
    - **Definition:** The process by which certain microorganisms (e.g., Rhizobia in legume root nodules, free-living bacteria like Azotobacter) convert atmospheric nitrogen gas (N2) into plant-available forms (e.g., ammonia).
    - **Significance:** A significant natural input of nitrogen, especially important in legume-based cropping systems, reducing the need for external N fertilizers.
- **Atmospheric Deposition (Wet and Dry)**
    - **Definition:** Nutrients (e.g., N, S, K, Ca) deposited onto the soil surface from the atmosphere, either dissolved in precipitation (wet deposition) or as dry particles (dry deposition).
    - **Significance:** Can be a notable input, particularly for nitrogen in areas with industrial or agricultural emissions (e.g., ammonia volatilization) or for sulfur near industrial areas. Sea spray can contribute to sodium and chloride inputs near coastal regions.
- **Sedimentation and Erosion Deposition**
    - **Definition:** Nutrients carried into the field with eroded soil particles from adjacent areas or by floodwaters.
    - **Significance:** Can be an input to depositional areas, though usually viewed as an environmental problem if coming from off-farm sources due to associated soil loss elsewhere.
- **Nutrient Release from Weathering of Soil Minerals**
    - **Definition:** The slow breakdown of parent rock material and primary minerals in the soil, releasing plant-available forms of nutrients (e.g., K, Ca, Mg, P, micronutrients).
    - **Significance:** A long-term, background source of nutrients, but often too slow to meet high crop demands, especially for highly mobile nutrients.

- **Seeds/Planting Material**
  - **Definition:** Nutrients contained within the seeds or planting material introduced into the system.
  - **Significance:** A relatively minor input, but technically part of the balance.

9.9.2 Components of Nutrient Outputs (Nutrients Leaving the System)

- **Crop Harvest/Product Removal**
  - **Definition:** Nutrients removed from the field in the harvested economic portion of the crop (e.g., grain, fruit, tubers, fodder).
  - **Significance:** The primary intended nutrient output. Quantifying nutrient removal by harvest is crucial for calculating replenishment needs.
- **Leaching**
  - **Definition:** The downward movement of soluble nutrients (e.g., nitrate-NO_3^-, sulfate-SO_4^{2-} , chloride-Cl^-) through the soil profile, below the root zone, often with percolating water.
  - **Significance:** A significant environmental concern, leading to groundwater contamination and nutrient loss from the system. Highly mobile nutrients like nitrate are most susceptible.
- **Gaseous Losses (Volatilization and Denitrification)**
  - **Definition**
    - **Volatilization:** Loss of nutrients as gases from the soil surface (e.g., ammonia-NH_3 from urea or manure application, especially in alkaline conditions).
    - **Denitrification:** Microbial conversion of nitrate (NO_3^-) to gaseous nitrogen (N_2) or nitrous oxide (N_2O) under anaerobic conditions (e.g., waterlogged soils), a potent greenhouse gas.
  - **Significance:** Major pathways for nitrogen loss from agricultural systems, contributing to air pollution and climate change.
- **Surface Runoff and Erosion**
  - **Definition:** Nutrients (dissolved or adsorbed to soil particles) carried away from the field by surface water flow.
  - **Significance:** Causes nutrient loss from the field and contributes to eutrophication of surface waters (e.g., P and N pollution in rivers and lakes).

- **Immobilization**
  - **Definition:** The conversion of plant-available inorganic nutrients into organic forms by soil microorganisms or plant residues, making them temporarily unavailable for plant uptake. While not a permanent loss from the total soil pool, it is a loss from the readily available pool for crops in the short term.
  - **Significance:** Can lead to temporary deficiencies if the mineralization rate is slower than crop demand. It's a critical part of internal nutrient cycling rather than a direct output from the system in a mass balance sense.

9.9.3 Calculating Nutrient Balance

The nutrient balance is typically calculated for key nutrients like Nitrogen (N), Phosphorus (P), and Potassium (K) for a specific field or farm over a growing season or a year.

- **Example (simplified for N)**

$$\Delta nsoil = Nfertilizer + Nmanure + NBNF + Natmospheric\_deposition - Ncrop\_removal - Nleaching - Ngaseous\_losses - Nerosion$$

- Understanding and monitoring these components allows for:
- **Identification of nutrient deficits or surpluses:** Guiding fertilizer recommendations.
- **Assessment of environmental impact:** Reducing nutrient losses to water and air.
- **Promotion of sustainable practices:** Encouraging practices like integrated nutrient management, cover cropping, and residue retention to improve nutrient cycling and reduce reliance on external inputs.

9.10 Conclusion

This chapter has provided a comprehensive overview of **crop production under moisture and nutrient limited conditions**, a reality for the majority of agricultural systems worldwide. We've explored how crucial factors like water and essential nutrients, when deficient, significantly impact a crop's ability to reach its full potential, leading to substantial yield gaps. The detailed examination of **indicators of moisture and nutrient stress**—from visible wilting and leaf discoloration to sophisticated physiological and analytical measurements—underscores the importance of early and accurate diagnosis. Furthermore, we delved into the stage-specific effects of moisture stress and nutrient deficiencies, highlighting why certain developmental phases are more vulnerable and how critical timely intervention can be for mitigating yield losses.

Understanding the intricate components of soil water balance and nutrient balance is fundamental. These balances reveal the constant fluxes of inputs and outputs, emphasizing that effective crop management requires not just adding resources, but optimizing their availability and minimizing losses within the soil-plant system. Finally, we discussed various methods for quantifying the effects of moisture stress and nutrient deficiencies, both individually and in their complex combined interactions. From precise field trials and laboratory analyses to advanced crop simulation models and remote sensing technologies, these tools are indispensable for diagnosing problems, understanding underlying mechanisms, and developing targeted solutions. As we strive to enhance global food security in the face of climate change and dwindling resources, integrating these insights and methodologies will be paramount to fostering resilient, productive, and sustainable agricultural systems that can thrive even under challenging environmental conditions.

Multiple Choice Questions (Bold means correct answer)

1. Which two factors are most frequently the primary drivers of the yield gap in agricultural systems?

   a) Weeds and diseases
   b) Temperature and light
   c) Moisture and nutrient limitations
   d) Soil pH and organic matter

2. What is the most common and easily observable visual symptom of moisture stress in plants?

   a) Interveinal chlorosis
   b) Purple discoloration
   c) Wilting
   d) Excessive branching

3. Leaf rolling or curling in crops like maize is an adaptive response to moisture stress that primarily aims to:

   a) Increase light absorption
   b) Minimize water loss through transpiration
   c) Enhance nutrient uptake
   d) Promote root growth

4. Which of the following is a physiological indicator of moisture stress?

   a) Stunted growth
   b) Reduced stomatal conductance
   c) Premature senescence of older leaves
   d) Shedding of flowers

5. When plants are well-watered, their leaf temperatures are typically:
   a) Higher than the surrounding air
   b) Lower than the surrounding air
   c) The same as the surrounding air
   d) Unrelated to air temperature
6. Which soil-based method provides objective, continuous data on soil water content?
   a) Hand-feel method
   b) Visual observation of cracks
   c) Soil moisture sensors
   d) Soil color changes
7. The Normalized Difference Moisture Index (NDMI) from remote sensing is specifically sensitive to:
   a) Plant nitrogen content
   b) Canopy temperature
   c) Plant water content
   d) Soil organic matter
8. Which growth stage is generally considered the most sensitive and critical to moisture stress for most crops, leading to the most significant yield losses?
   a) Seed germination
   b) Vegetative growth
   c) Reproductive growth (flowering, pollination)
   d) Maturity/ripening
9. Moisture stress during the grain filling stage primarily affects:
   a) The number of grains per ear
   b) The viability of pollen
   c) The size and weight of individual grains
   d) The onset of flowering
10. What is the primary cause of flower abortion or shedding in plants under moisture stress during the reproductive stage?
    a) Increased nutrient availability
    b) To conserve water and energy for survival
    c) Enhanced photosynthesis
    d) Accelerated root growth
11. Nitrogen (N) deficiency symptoms typically appear first on:
    a) Younger leaves
    b) Root tips
    c) Older leaves
    d) Flowers
12. What is the characteristic visual symptom of Phosphorus (P) deficiency?
    a) Yellowing along leaf margins
    b) Interveinal chlorosis on older leaves

c) Dark green or bluish-green leaves, often with purple tints

d) Death of terminal buds

13. Potassium (K) deficiency is often characterized by:

    a) Pronounced interveinal chlorosis on young leaves

    b) Dark green, stunted growth

    c) Yellowing and scorching along the margins of older leaves

    d) Little leaf symptom

14. Which nutrient deficiency primarily affects younger leaves and growing points due to its immobility in the plant?

    a) Nitrogen
    b) Phosphorus
    c) Potassium
    d) Calcium

15. What is the central atom in the chlorophyll molecule, whose deficiency leads to interveinal chlorosis on older leaves?

    a) Iron
    b) Magnesium
    c) Manganese
    d) Copper

16. Sulfur (S) deficiency symptoms resemble Nitrogen deficiency but typically appear first on:

    a) Older leaves
    b) Younger leaves
    c) Roots
    d) Stems

17. Which micronutrient deficiency is characterized by pronounced interveinal chlorosis on younger leaves with veins remaining distinctly green?

    a) Zinc
    b) Manganese
    c) Iron
    d) Boron

18. The little leaf symptom and stunted growth with shortened internodes are characteristic of which micronutrient deficiency?

    a) Iron
    b) Boron
    c) Zinc
    d) Molybdenum

19. What is the primary purpose of soil testing in nutrient management?

    a) To measure the exact amount of nutrient taken up by the plant.

    b) To determine the genetic potential of the crop.

    c) To assess available nutrient concentrations in the soil before planting.

    d) To identify hidden hunger during the reproductive stage.

20 What is the main advantage of plant tissue analysis over soil testing for diagnosing nutrient deficiencies?

a) It's less expensive.

b) It provides a direct measure of what nutrients the plant has actually taken up.

c) It can predict nutrient needs years in advance.

d) It identifies specific soil pH issues.

21. The general soil water balance equation starts with ΔS, which represents:

a) Deep percolation

b) Surface runoff

c) Change in Soil Moisture Storage

d) Evapotranspiration

22. Which component of the soil water balance represents water artificially applied to the soil?

a) Precipitation (P) **b) Irrigation (I)**

c) Capillary Rise (CR) d) Surface Runoff (R)

23. The combined process of water loss from the soil surface and plant leaves to the atmosphere is called:

a) Infiltration b) Percolation

c) Evapotranspiration d) Runoff

24. What is the upward movement of water from a deeper water table into the root zone called?

a) Deep percolation **b) Capillary rise**

c) Surface runoff d) Transpiration

25. Water that flows over the land surface when infiltration capacity is exceeded is known as:

a) Leaching b) Evaporation

c) Surface runoff d) Capillary action

26. Which component of soil water balance contributes to groundwater recharge?

a) Evaporation b) Transpiration

c) Deep percolation d) Surface runoff

27. Which of the following is considered the primary intended nutrient output from a field?
    a) Leaching
    b) Gaseous losses
    c) Crop Harvest/Product Removal
    d) Surface Runoff
28. Biological Nitrogen Fixation (BNF) is an example of a nutrient:
    a) Output
    b) Loss pathway
    c) Input
    d) Storage component
29. The conversion of plant-available inorganic nutrients into organic forms by soil microorganisms, making them temporarily unavailable, is known as:
    a) Mineralization
    b) Immobilization
    c) Volatilization
    d) Denitrification
30. Which nutrient loss pathway is a significant environmental concern for groundwater contamination?
    a) Ammonia volatilization
    b) Crop harvest
    c) Nitrate leaching
    d) Atmospheric deposition
31. Which of the following is a key input of nutrients in intensive agricultural systems that is externally controlled?
    a) Weathering of soil minerals
    b) Biological Nitrogen Fixation
    c) Mineral Fertilizers
    d) Atmospheric Deposition
31. The Yield Response Factor (Ky) quantifies the sensitivity of crop yield to:
    a) Soil pH
    b) Water deficits
    c) Nitrogen content
    d) Temperature fluctuations
32. When Ky > 1, it indicates that the crop is:
    a) Highly tolerant to water stress
    b) Very sensitive to water stress
    c) Not affected by water stress
    d) Efficient in water use
33. Which method for quantifying water deficit uses infrared thermometers to measure canopy temperature?
    a) Soil moisture sensors
    b) Lysimeters
    c) Crop Water Stress Index (CWSI)
    d) Evapotranspiration (ET) measurement via Penman-Monteith

34. The Crop Water Stress Index (CWSI) works on the principle that under water stress, reduced transpiration leads to:
    a) Lower canopy temperature
    b) Higher canopy temperature
    c) Increased stomatal conductance
    d) Enhanced photosynthesis
35. Which method is considered the gold standard for directly quantifying the combined effects of moisture stress and nutrient deficiencies?
    a) Remote sensing
    b) Crop simulation models
    c) Factorial Field Experiments
    d) Plant tissue analysis
36. If a plant shows reduced photosynthesis and stomatal closure, it is most likely experiencing:
    a) Nutrient toxicity
    b) Optimal conditions
    c) Water stress
    d) Excessive branching
37. Which type of crop model is particularly useful for simulating complex interactions between water and nutrient dynamics?
    a) Statistical models
    b) Empirical models
    c) Mechanistic crop models (e.g., DSSAT, APSIM)
    d) Economic models
38. What is a key limitation of visual symptom scoring for nutrient deficiencies?
    a) It is too expensive.
    b) Symptoms often appear only after significant yield loss has occurred (hidden hunger).
    c) It requires complex laboratory equipment.
    d) It cannot be used in the field.
39. Nitrogen deficiency in crops like cereals can significantly reduce the number of:
    a) Roots
    b) Tillers/branches
    c) Leaves
    d) Flowers in fruit crops
40. A negative nutrient balance in a soil-plant system typically leads to:
    a) Nutrient accumulation
    b) Improved soil structure

c) Nutrient depletion and declining soil fertility

d) Increased greenhouse gas emissions

41. What is the role of Boron (B) in plants, making its deficiency critical for reproductive success?

    a) Component of chlorophyll

    b) Essential for cell wall integrity and pollen tube growth

    c) Regulates stomatal opening

    d) Forms ATP

42. In a highly waterlogged soil, which nutrient loss pathway is most likely to be accelerated?

    a) Volatilization b) Leaching

    c) Denitrification d) Crop harvest

43. Which factor affecting achievable production is often constrained by a farmer's economic resources or knowledge?

    a) Genetic potential of the crop b) Climatic conditions

    c) Management decisions d) Soil texture

44. The "rosette" appearance in plants, characterized by shortened internodes and stunted growth, is a symptom of severe deficiency in:

    a) Calcium b) Boron

    c) Zinc d) Copper

45. When comparing WUE (Water Use Efficiency) and NUE (Nutrient Use Efficiency) under combined stress, reduced nutrient availability might lead to:

    a) Increased WUE

    b) Reduced WUE by limiting photosynthetic capacity

    c) No change in WUE

    d) Reduced transpiration only

46. What is the main benefit of conducting nutrient omission trials in the field?

    a) To determine the global demand for a specific nutrient.

    b) To directly quantify the yield reduction caused by a specific nutrient deficiency at that site.

    c) To predict future climate impacts on nutrient cycling.

    d) To identify areas suitable for organic farming.

47. Thermal imaging using drones can detect water stress by identifying:
    a) Variations in plant height
    b) Differences in canopy temperature
    c) Nutrient content of leaves
    d) Soil organic matter content

48 Which of the following is a common visual symptom of Calcium (Ca) deficiency in fruits like tomatoes?
    a) Blossom-end rot
    b) Purple discoloration
    c) Marginal scorching
    d) Interveinal chlorosis on older leaves

49. If atmospheric CO_2 concentration were to significantly increase, which type of crop would generally show a more significant increase in photosynthetic efficiency and yield under otherwise optimal conditions?
    a) C4 crops (e.g., maize)
    b) C3 crops (e.g., wheat)
    c) Both equally
    d) Neither would show an increase

10

Weather Forecasting

10.1 Introduction

Weather forecasting stands as a testament to humanity's enduring quest to understand and anticipate the natural world. Far from being a mere casual prediction, it is a highly sophisticated scientific and technological discipline dedicated to projecting the future state of the Earth's atmosphere for specific locations and times. This complex endeavor begins with the meticulous collection of vast quantities of real-time data, encompassing everything from surface observations of temperature, pressure, wind, and humidity to upper-air measurements gathered by weather balloons, aircraft, radar, and advanced satellite systems.

The core of modern weather forecasting lies in numerical weather prediction (NWP) models. These are intricate computer programs that embody the fundamental laws of physics governing atmospheric processes, such as fluid dynamics and thermodynamics. By feeding the enormous datasets of current atmospheric conditions into these models, supercomputers solve complex equations to simulate how the atmosphere will evolve over minutes, hours, and days. The raw output from these models is then further refined by human meteorologists who apply their expertise, integrating local geographical influences, historical patterns (climatology), and statistical techniques to produce the most accurate and actionable forecasts.

The importance of accurate weather forecasting cannot be overstated. It serves as a vital tool across countless sectors, guiding daily personal decisions, from what to wear to whether to carry an umbrella. Crucially, for agriculture, precise weather forecasts are indispensable. Farmers rely heavily on these predictions to make informed decisions that directly impact crop yield and livestock health. Knowing when to plant seeds, schedule irrigation, apply fertilizers and pesticides, or harvest crops is directly influenced by anticipated rainfall, temperature extremes, and wind conditions. Timely warnings of frost, hailstorms, or prolonged droughts can enable farmers to take protective measures, minimizing losses and ensuring food security. For example, in regions like Telangana, where agriculture is a dominant sector, understanding the monsoon's progression and anticipating dry spells or heavy downpours is critical for the success of paddy and other major crops.

Beyond agriculture, weather forecasting provides crucial information for transportation, especially aviation and shipping, where precise forecasts are indispensable for safety and efficiency. Furthermore, weather warnings, generated through forecasting, are absolutely critical for protecting lives and property from the devastating impacts of severe weather events like hurricanes, floods, and blizzards. Despite the inherent chaotic nature of the atmosphere, continuous advancements in observational technology, computational power, and meteorological understanding are constantly pushing the boundaries of forecasting skill, leading to ever more reliable predictions that underpin societal resilience and economic well-being.

Weather forecasting is a highly sophisticated scientific and technological discipline, a testament to humanity's ongoing endeavor to understand and anticipate the natural world. It involves the meticulous collection of vast quantities of real-time atmospheric data – from surface observations of temperature, pressure, wind, and humidity to upper-air measurements gathered by weather balloons, aircraft, radar, and advanced satellite systems. This deluge of information is then fed into cutting-edge numerical weather prediction (NWP) models. These complex computer programs, built upon the fundamental laws of atmospheric physics, simulate how the atmosphere will evolve over various timescales. The raw output from these models is then expertly interpreted and refined by human meteorologists, who integrate local geographical influences, historical climatic patterns, and statistical techniques to generate the most accurate and actionable forecasts.

The importance of accurate weather forecasting permeates nearly every facet of modern life, extending far beyond daily planning. Its critical role is particularly pronounced in agriculture, where the very livelihood of farmers and global food security hinge on atmospheric conditions. Precise weather forecasts are indispensable for optimizing crucial farm operations. For instance, knowing when rain is expected allows farmers to strategically time planting for optimal seed germination or to delay harvesting to prevent spoilage of mature crops. Conversely, predictions of dry spells enable efficient irrigation planning, conserving precious water resources and reducing costs.

Beyond these immediate decisions, weather forecasts are vital for managing fertilizer application, ensuring nutrients are absorbed effectively and not washed away by heavy rains. They also play a pivotal role in pest and disease management, as certain weather conditions can trigger outbreaks; forewarning allows for timely preventative measures, minimizing crop losses and the need for chemical interventions. Furthermore, forecasts of extreme weather events like frost, hail, or strong winds empower farmers to take protective actions, such as covering sensitive crops or securing infrastructure, thereby mitigating potential devastation. In a world increasingly affected by climate variability and extreme weather, reliable weather forecasting transforms weather from an unpredictable risk into a strategic advantage, enabling farmers to enhance productivity, conserve resources, and build resilience in their operations, ultimately contributing to a more sustainable and secure food supply.

10.2 Types of Weather Forecasting

Weather forecasting is broadly categorized by the lead time of the prediction, which directly correlates with the methods used and the level of detail and accuracy achievable. These categories serve different purposes for various users, from individuals planning their day to large industries making strategic decisions.

Here are the main types of weather forecasting based on their timescale:

- **Nowcasting (Very Short Range Forecast):** This type of forecast covers a period from a few minutes up to 6-12 hours, sometimes extending to 2-3 hours for very precise details. Nowcasting focuses on immediate, rapidly evolving weather phenomena like thunderstorms, fog, or sudden heavy rainfall. It relies heavily on real-time observations from radar, satellites, and ground-based sensors, using extrapolation techniques to predict movement and development. Its high accuracy for a very limited timeframe makes it invaluable for aviation, emergency services, and early warnings for localized severe weather.
- **Short-Range Forecast:** Typically covering a period of 1 to 3 days (or up to 72 hours), short-range forecasts are the most common type encountered by the general public. These forecasts are largely driven by numerical weather prediction (NWP) models, which simulate atmospheric processes. While still very accurate, especially for the first 24-48 hours, the skill gradually decreases with increasing lead time due to the chaotic nature of the atmosphere. They provide essential information for daily activities, transportation, and initial planning for events.
- **Medium-Range Forecast:** This category extends from 3 days to about 10 days. Medium-range forecasts also rely heavily on NWP models, but their accuracy is lower than short-range forecasts. They provide a general outlook on weather patterns, indicating trends in temperature, precipitation, and significant weather systems. These forecasts are

crucial for various sectors, including agriculture (for planning planting, irrigation, and harvesting weeks in advance), construction, and energy management.

- **Subseasonal Forecasts:** Bridging the gap between traditional medium-range weather forecasts and seasonal climate predictions, subseasonal forecasts typically cover a period of 2 weeks to 2 months (sometimes up to 3 months). This is often referred to as the predictability desert" due to the complex interplay of atmospheric and oceanic influences at this timescale. Unlike shorter-range forecasts that primarily depend on the initial state of the atmosphere, subseasonal forecasts draw predictability from slowly evolving boundary conditions, such as sea surface temperatures and soil moisture, and atmospheric oscillations like the Madden-Julian Oscillation (MJO).
  - **Importance in Agriculture:** Subseasonal forecasts hold immense potential for agriculture. They can provide crucial lead time for decisions that transcend daily operations but fall short of long-term climate strategies. For example, a subseasonal forecast might indicate a higher probability of extended dry spells or unusual heat waves in the coming weeks. This allows farmers to:
  - **Adjust irrigation schedules** more proactively to conserve water or prepare for increased demand.
  - **Optimize planting times** for specific crop varieties that might be more resilient to anticipated conditions.
  - **Plan for pest and disease outbreaks** that are often linked to specific temperature and humidity patterns over several weeks.
  - **Make informed decisions about crop insurance** or market strategies based on expected yield impacts from prolonged weather anomalies.
  - **Manage livestock** by preparing for extended periods of heat stress or cold. By providing this intermediate-term outlook, subseasonal forecasts empower agricultural stakeholders to make more resilient and profitable decisions, reducing risks associated with weather variability.
- **Long Range Forecasts:** These forecasts predict average weather patterns for a season (e.g., 3 months) or even up to a year in advance. They do not predict daily weather but rather the likelihood of above, below, or near-normal conditions for temperature and precipitation. Seasonal forecasts are largely based on the behavior of large-scale climate phenomena like El Niño-Southern Oscillation (ENSO), Indian Ocean Dipole (IOD), and

other teleconnections that influence global weather patterns over longer periods. They are vital for long-term planning in sectors like water resource management, energy, disaster preparedness, and agricultural policy at a macro level, helping in crop selection and water allocation for an entire growing season.

- **Climate Forecasts (or Climate Outlooks):** Extending beyond a season, these forecasts can cover periods of several months to even a year or more. They provide very general indications of climate trends and anomalies, primarily useful for strategic planning, resource allocation, and understanding potential shifts in long-term weather patterns related to climate change (Fig.10.1). Their accuracy is inherently lower, focusing on probabilities and broad trends rather than specific weather events.

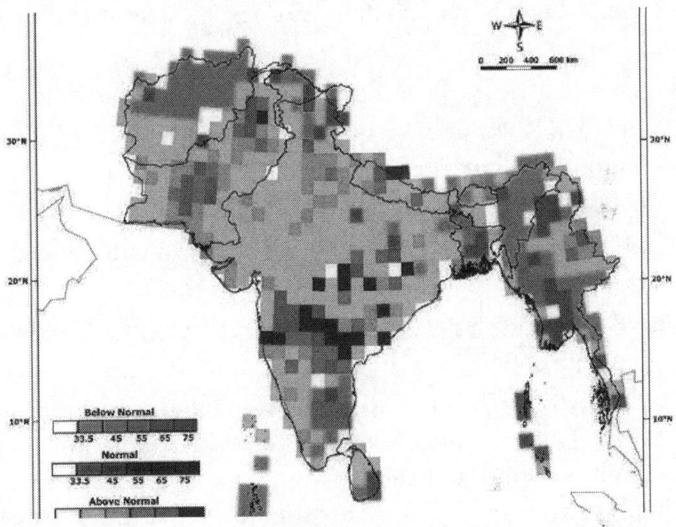

Fig. 10.1: Climate outlook (Monsoon-2025) in South Asia

10.3 Tools Used for Weather Forecasting

Weather forecasting, a cornerstone of modern society, is an intricate blend of scientific observation, advanced computational power, and expert human analysis. To achieve its goal of predicting future atmospheric conditions, the field employs an extensive toolkit, constantly evolving with technological advancements. These tools can be broadly categorized into systems for data collection, and powerful platforms for data processing and analysis.

10.3.1 Observation Systems -The Eyes and Ears of Weather Forecasting

The accuracy of any forecast hinges on the quality and quantity of initial data describing the current state of the atmosphere. A diverse array of instruments gathers this essential information:

- **Ground-Based Weather Stations (Automated and Manual):** These form the foundational layer of atmospheric observation, providing crucial surface-level data.
  - **Thermometers** precisely measure air temperature, often using highly accurate electronic sensors like Platinum Resistance Thermometers.
  - **Barometers** detect atmospheric pressure, a key indicator of approaching high or low-pressure systems that dictate weather changes. Digital barometers are now commonplace, offering continuous readings.
  - **Hygrometers** quantify the amount of water vapor in the air (humidity), which is vital for predicting fog, clouds, and precipitation.
  - **Anemometers** measure wind speed, with cup anemometers and the more advanced ultrasonic anemometers being standard.
  - **Wind Vanes** (or wind socks) indicate wind direction, providing essential context for wind speed data.
  - **Rain Gauges** accurately measure liquid precipitation, with automated tipping bucket gauges providing real-time data.
  - Specialized instruments like **Snow Gauges** measure snowfall's water equivalent, **Pyranometers**assess solar radiation, **Ceilometers** determine cloud base height using lasers, and **Disdrometers**analyze raindrop size distribution, all contributing to a comprehensive surface picture.
- **Weather Balloons (Radiosondes/Dropsondes):** These are vital for probing the vertical structure of the atmosphere.
  - **Radiosondes** are small, instrument packages attached to balloons that ascend through the atmosphere, measuring and wirelessly transmitting data on temperature, humidity, and pressure up to altitudes of 30 kilometers. Their ascent also provides wind speed and direction data (called a rawinsonde observation).
  - **Dropsondes** are similar instruments released from aircraft (often specialized hurricane hunter" planes) into severe weather systems, providing crucial in-situ data from within storms.

- **Weather Radar (e.g., Doppler Radar, NEXRAD):** These systems are indispensable for observing precipitation and storm dynamics.
  - Radar units emit microwave pulses that reflect off hydrometeors (rain, snow, hail) in the atmosphere. The time it takes for the pulse to return indicates the distance to the precipitation, and the strength of the return signal indicates its intensity.
  - **Doppler Radar** takes this a step further by utilizing the Doppler effect: it measures the shift in frequency of the returned pulse, which reveals the motion of the precipitation towards or away from the radar. This allows meteorologists to detect wind shear, rotation within thunderstorms (indicative of tornadoes), and the speed of rain bands, making it critical for severe weather warnings.
- **Weather Satellites:** Orbiting high above Earth, satellites offer a synoptic, large-scale view of the atmosphere that ground-based observations cannot provide.
  - They are equipped with various sensors to track cloud cover and movement (both visible light and infrared imagery), measure atmospheric temperature profiles, detect water vapor even in clear skies, and monitor sea surface temperatures (SSTs) – a crucial factor for tropical cyclone formation and long-range forecasts (Fig.10.2).

Fig. 10.2: Weather satellite and imagery

  - **Geostationary satellites** remain over a fixed point on the equator, providing continuous updates for a specific region, while polar-orbiting satellites circle the Earth, offering global coverage over time.
- **Aircraft Observations (AMDAR - Aircraft Meteorological Data Relay):** Commercial aircraft, equipped with sophisticated sensors, automatically collect and transmit meteorological data (temperature,

wind speed/direction, turbulence) along their flight paths. This provides a constant stream of valuable upper-air observations, particularly over data-sparse regions like oceans.

- **Ocean Buoys (Fixed and Drifting):** These automated stations deployed across oceans collect vital data on sea surface temperature, air temperature, wind speed and direction, and wave height. This oceanic data is paramount for marine weather forecasting, understanding large-scale climate drivers like El Niño, and improving hurricane predictions.
- **Lidar (Light Detection and Ranging) and Sodar (Sound Detection and Ranging):** These specialized remote sensing tools use light (laser pulses) or sound waves, respectively, to measure atmospheric properties.
  - **Lidar** can profile aerosols, measure cloud properties, and determine wind speed and direction at various altitudes.
  - **Sodar** is particularly useful for measuring wind profiles in the lower atmosphere, important for air quality and wind energy applications.

10.3.2 Data Processing and Analysis Tools - Making Sense of the Data Deluge

Collecting data is only the first step. The real power of modern weather forecasting lies in how this immense volume of information is processed, analyzed, and transformed into actionable predictions.

- **Supercomputers:** These incredibly powerful machines are the central nervous system of weather forecasting. Their immense computational capacity is essential for running Numerical Weather Prediction (NWP) models. These models are essentially vast sets of mathematical equations representing the fundamental laws of fluid dynamics, thermodynamics, and radiation that govern atmospheric behavior. Supercomputers solve these complex equations, simulating how the atmosphere will evolve over time.
- **Numerical Weather Prediction (NWP) Models:** These are the heart of modern forecasting.
  - **Global Models** (e.g., the Global Forecast System (GFS) from the US, the European Centre for Medium-Range Weather Forecasts (ECMWF) model) simulate the entire Earth's atmosphere, providing broad weather patterns.
  - **Regional/Mesoscale Models** (e.g., Weather Research and Forecasting (WRF) model, High-Resolution Rapid Refresh (HRRR)) focus on smaller geographic areas with higher spatial resolution, allowing

for more detailed and accurate short-range forecasts of localized phenomena.

- **Ensemble Models** address the inherent chaos of the atmosphere. Instead of running a single forecast, they run multiple versions of the same model with slightly perturbed initial conditions or different physical parameterizations. This produces a range of possible outcomes, allowing forecasters to assess the probability of certain events and the confidence level of a forecast.

- **Data Assimilation Systems:** These sophisticated algorithms are the bridge between observations and NWP models. They intelligently combine disparate observational data (from all the sources mentioned above) with a previous forecast to create the most accurate and physically consistent initial conditions for the next model run. This is a crucial step, as even tiny errors in the initial state can lead to significant forecast deviations due to the atmosphere's chaotic nature.
- **Geographic Information Systems (GIS):** GIS tools are used to visualize, analyze, and manage geographical weather data. They allow meteorologists to overlay different data layers (e.g., radar reflectivity, satellite imagery, model output, terrain features) to identify patterns, understand spatial relationships, and gain a more comprehensive understanding of weather phenomena.
- **Visualization Software and Workstations:** Advanced meteorological workstations and specialized software provide meteorologists with powerful graphical interfaces to view, interpret, and interact with the massive amounts of data generated by observation systems and NWP models. This allows them to quickly identify critical features, analyze trends, detect potential model biases, and ultimately refine the raw model output into a user-friendly and accurate forecast.
- **Artificial Intelligence (AI) and Machine Learning (ML):** These cutting-edge technologies are rapidly transforming weather forecasting:
  - They are being used to improve data assimilation, extracting more signal from noisy observations.
  - AI/ML can enhance the accuracy of NWP models by learning complex relationships in past data that traditional physics-based models might struggle with.
  - They are adept at post-processing model output, refining global model predictions to highly localized areas (downscaling) and correcting for known model biases.

- AI/ML excels at identifying subtle patterns in vast datasets that might be imperceptible to human eyes, leading to better predictions of extreme events.
- They can also be used for personalized forecasting, tailoring weather information to specific user needs.

By harnessing this sophisticated suite of tools, meteorologists can observe the atmosphere comprehensively, process an immense volume of data, simulate future conditions, and ultimately produce the increasingly accurate and timely weather forecasts that are indispensable for everything from daily personal planning to critical disaster preparedness and economic decision-making across the globe.

10.4 Principles of Weather Forecasting

The principles of weather forecasting vary significantly depending on the forecast's lead time. This is because different atmospheric phenomena dominate predictability at different scales. While all types rely on observations and an understanding of atmospheric physics, the emphasis shifts from detailed initial conditions to large-scale climate drivers as the forecast horizon extends.

Here's an elaboration on the principles for each type:

10.4.1 Nowcasting (Very Short Range Forecast: 0-6 hours)

- **Principle: Extrapolation and Rapid Assimilation of Real-time Observations**

Nowcasting focuses on highly localized, rapidly evolving weather events that are often too small or fast-moving for traditional Numerical Weather Prediction (NWP) models to resolve accurately in their early stages.

- **Observation-Driven:** The primary principle is to heavily rely on the *most current and high-resolution observations*. This means immediate access to radar data (showing precipitation type, intensity, and movement, including storm rotation), high-frequency satellite imagery (tracking cloud development and movement), and dense networks of surface weather stations.
- **Extrapolation:** The core technique is extrapolation, which involves projecting the observed movement and evolution of current weather features (e.g., a thunderstorm, a fog bank) forward in time. This is often done using kinematic methods, assuming continuity of motion and development.

- **Rapid Refresh and Assimilation:** Nowcasting systems require extremely rapid data assimilation cycles (often every few minutes) to incorporate the latest observations and update the forecast in near real-time. This allows for quick adjustments as conditions change.
- **Identification of Mesoscale Phenomena:** The emphasis is on identifying and tracking mesoscale (small to medium scale) weather phenomena that have significant immediate impacts, such as individual thunderstorms, squall lines, localized heavy rainfall, or sudden wind gusts.
- **High Spatial and Temporal Resolution:** To capture these small-scale features, nowcasting requires observations and forecast outputs at very high spatial resolution (e.g., hundreds of meters to a few kilometers) and high temporal frequency (e.g., every 5-15 minutes).
- **Increasing Role of Convection-Permitting Models:** While traditionally extrapolation-based, modern nowcasting increasingly integrates rapidly updating, high-resolution NWP models that explicitly resolve convective processes, allowing for better prediction of storm initiation and evolution.

10.4.2 Short-Range Forecast (1-3 Days)

- **Principle: Initial Value Problem Solved by Numerical Weather Prediction (NWP)**

Short-range forecasting is fundamentally treated as an initial value problem, where the current state of the atmosphere is precisely defined, and then the laws of physics are used to predict its evolution.

- **Dominance of Atmospheric Dynamics:** The forecast is primarily driven by the large-scale and synoptic-scale (hundreds to thousands of kilometers) atmospheric motions and thermodynamic processes governed by the fundamental equations of fluid dynamics, thermodynamics, and radiative transfer.
- **Numerical Weather Prediction (NWP):** This is the cornerstone. Sophisticated NWP models, running on supercomputers, take the observed initial state of the atmosphere (via data assimilation) and numerically integrate the governing equations forward in time.
- **Accurate Initial Conditions:** The principle of data assimilation is crucial. The quality of the short-range forecast is highly dependent on how accurately the initial state of the atmosphere is represented in the model. Errors in the initial conditions grow rapidly due to the chaotic nature of the atmosphere.

- **Parameterization of Sub-Grid Scale Processes:** Processes too small to be explicitly resolved by the model's grid (e.g., clouds, turbulence, convection) are *parameterized* – their effects are represented by simplified mathematical relationships based on the larger-scale variables the model *does* resolve.
- **Deterministic Forecasts (with Ensemble Guidance):** While the aim is to produce a single, most likely forecast (deterministic), the inherent chaotic nature means that small initial errors can lead to different outcomes. Therefore, ensemble forecasting is used: running multiple model simulations with slightly varied initial conditions or model physics to provide a range of possible outcomes and quantify forecast uncertainty.

10.4.3 Medium-Range Forecast (3-10 Days)

• Principle: Continued Dominance of NWP with Increasing Emphasis on Ensemble Forecasting and Model Physics

Medium-range forecasting extends the principles of short-range forecasting, but the challenge of atmospheric chaos becomes more pronounced.

- **Numerical Weather Prediction (NWP) Remains Central:** NWP models are still the primary tool, but the longer lead time means that initial condition errors have more time to amplify.
- **Increased Reliance on Ensemble Forecasting:** The principle of probabilistic forecasting becomes much more significant. Ensemble prediction systems are critical for assessing the uncertainty inherent in the forecast. A wider spread in ensemble members indicates lower confidence, while a tighter clustering suggests higher confidence.
- **Importance of Model Physics and Resolution:** Accurate representation of physical processes (e.g., cloud formation, precipitation, land-surface interactions, atmospheric boundary layer) within the NWP models becomes even more crucial, as subtle biases can accumulate over longer periods. Higher resolution global models are also beneficial.
- **Teleconnections and Global Circulation:** While initial conditions are still important, large-scale atmospheric circulation patterns and their interactions with other parts of the climate system (e.g., ocean temperatures) begin to exert a more noticeable influence on predictability.

10.4.4 Subseasonal Forecasts (2 Weeks - 2 Months)

• Principle: Identifying and Leveraging Sources of Predictability Beyond Initial Conditions (Boundary Forcing)

Often called the predictability desert," subseasonal forecasting transitions from being purely an initial value problem to incorporating slowly varying boundary conditions and large-scale atmospheric oscillations.

- **Bridging the Gap:** The core principle is to bridge the predictability gap between the deterministic skill of medium-range forecasts and the statistical predictability of seasonal forecasts.
- **Sources of Predictability (Boundary Forcing):** Unlike shorter-range forecasts, skill here comes not just from the exact initial state of the atmosphere but increasingly from persistent anomalies in underlying slower-varying components of the Earth system. Key sources include:
  - **Sea Surface Temperatures (SSTs):** Anomalies in SSTs (e.g., from El Niño-Southern Oscillation - ENSO, or the Indian Ocean Dipole - IOD) can influence atmospheric circulation patterns for weeks to months.
  - **Soil Moisture:** Persistent soil moisture anomalies (very dry or very wet) can impact local temperature and precipitation patterns by affecting evaporation and heat transfer.
  - **Snow Cover and Sea Ice:** The extent of snow and ice cover can influence local and regional atmospheric conditions through albedo and heat exchange.
  - **Intraseasonal Oscillations:** Phenomena like the Madden-Julian Oscillation (MJO), a propagating area of enhanced convection and rainfall in the tropics, strongly influence global weather patterns on subseasonal timescales. Predicting the MJO's phase and amplitude is a major principle here.
- **Coupled Models:** Forecast systems for subseasonal ranges often use coupled atmosphere-ocean-land models to capture the interactions between these components.
- **Ensemble Forecasting (Probabilistic):** Due to high uncertainty, forecasts are inherently probabilistic, providing probabilities of above/below normal conditions rather than precise daily forecasts. Ensemble prediction systems are indispensable.

- **Statistical and Machine Learning Methods:** Statistical models that identify historical relationships between these slow-varying boundary conditions and subsequent weather, as well as modern machine learning techniques, play a significant role in extracting and enhancing predictability.

10.4.5. Long range Forecasts (3-6 Months)

- **Principle: Relying on Slowly Varying Boundary Conditions and Low-Frequency Climate Variability**

Seasonal forecasts abandon attempts at daily weather prediction and focus on the likelihood of average conditions (e.g., mean temperature, total precipitation) deviating from climatological norms over an entire season.

- **Boundary Value Problem:** The forecast is treated more as a boundary value problem, where the primary drivers are the slowly changing boundary conditions of the Earth system rather than the rapidly evolving atmospheric initial state.
- **Dominance of Ocean-Atmosphere Interaction:** The most critical principle is the influence of large-scale ocean phenomena, particularly:
  - **El Niño-Southern Oscillation (ENSO):** The most significant driver of interannual climate variability globally. Predicting the phase (El Niño, La Niña, Neutral) of ENSO is paramount.
  - **Indian Ocean Dipole (IOD):** Influences weather patterns, especially in the Indian Ocean region.
  - **Atlantic Multi-decadal Oscillation (AMO):** A longer-term oscillation in Atlantic Ocean temperatures.
- **Coupled Climate Models:** Advanced coupled ocean-atmosphere-land-ice models are used, designed to simulate the interactions between these components over long periods.
- **Probabilistic Nature:** Seasonal forecasts are almost entirely probabilistic. They do not predict specific days of rain or sunshine but rather the probability of a season being warmer/colder or wetter/drier than average.
- **Climatological Basis:** The forecasts are always compared to the long-term climatology (average conditions for that season over many years) to determine anomalies.

10.4.6 Climate Forecasts (Beyond 6 Months to Years)

- **Principle: Analysis of Climate Trends, Decadal Variability, and Climate Change Projections**

Long-range forecasts move into the domain of climate prediction, focusing on very broad trends and potential shifts rather than specific weather anomalies.

- **Climate Drivers:** The principles are based on understanding and projecting long-term climate drivers, including:
  - **Greenhouse Gas Concentrations:** The most significant human-induced factor influencing long-term climate trends.
  - **Solar Variability:** Changes in solar radiation.
  - **Volcanic Activity:** Large volcanic eruptions can have a temporary cooling effect.
  - **Oceanic Cycles (Decadal to Multi-Decadal):** Longer-term oscillations in ocean temperatures that influence global climate patterns.
- **Climate Models (Earth System Models):** Highly complex Earth System Models (ESMs) are used, which integrate atmospheric, oceanic, land, and cryospheric components, as well as biogeochemical cycles (e.g., carbon cycle).
- **Scenario-Based Projections:** Forecasts are often presented as projections based on different emission scenarios (e.g., how temperatures might change if greenhouse gas emissions follow a high or low trajectory).
- **Focus on Averages and Extremes:** Predictions are not about specific weather events but about long-term averages, changes in the frequency or intensity of extreme events (e.g., heatwaves, heavy rainfall), and regional climate shifts over decades.
- **Less Detailed and More Uncertain:** Due to the long lead time and the complexities of the Earth system, long-range forecasts have the lowest spatial and temporal detail and the highest uncertainty among all forecast types.

In essence, as the forecast lead time increases, the primary source of predictability shifts from the exact initial state of the fast-moving atmosphere to the more slowly evolving components of the Earth system (oceans, land surface) and the external forcings like greenhouse gas concentrations. This transition necessitates a change in the models used, the type of information provided, and the confidence levels associated with the predictions.

10.5 Forecasting Long Range Rainfall

The Southwest Monsoon (SWM) rainfall, which graces India from June to September, is unequivocally the nation's lifeline, delivering 70-80% of its annual precipitation. Its timely onset, healthy progression, and equitable distribution are paramount for India's largely agrarian economy, water security, hydroelectric power generation, and overall national prosperity. Consequently, the Long Range Forecast **(LRF)** for the Southwest Monsoon is a cornerstone of strategic planning for the country.

• How the Long-Range Forecast is Made? A Multi-Pronged Approach

The India Meteorological Department (IMD), the nodal agency for weather and climate services in India, employs a sophisticated and evolving methodology for its operational LRF. This methodology has progressed from rudimentary statistical correlations to a highly integrated system combining advanced numerical models with refined statistical techniques, often within a Multi-Model Ensemble (MME) framework.

10.5.1 Statistical (Empirical) Models

- **Foundation in Teleconnections:** These models leverage the principle of teleconnections," identifying robust statistical relationships between global atmospheric and oceanic parameters (predictors) and the subsequent Indian monsoon rainfall over decades. Sir Gilbert Walker's pioneering work in the early 20th century, which linked the Southern Oscillation (later understood as part of ENSO) to the monsoon, laid the groundwork.
- **Refined Predictor Selection:** IMD continuously refines its statistical ensemble forecasting system (SEFS), optimizing the selection of predictors based on their consistency and predictive skill. These models are built on the premise that certain global climate patterns occurring months before the monsoon can subtly force" the Indian monsoon's behavior.

10.5.2 Dynamical Models (Numerical Weather Prediction - NWP)

- **Coupled Global Climate Models (CGCMs):** These are the workhorses of modern long-range forecasting. Running on powerful supercomputers, CGCMs numerically simulate the complex physical interactions within the Earth's climate system – atmosphere, ocean, land, and cryosphere. IMD's **Monsoon Mission Climate Forecasting System (MMCFS)** is a key indigenous CGCM used for this purpose.

- **Multi-Model Ensemble (MME) Approach:** Recognizing the inherent uncertainties in any single climate model, the MME approach is the preferred method. It combines forecasts from several CGCMs developed by various global climate prediction and research centers (including IMD's own). By averaging or weighting the outputs of these diverse models, the MME often yields a more robust, skillful, and reliable forecast, effectively reducing the individual biases and limitations of a single model.
- **Data Assimilation:** Both statistical and dynamical forecasting processes are underpinned by rigorous data assimilation. Vast amounts of real-time and historical observational data from a global network of satellites, ocean buoys, land-based stations, and other sources are continuously fed into the systems to establish the most accurate initial conditions for the models and to inform the state of predictor parameters.
- **Two-Stage Forecasts:** IMD typically issues its LRF in two phases to provide updated information as the monsoon season approaches:
- **First Stage (Mid-April):** Provides an initial quantitative and probabilistic forecast for the total seasonal (June-September) rainfall over India as a whole.
- **Second Stage (End of May):** An updated forecast based on the latest observations and model runs, often including probabilistic forecasts for rainfall over specific homogenous regions of India (e.g., Northwest India, Central India) and a specific forecast for June rainfall.

10.5.3 Emerging Technologies

Research and operational integration of Artificial Intelligence (AI) and Machine Learning (ML) techniques are ongoing. These technologies hold promise for identifying more subtle patterns in vast climate datasets, improving predictor selection, refining model output, and potentially enhancing the overall accuracy of monsoon forecasts.

10.5.4 Significance of the Long-Range Monsoon Forecast

The LRF of the Southwest Monsoon is not just an academic exercise; it is a critical input for various sectors, driving significant socio-economic decisions across India.

- **Agriculture (The Primary Beneficiary):** With a large portion of Indian agriculture being rain-fed, an accurate monsoon forecast is paramount for farmers. It influences decisions on:

  - **Crop Choice:** Selecting drought-resistant varieties during anticipated weak monsoons or water-intensive crops during expected normal/excess rainfall.
  - **Sowing and Harvesting Schedules:** Optimizing these crucial activities to prevent crop damage and maximize yields.
  - **Input Management:** Efficient application of fertilizers and pesticides, preventing wastage or ensuring their efficacy.
  - **Irrigation Strategies:** Planning for supplementary irrigation, or conversely, conserving water resources.
- **Water Resource Management:** The monsoon is the primary source of replenishment for India's reservoirs, lakes, and groundwater. Forecasts aid in:
  - **Reservoir Operations:** Managing water levels for drinking water supply, irrigation, and hydroelectric power generation.
  - **Flood and Drought Preparedness:** Enabling early warnings, deploying resources, and implementing mitigation plans.
- **Economy:** A robust monsoon invariably leads to higher agricultural output, which stimulates rural demand, boosts consumer spending, and contributes significantly to GDP growth. Conversely, a poor monsoon can trigger food inflation, rural distress, and an overall economic slowdown, making the forecast a key economic indicator.
- **Disaster Preparedness:** Timely forecasts of prolonged dry spells or intense rainfall events are vital for disaster management agencies to prepare for potential droughts or floods, protecting lives and property.
- **Policy Making:** Government bodies at national and state levels utilize these forecasts for crucial policy decisions related to food security, agricultural subsidies, import/export policies for commodities, and resource allocation.

10.5.5 Role of Key Global Climatic Parameters (Predictors)

The Indian Southwest Monsoon's variability is profoundly influenced by large-scale global climate phenomena, known as teleconnections. Understanding and predicting the state of these parameters is central to LRF.

- **El Niño-Southern Oscillation (ENSO)**
  - **What it is:** ENSO is the most dominant mode of interannual climate variability on Earth, involving periodic fluctuations in sea surface temperatures (SSTs) and atmospheric pressure across the equatorial Pacific Ocean.

- **El Niño:** Characterized by warmer-than-average SSTs in the central and eastern equatorial Pacific. This warming often shifts rainfall patterns eastward in the Pacific, affecting global atmospheric circulation.
- **La Niña:** Characterized by cooler-than-average SSTs in the same region. This typically enhances the normal Walker Circulation, intensifying convection over the western Pacific and Indian Ocean.
- **Neutral:** Conditions are neither significantly warm nor cold.

- **Role in Monsoon:** ENSO has a strong, though not exclusive, influence on the Indian Monsoon.
  - **El Niño (Warm Phase):** Generally correlates with suppressed rainfall over India. The anomalous warming in the Pacific often leads to a weakening of the monsoon trough over India and a reduction in moisture convergence, increasing the probability of a deficient monsoon or drought years in India.
  - **La Niña (Cold Phase):** Generally correlates with **enhanced rainfall** over India. The cooler Pacific waters tend to strengthen the normal monsoon circulation, increasing the likelihood of above-normal rainfall and even floods.
- **Current Outlook (as of June 2025):** The latest assessment indicates that ENSO is currently in a Neutral phase and is highly likely to remain so during the upcoming Southwest Monsoon season. This neutral" influence is generally considered favorable, as it removes the strong suppressive or enhancing signal that El Niño or La Niña could impart, allowing other factors to play a more prominent role.

- **Indian Ocean Dipole (IOD)**
  - **What it is:** The IOD describes the oscillation of sea surface temperatures across the tropical Indian Ocean, creating a difference between its western and eastern poles.
    - **Positive IOD:** Warmer-than-average SSTs in the western Indian Ocean and cooler-than-average in the eastern part. This often strengthens convection over the western basin.
    - **Negative IOD:** Cooler-than-average SSTs in the western Indian Ocean and warmer-than-average in the eastern part.
    - **Neutral IOD:** No significant temperature difference across the basin.
  - **Role in Monsoon:** The IOD often acts as a significant modulator of the ENSO impact on the Indian monsoon.

- **A positive IOD** typically favors good monsoon rainfall over India, often mitigating the negative impact of an El Niño or enhancing a normal monsoon. It increases moisture convergence over the subcontinent.
- A **negative IOD** can have an adverse impact, potentially reducing monsoon rainfall.

- **Current Outlook (as of June 2025):** Current conditions indicate a **Neutral IOD**, with IMD projections suggesting its continuation through the monsoon season. This means it is unlikely to provide a strong enhancing or suppressing signal this year.

- **Madden-Julian Oscillation (MJO)**
  - **What it is:** The MJO is the largest coherent intraseasonal (30-90 day) fluctuation in tropical weather. It is an eastward-propagating pulse" of clouds and rainfall that moves around the equator, primarily over the Indian and Pacific Oceans. It has distinct phases (typically 8 phases) corresponding to its position and intensity.
  - **Role in Monsoon:** The MJO significantly influences the day-to-day and week-to-week variability of the monsoon.
    - When the MJO's convective (wet) phase is over the Indian Ocean, it tends to **enhance** monsoon rainfall over India, contributing to active spells.
    - When its convective phase moves eastward into the Pacific, it can lead to a **break in monsoon**activity over India.
  - **Role in LRF:** While not a long-term predictor of the *total seasonal rainfall* like ENSO or IOD, the expected evolution of MJO (predicted on subseasonal timescales) is crucial for refining the **within-season distribution** of rainfall, identifying active and break phases, and providing guidance for short-to-medium range forecasts during the monsoon season itself. Its signal is often incorporated into subseasonal forecast models that feed into the LRF process.

- **Eurasian Snow Cover (including Himalayan Snow Cover)**
  - **What it is:** Refers to the extent and depth of snow cover over the vast Eurasian landmass and specifically the Himalayas during the preceding winter and spring months.
  - **Role in Monsoon:** There is generally an **inverse relationship** between the extensive winter/spring snow cover over Eurasia and the subsequent Indian Summer Monsoon Rainfall.

- **Mechanism:** Extensive and prolonged snow cover leads to increased reflection of solar radiation (high albedo) and a larger amount of energy being used for melting snow rather than heating the land surface. This results in the Asian landmass remaining cooler than usual during spring, weakening the crucial land-ocean temperature gradient that drives the monsoon circulation. A weaker gradient can translate to weaker monsoon winds and reduced moisture advection, leading to a poorer monsoon.
- **Current Outlook (as of June 2025):** IMD noted **below-normal snow cover** over the Northern Hemisphere and Eurasia during January-March 2025. This is generally considered a positive factor for the upcoming monsoon, as it implies stronger spring heating of the landmass.

The final LRF for the Southwest Monsoon is a carefully weighted synthesis of the expected influences from these various global and regional climatic parameters, integrated through sophisticated statistical models and multi-model ensemble dynamical forecasts. It represents the most comprehensive scientific assessment available to guide India's preparations for its most vital weather season.

Long-range forecasting (LRF) of monsoon rainfall in India, particularly for the crucial Southwest Monsoon, holds immense promise for the agricultural sector. However, it also faces significant inherent limitations. Understanding both aspects is key to maximizing its utility for farmers.

10.6 Present Limitations of Long-Range Monsoon Forecasting in Agriculture

Despite significant advancements, LRF for monsoon rainfall, especially for agricultural applications, currently faces several challenges:

10.6.1 Inherent Uncertainty and Margin of Error

- **Chaotic Nature of the Atmosphere:** The Earth's atmosphere is a chaotic system. While long-term drivers like ENSO provide a signal, the exact trajectory of the monsoon is influenced by numerous complex, non-linear interactions. Small errors in initial conditions or model physics can amplify significantly over months, leading to substantial forecast errors.
- **Probabilistic vs. Deterministic:** LRFs are inherently probabilistic (e.g., 70% chance of normal rainfall") rather than deterministic (It will rain exactly 88 cm"). Farmers, however, often need more precise, actionable information for specific decisions. The IMD's forecasts typically come

with a $\pm$5% error margin for the all-India average. While this has improved, it still leaves room for significant deviation.

- **Predictability Desert":** The subseasonal-to-seasonal range (2 weeks to 3 months) is often termed the predictability desert" because it's too long for detailed daily forecasts but too short for the strong, consistent signals of long-term climate drivers to fully dominate.

10.6.2 Limited Spatial and Temporal Detail (Granularity)

- **All-India vs. Regional Variability:** An all-India normal monsoon" forecast can mask significant regional disparities. One region might experience floods while another faces drought in the same year. LRFs struggle to provide reliable district-level or even block-level rainfall predictions, which are what individual farmers truly need.
- **Intra-seasonal Variability:** Even if the overall seasonal rainfall is predicted accurately, the LRF cannot reliably predict crucial intra-seasonal variations, such as:
- **Onset and Withdrawal Dates:** Small delays or early withdrawals can severely impact crop cycles.
- **Active and Break Spells:** Long dry spells (breaks") during the monsoon can devastate crops even in a year with overall normal rainfall. Current LRFs cannot reliably predict the duration or timing of these breaks months in advance.
- **Extreme Events:** Predicting localized heavy rainfall events, cloudbursts, or prolonged heatwaves within the monsoon season remains a challenge for long-range models.

10.6.3 Model and Data Constraints

Model Biases: All numerical models have inherent biases and systematic errors in how they represent complex atmospheric processes. These biases can accumulate over long simulation periods, leading to inaccuracies.

Incomplete Observations: Despite advances, observational networks (especially over oceans and remote areas) still have gaps. Lack of comprehensive, high-resolution data limits the accuracy of initial conditions for models.

Changing Predictor Relationships: Climate change is altering established teleconnections. Some historically reliable statistical predictors may lose their relevance or the strength of their correlation, requiring continuous recalibration and research.

Computational Intensity: Running high-resolution dynamical models for long periods requires immense computational resources, which can be a limiting factor.

10.6.4 Communication and Interpretation Gaps (Last-Mile Connectivity")

- **Understanding Probabilistic Forecasts:** Farmers may find it challenging to interpret probabilistic forecasts (e.g., a 33% chance of above-normal rainfall) and translate them into concrete farming actions.
- **Accessibility:** Even if accurate, forecasts may not reach all farmers in a timely, understandable, and actionable format, especially in remote or less digitally connected areas. Many farmers still rely on informal sources for weather information.

10.7 Potential Applications in Agriculture

Despite the limitations, the LRF of monsoon rainfall offers immense potential for strategic decision-making in Indian agriculture, particularly as forecasting skill continues to improve.

10.7.1 Strategic Crop Planning and Diversification:

- **Crop Choice:** If a forecast indicates a higher probability of a deficient monsoon, farmers can opt for drought-resistant crops (e.g., certain millets, pulses) over water-intensive ones (like paddy in some regions). Conversely, an above normal" forecast could encourage cultivation of higher-value, water-demanding crops.
- **Varietal Selection:** Choosing crop varieties with shorter growth cycles if a delayed or shorter monsoon is anticipated.
- **Cropping Patterns:** Advising shifts in cropping patterns at regional levels based on likely seasonal rainfall scenarios.

10.7.2 Optimizing Input Management

- **Sowing Dates:** LRFs can provide broad guidance on optimal sowing windows, helping farmers decide whether to delay or advance planting based on the likely monsoon onset, thereby maximizing germination and reducing risk.
- **Water Management:** Anticipating a deficient monsoon encourages water conservation strategies (e.g., drip irrigation adoption, rainwater harvesting, building check dams) at both farm and policy levels. Conversely, forecasts for excess rainfall can inform preparations for drainage and flood management.

- **Fertilizer and Pesticide Use:** Knowing the general rainfall outlook can help optimize the timing and quantity of fertilizer and pesticide application, preventing wash-offs and ensuring efficacy, thus reducing costs and environmental impact.

10.7.3 Risk Mitigation and Financial Planning

- **Crop Insurance:** Long-range forecasts can help farmers make informed decisions about purchasing crop insurance, and also assist insurance companies in pricing policies and managing risk portfolios.
- **Credit and Investment:** Banks and financial institutions can use LRFs to assess agricultural credit risk and adjust lending policies, while farmers can make better investment decisions in farm machinery or new technologies based on expected profitability.
- **Contingency Planning:** State agricultural departments can develop district-level contingency plans for drought or flood scenarios, stocking appropriate seeds, fertilizers, and relief materials in anticipation.

10.7.4 Market and Price Management

- **Government Policy:** The government uses LRFs to formulate agricultural policies, including decisions on minimum support prices (MSPs), import/export of food grains, and buffer stock management, ensuring food security and price stability.
- **Supply Chain Planning:** Agri-businesses, food processors, and commodity traders can use LRFs to anticipate supply fluctuations, manage logistics, and plan sourcing and distribution strategies.

10.7.5 Long-Term Adaptation to Climate Change

- **Resilience Building:** As climate change introduces greater monsoon variability and extreme events, LRFs, even with their current limitations, become crucial for long-term adaptation strategies, enabling a more proactive rather than reactive approach to agricultural planning.
- **Research and Development:** LRFs highlight critical areas for agricultural research, such as developing climate-resilient crop varieties suitable for changing rainfall patterns.

In conclusion, while the long-range forecast for monsoon rainfall is not a perfect crystal ball, its value in agriculture is steadily increasing. As forecasting science and communication infrastructure improve, the ability of farmers and policymakers to translate these broad outlooks into actionable strategies will significantly enhance India's agricultural resilience and food security in the face of a changing climate. The key lies in not expecting daily weather from a seasonal forecast but in using it for strategic, probabilistic decision-making.

10.8 Sub-Seasonal Forecasts

Subseasonal forecasting, sometimes referred to as extended-range forecasting," bridges the critical gap between traditional medium-range weather forecasts (up to 10-15 days) and long-range/seasonal climate outlooks (3 months or more). Covering a timeframe of roughly 2 weeks to 2 months, this forecasting type provides insights into potential deviations from average conditions that are crucial for decision-making across various sectors, especially agriculture. It's considered a predictability desert" because it's too long for the atmospheric initial conditions to dominate entirely and too short for slowly evolving ocean signals (like ENSO) to fully dictate the outcome.

• How Sub seasonal Forecasts Are Made?

Subseasonal forecasting relies on a blend of approaches, drawing from both numerical modeling and the identification of predictable atmospheric and oceanic oscillations.

10.8.1 Dynamical Models (Coupled Climate Models)

- **High-Resolution NWP Models with Extended Runs:** Subseasonal forecasts typically utilize the same sophisticated Numerical Weather Prediction (NWP) models (often global coupled atmosphere-ocean models) used for medium-range forecasts, but they are run for longer periods (e.g., 45-60 days).
- **Ensemble Forecasting is Essential:** Due to the increasing influence of atmospheric chaos at these lead times, **ensemble forecasting** is absolutely critical. Instead of a single model run, multiple simulations are performed with slightly perturbed initial conditions or different model physics. The spread of these ensemble members provides a measure of forecast uncertainty, and the ensemble mean often provides the most skillful prediction.
- **Data Assimilation:** Just like shorter-range forecasts, accurate initial conditions are vital. Sophisticated data assimilation techniques are used to blend observations from various sources (satellites, radar, surface stations, buoys) into the model's starting point.

- **Sources of Predictability (Boundary Forcing & Oscillations)**
  - While initial atmospheric conditions become less important after about two weeks, subseasonal predictability arises from slowly varying boundary conditions and large-scale atmospheric/oceanic oscillations that have memory on these timescales. Key factors include:
    - **Sea Surface Temperatures (SSTs):** Regional SST anomalies, even outside the main ENSO regions, can influence overlying atmospheric convection and circulation, impacting rainfall patterns weeks later.
    - **Soil Moisture:** Anomalies in soil moisture (very wet or very dry) can influence surface energy balance, impacting local temperature and rainfall through land-atmosphere feedback.
    - **Snow Cover:** The extent and persistence of snow cover can influence regional temperature and atmospheric stability.
    - **Madden-Julian Oscillation (MJO):** This is a critical source of subseasonal predictability, especially for the tropics.The MJO is an eastward-propagating pulse" of enhanced cloudiness and rainfall that moves around the equator. Its location and phase (typically 8 phases) strongly influence global weather patterns, including the Indian monsoon's active and break spells. Predicting the MJO's future phase and amplitude is a significant component of subseasonal forecasting.
    - **Other Intraseasonal Oscillations:** Besides MJO, other modes of variability, though weaker, can also contribute.

10.8.2 Statistical and Hybrid Methods

- **Bias Correction and Calibration:** Raw output from dynamical models often has systematic biases.Statistical methods, often based on **hindcasts (reforecasts)** – historical runs of the same model – are used to bias-correct and calibrate the model output against observed data, enhancing skill.
- **Machine Learning/AI:** Increasingly, AI and ML algorithms are being developed to identify complex patterns and relationships in historical data and model outputs, further improving skill in this challenging timescale.

- **Probabilistic Outputs (Weekly/Bi-weekly Averages)**

Subseasonal forecasts are typically expressed as **probabilities** of exceeding or falling below certain thresholds (e.g., probability of above-normal rainfall) for specific **weekly or bi-weekly periods**. They do not provide daily weather details but rather anomalies from climatological averages.

10.8.3 Present Status of Subseasonal Forecasting in India

India, through the **India Meteorological Department (IMD)** and its associated institutions like the **Indian Institute of Tropical Meteorology (IITM)**, has made significant strides in subseasonal forecasting.

- **Operational Products:** IMD issues operational Extended Range Forecasts" which typically provide a probabilistic outlook for rainfall and temperature for the next two weeks, sometimes extending up to four weeks, usually updated weekly. These forecasts often focus on weekly averages or anomalies.
- **Monsoon Mission Climate Forecasting System (MMCFS):** IMD's state-of-the-art coupled global climate model (MMCFS) is used for both seasonal and subseasonal predictions. Continuous efforts are made to improve its representation of the MJO and other intraseasonal oscillations, which are key to subseasonal skill over the Indian subcontinent.
- **Focus on Monsoon Intraseasonal Oscillations (MISOs):** A significant research and operational focus in India is on improving the prediction of **Monsoon Intraseasonal Oscillations (MISOs)**, which are the northward-propagating active and break spells of the monsoon. Accurate MISO prediction is crucial for subseasonal rainfall forecasts over India.
- **Multi-Model Ensemble Approach:** India also leverages outputs from international Subseasonal-to-Seasonal (S2S) prediction projects, participating in and contributing to global multi-model ensemble efforts to enhance predictability.
- **Challenges:** While progress is evident, challenges remain. Models still have biases in representing complex processes like cloud formation and convection, and the skill can vary significantly depending on the region and the time of year. Predicting the exact timing and intensity of MISO phases remains complex.

10.8.4 Future Applications in Agriculture

Subseasonal forecasts hold tremendous untapped potential for Indian agriculture, offering a decision-making window that is more immediate than seasonal outlooks but provides more lead time than short-range forecasts. This sweet spot" allows for proactive adjustments rather than reactive measures.

- **Optimizing Crop Management**
    - **Refined Sowing and Transplanting:** If a forecast indicates a likely dry spell in 3-4 weeks, farmers can delay sowing or transplanting paddy seedlings, or choose shorter-duration varieties. Conversely, anticipating good rain can ensure timely planting.
    - **Irrigation Scheduling:** Forewarning of a prolonged dry spell can prompt farmers to plan for supplementary irrigation, conserve water, or utilize available resources more efficiently. Similarly, anticipating heavy rainfall can help prevent over-irrigation.
    - **Fertilizer and Pesticide Application:** Knowing the likelihood of heavy rain in the coming weeks can help farmers avoid applying fertilizers just before they might be washed away, improving efficiency and reducing costs. Similarly, anticipating extended wet periods might signal a higher risk of pest or disease outbreaks, allowing for proactive spraying.
- **Risk Mitigation and Contingency Planning**
    - **Drought/Flood Preparedness:** Subseasonal forecasts can provide crucial lead time (e.g., 2-4 weeks) for regional agricultural departments and farmers to prepare for potential agricultural droughts (soil moisture deficits) or localized floods. This could involve identifying alternative crops, preparing drainage systems, or arranging for emergency water supplies.
    - **Crop Harvesting and Storage:** A forecast for heavy rains during harvesting can prompt farmers to accelerate harvest, protect harvested produce, or arrange for proper storage to prevent spoilage.
    - **Livestock Management:** Anticipating heatwaves or prolonged cold snaps can help farmers plan for shelter, feed, and water for livestock, reducing stress and mortality.
- **Input and Market Decisions**
    - **Input Procurement:** Farmers can make more informed decisions about purchasing seeds, fertilizers, and other inputs if they have a clearer picture of likely conditions in the coming weeks.
    - **Market Strategies:** Traders and agri-businesses can use subseasonal outlooks to anticipate regional supply fluctuations, aiding in procurement, storage, and pricing decisions.

• Tailored Agrometeorological Advisories (AAS)

The true power of subseasonal forecasts lies in their integration into actionable **Agrometeorological Advisory Services (AAS)**. Instead of generic weather updates, AAS can provide specific, localized advice to farmers based on the expected conditions for the next few weeks (e.g., Given the likelihood of below-normal rainfall in the next 3 weeks, consider delaying paddy transplanting by 10 days and opt for short-duration maize varieties").

This requires robust communication channels and extension services to translate complex probabilistic forecasts into simple, practical recommendations for diverse farming communities.

10.9 Medium Range Weather Forecasting

Medium-range weather forecasting (MRWF) typically covers a period of 3 to 10 days. This timescale is crucial because it provides enough lead time for planning significant activities, yet the forecasts still retain a reasonable level of accuracy for general weather patterns.

• Methods Used in Medium-Range Weather Forecasting

The primary method for medium-range forecasting is **Numerical Weather Prediction (NWP)**, but it relies on a sophisticated chain of processes:

10.9.1 Observation Systems (Data Collection):

- **Global Networks:** A vast network of observations from around the world is essential. This includes surface weather stations (automated and manned), weather balloons (radiosondes), commercial aircraft (AMDAR), ocean buoys, and ships.
- **Satellite Data:** Satellites (both geostationary and polar-orbiting) provide crucial global coverage, especially over data-sparse regions like oceans. They measure temperature, humidity, cloud cover, wind, and sea surface temperatures.
- **Radar Data:** While primarily used for short-range and nowcasting, radar data provides valuable information on precipitation and storm structures that can be assimilated into models to improve their initial state.

• Data Assimilation

This is a critical step where all the diverse observational data, which are often irregularly spaced and collected at different times, are combined with a previous forecast (a first guess") to create the most accurate and dynamically consistent initial conditions" for the NWP model. This process involves complex mathematical algorithms (e.g., 4D-Var, Ensemble Kalman Filters) that aim to minimize errors in the model's starting point.

10.9.2 Numerical Weather Prediction (NWP) Models

- **High-Resolution Global Models:** Medium-range forecasts are primarily generated by sophisticated global NWP models (e.g., ECMWF's IFS, US's GFS, IMD's NCUM/MMCFS). These models use a grid system to divide the atmosphere into thousands of points and layers. At each point, they solve complex partial differential equations that represent the fundamental laws of atmospheric physics (conservation of mass, momentum, energy, and water vapor).
- **Parameterization:** Processes that occur on scales smaller than the model's grid resolution (e.g., cloud formation, precipitation, turbulence, radiation, land-surface interactions) cannot be explicitly resolved. Their effects are represented through parameterization schemes"– simplified mathematical formulas that approximate their collective impact on the larger-scale model variables.
- **Computational Power:** Running these models requires immense supercomputing power, as millions of calculations are performed for each time step.

- **Ensemble Forecasting**
  - **Addressing Chaos:** Due to the inherent chaotic nature of the atmosphere (the butterfly effect"), tiny uncertainties in the initial conditions or model physics can lead to significantly different outcomes over time. To address this, **ensemble forecasting** is widely used.
  - **Multiple Runs:** Instead of a single deterministic" forecast, an ensemble system runs multiple (e.g., 20 to 50 or more) slightly different versions of the same NWP model. The variations can come from perturbing the initial conditions within the observational uncertainty range or by slightly altering the model's physical parameterization schemes.
  - **Probabilistic Output:** The output of an ensemble is a range of possible future atmospheric states. This allows forecasters to assess the probability of certain events (e.g., a 60% chance of heavy rain, a 30% chance of temperatures exceeding 40^{0}C) and the confidence level in the forecast. A wide spread among ensemble members indicates high uncertainty, while a tight clustering suggests higher confidence.

- **Post-processing and Human Interpretation**

Raw model output needs to be refined. **Statistical post-processing** (e.g., Model Output Statistics - MOS, Kalman filtering) corrects for systematic model biases based on past performance.

- **Human Meteorologists:** Experienced meteorologists play a crucial role. They interpret the model outputs, considering local geographical features, known model biases, and recent atmospheric trends. They synthesize information from various models (multi-model ensemble approach) and apply their expertise to produce the final forecast.

- **Forecast Output**

The output of medium-range forecasts is typically presented in various forms, offering both deterministic and probabilistic information:

- **Deterministic Forecasts (for the initial 3-5 days)**
  - Daily maximum and minimum temperatures.
  - Daily precipitation amounts (often with probabilities: e.g., isolated showers,""scattered rainfall,""heavy rain").
  - Wind speed and direction.
  - Humidity (Relative Humidity).
  - Cloud cover.
  - Specific phenomena like fog, frost, or thunderstorms (with likelihood).
- **Probabilistic Forecasts (for the full 3-10/15 day period)**
  - Probabilities of temperature or precipitation being above, below, or near normal for a specific future day or a 3-day period.
  - Probability of extreme events (e.g., chance of a heatwave, probability of heavy rainfall exceeding a certain threshold).
  - Ensemble spread: Visualizations showing the range of possible outcomes from the ensemble members for a given parameter (e.g., a plume diagram for temperature).
  - **Synoptic Charts:** Maps showing pressure systems, fronts, and overall circulation patterns for several days ahead.
  - **Derived Products:** Forecasts for specialized parameters like:
    - **Crop-specific variables:** Growing degree days, evapo-transpiration rates.

- **Weather warnings:** For potential heavy rainfall, strong winds, or cold waves within the forecast period.

10.9.3 Present Limitations

Despite remarkable progress, medium-range forecasting still faces inherent limitations:

- **Decreasing Skill with Lead Time:** While highly skillful for the first 3-5 days, the accuracy of forecasts diminishes significantly beyond 7 days due to the chaotic nature of the atmosphere. Small initial errors amplify rapidly.
- **Resolution vs. Phenomena:** Even high-resolution global models struggle to fully resolve smaller-scale, localized phenomena (e.g., individual thunderstorms, mesoscale convective systems) that can have significant impacts. Parameterization schemes, though improving, are still approximations.
- **Complex Interactions in Tropics:** Forecasting in the tropics (like India) is particularly challenging. Atmospheric dynamics are weaker, and convection (thunderstorms) plays a dominant role, releasing latent heat that fuels weather systems. Accurately parameterizing tropical convection is a major hurdle.
- **Data Gaps:** Despite advancements, observational data gaps persist, especially over oceans, mountainous regions, and remote areas. This can lead to less accurate initial conditions for models.
- **Model Biases:** All NWP models have systematic biases, which can affect the forecast skill, especially as lead time increases. Continuous research is needed to identify and correct these biases through model improvements and statistical post-processing.
- **Uncertainty in Forecasts of Extreme Events:** While models can indicate the *potential* for extreme events, predicting their exact location, timing, and intensity with high confidence several days in advance remains difficult.

10.9.4 Applications in Agriculture (India)

Medium-range weather forecasts are arguably the most impactful for day-to-day and week-to-week agricultural planning in India, offering a practical window for decision-making. The National Centre for Medium Range Weather Forecasting (NCMRWF) and IMD's Agrometeorological Advisory Services (AAS) are built upon these forecasts.

- **Optimizing Field Operations**
  - **Sowing and Planting:** Advisories based on predicted rainfall and soil moisture for the next week can guide farmers on optimal sowing dates for various crops, minimizing the risk of crop failure due to early dry spells or unexpected heavy rain.
  - **Irrigation Scheduling:** Farmers can plan irrigation more efficiently, avoiding unnecessary watering before predicted rainfall or ensuring adequate water supply during expected dry periods. This conserves water and energy.
  - **Fertilizer and Pesticide Application:** Advisories can recommend applying fertilizers or pesticides to avoid wash-off from heavy rain or to optimize absorption during dry spells. This maximizes efficacy and reduces input costs and environmental pollution.
  - **Weeding and Intercultural Operations:** Planning these activities when dry weather is expected can prevent soil compaction and ensure effective weed control.
- **Pest and Disease Management**
  - Many pest and disease outbreaks are highly sensitive to weather conditions (temperature, humidity, rainfall). Medium-range forecasts can predict conducive conditions for outbreaks (e.g., prolonged humidity for fungal diseases or specific temperatures for insect multiplication), allowing farmers to implement preventative measures or timely treatments, reducing crop losses.
- **Harvesting and Post-Harvest Management**
  - Predicting dry windows allows farmers to plan harvesting operations effectively, ensuring crops are harvested under optimal conditions to prevent spoilage (e.g., drying grains before storing).
  - Forecasts of heavy rain or strong winds can prompt farmers to accelerate harvesting or take protective measures for standing crops, minimizing losses.
- **Protection from Extreme Weather**
  - Warnings for potential hailstorms, severe thunderstorms, or cold waves can enable farmers to take protective measures for crops and livestock (e.g., covering sensitive crops, moving livestock to shelter).
  - In coastal areas, forecasts of cyclonic weather allow for timely evacuation and protection of agricultural assets.

- **Resource Allocation and Policy Decisions**
  - District and state agricultural departments can use medium-range forecasts to make short-term resource allocation decisions, such as distributing specific seeds, fertilizers, or providing extension services based on anticipated regional weather.
  - These forecasts inform local advisories that are disseminated to farmers, often in local languages, through various channels like radio, mobile apps (e.g., Meghdoot app by IMD), and community meetings.
  - In summary, medium-range weather forecasting provides a crucial link between daily operations and broader seasonal planning in agriculture. Its continued improvement in terms of accuracy, spatial resolution, and communication mechanisms is vital for building a climate-resilient and productive agricultural sector in India.

10.10 Short Range Forecasting

10.10.1 Short-Range Weather Forecasting (SRWF): 1 to 3 Days

Short-range weather forecasting (SRWF) focuses on predicting atmospheric conditions for the immediate future, typically 1 to 3 days (24 to 72 hours) ahead. This is the most common and generally the most accurate type of forecast, directly impacting daily planning and operational decisions across virtually all sectors, critically so for agriculture.

- **Methods Used in Short-Range Weather Forecasting**

SRWF employs a highly integrated and dynamic approach, combining vast amounts of real-time observations, powerful numerical models, and expert human judgment. Dense Observational Networks (Data Collection via Remote Sensing & In-Situ):

- The foundation of SRWF is an extensive and continuously updated observational database. This involves both in-situ (direct contact) and remote sensing (indirect measurement) tools.

- **In-Situ Observations**
  - **Surface Weather Stations:** Thousands of Automated Weather Stations (AWS) and manned observatories provide continuous, high-frequency data on temperature, humidity, atmospheric pressure, wind speed/direction, and precipitation from the ground level.
  - **Upper-Air Observations (Radiosondes & AMDAR):Radiosondes**, small instrument packages carried by weather balloons, are launched

twice daily globally, providing crucial vertical profiles of temperature, humidity, and wind up to high altitudes. Commercial aircraft equipped with **Aircraft Meteorological Data Relay (AMDAR)** systems also contribute valuable real-time upper-air data along their flight paths.

- **Oceanic Observations:** Network of **ocean buoys** (fixed and drifting) and data from ships provide essential sea surface temperature, wind, and wave information, which are critical for marine forecasts and influencing coastal weather.

- **Remote Sensing Technologies:** These tools measure atmospheric properties from a distance, providing broad coverage and high temporal resolution.
  - **Weather Radar (e.g., Doppler Radar):** These ground-based systems are indispensable. They emit microwave pulses that bounce off precipitation particles. By analyzing the returned signal, radar detects the **location, intensity, and type of precipitation**. Crucially, **Doppler Radar**utilizes the Doppler effect to measure the motion of precipitation towards or away from the radar, providing vital information on wind shear, rotation within thunderstorms (indicative of tornadoes), and the dynamics of severe weather events. Radar data is updated every 5-10 minutes, making it ideal for tracking rapidly evolving storms.
  - **Weather Satellites (Satellite Pictures):** Satellites are perhaps the most vital remote sensing tool for SRWF, providing a synoptic (large-scale, comprehensive) view of the Earth's atmosphere.
  - **Geostationary Satellites:** Orbit at a fixed position above the equator (e.g., INSAT series for India). They provide continuous, frequent updates (every 15-30 minutes) of visible light (for cloud cover, fog, snow), infrared (for cloud top temperature, which indicates cloud height and intensity, and land/sea surface temperature), and water vapor imagery (showing moisture content in the middle and upper troposphere, even in cloud-free areas). This allows meteorologists to track cloud evolution, storm development, and atmospheric flow patterns in real-time.
  - **Polar-Orbiting Satellites:** Orbit over the poles, providing global coverage (passing over any given point twice a day). They carry advanced sensors (sounders) that can penetrate clouds to retrieve more detailed vertical profiles of atmospheric temperature and humidity.

- **Lightning Detection Networks:** Detect lightning strikes, which are indicators of active thunderstorms and aid in tracking their intensity and movement.
- **Lidar and Sodar:** Ground-based remote sensing tools that use light (laser) and sound waves, respectively, to measure wind profiles, aerosols, and boundary layer characteristics.

- **Data Assimilation:**This is a highly sophisticated process where all the heterogeneous observational data (from all the sources above, often with different accuracies, resolutions, and times) are optimally combined with a first guess" (a short-range forecast from a previous model run). Advanced mathematical algorithms (e.g., 3D-Var, 4D-Var, Ensemble Kalman Filters) continuously integrate these new observations to create the most accurate, dynamically consistent, and comprehensive initial conditions" for the NWP models. The quality of these initial conditions is paramount, as even small errors can rapidly grow and affect forecast accuracy over short periods.

10.10.2 Numerical Weather Prediction (NWP) Models

- **High-Resolution Regional Models (Mesoscale Models):** For short-range forecasts, these models (e.g., WRF, HRRR, IMD's NCUM-R for India) are critical. They divide the atmosphere into a very fine grid (typically 1-10 km horizontal resolution) and numerous vertical layers. They numerically solve complex partial differential equations that describe the fundamental physical laws governing atmospheric behavior (conservation of mass, momentum, energy, and water vapor). The finer resolution allows them to explicitly resolve or better parameterize smaller-scale atmospheric features that directly impact local weather.
- **Global Models:** Global models (e.g., GFS, ECMWF's IFS) provide the broader synoptic-scale (large-scale) atmospheric patterns that provide boundary conditions for the regional models.
- **Rapid Update Cycles (RUCs):** Many short-range NWP systems operate in rapid update cycles (e.g., every 1-3 hours), quickly ingesting the latest observations to provide continuously updated forecasts.
- **Physical Parameterization:** Processes occurring at scales smaller than the model's grid (e.g., cloud microphysics, boundary layer turbulence, cumulus convection in coarser models) are approximated through parameterization schemes."
- **Ensemble Forecasting:** While deterministic models provide a single most likely outcome, ensemble forecasting is increasingly integral even

for short-range forecasts. By running multiple model simulations with slightly varied initial conditions or physics, ensembles provide a range of possible scenarios and quantify the forecast uncertainty. This allows meteorologists to communicate probabilities (e.g., 70% chance of heavy rain") and confidence levels.

10.10.3 Synoptic Methods and Human Interpretation

- **Synoptic Meteorology:** This is the traditional method of weather analysis, where meteorologists manually (or with computer assistance) plot and analyze real-time observed weather data (pressure, temperature, wind, clouds) on **synoptic charts**. This allows them to identify and track large-scale weather systems like high/low-pressure areas, fronts, and troughs, understanding their current state and likely evolution. While NWP models are dominant, the understanding derived from synoptic analysis guides forecasters' interpretation of model output and helps identify potential model errors.
- **Post-processing and Forecaster Expertise:** Raw model output is refined using statistical post-processing (e.g., Model Output Statistics - MOS) to correct systematic biases. Ultimately, experienced human meteorologists play a crucial role. They synthesize information from all sources – radar loops, satellite imagery, model outputs (including ensemble guidance), and their synoptic understanding – to produce the final, highly detailed, and actionable forecast. They are also responsible for issuing timely and accurate severe weather warnings.

10.10.4 Tools Used

The tools" in SRWF encompass the entire technological infrastructure:

- **Observational Instruments:** AWS, Radiosonde launchers, Doppler Weather Radar (DWR) systems, Weather Satellites (receiving stations and processing software), AMDAR data receivers, Ocean buoys, Lightning detection networks, Lidars, Sodars.
- **Computational Hardware:** High-performance Supercomputers are essential for running complex NWP models and data assimilation systems quickly.
- **Modeling Software:** Sophisticated NWP models (e.g., WRF, HRRR, NCUM-R), and specialized data assimilation software.
- **Meteorological Workstations and Visualization Software:** Advanced graphical interfaces (e.g., IMD's dedicated workstations, AWIPS in the US) allow meteorologists to view, analyze, and overlay various data

layers (radar, satellite, model output, observations) in real-time. This enables rapid identification of developing weather, assessment of model performance, and timely issuance of warnings.

- **Communication Systems:** Robust infrastructure for rapid dissemination of forecasts and warnings to the public and stakeholders (internet, mobile apps like IMD's Meghdoot, radio, television, SMS alerts, specialized dedicated lines).

10.10.5 Importance and Applications in Agriculture

Short-range forecasts are paramount for day-to-day tactical decision-making at the farm level, directly influencing efficiency, productivity, and immediate risk management. They bridge the gap between knowing general trends and acting precisely.

- **Precise Timing of Farm Operations**
    - **Sowing/Planting:** Farmers can precisely time sowing or transplanting based on the forecast for the next 24-72 hours, ensuring optimal soil moisture for seed germination or successful establishment of seedlings.
    - **Irrigation Scheduling:** The most direct and significant application. Farmers can avoid irrigating if significant rain is expected soon, conserving water and energy. Conversely, they can plan for immediate irrigation if dry conditions are forecast, preventing crop stress.
    - **Fertilizer and Pesticide Application:** Crucial for efficacy and cost-effectiveness. Advisories based on predicted wind speed/direction prevent spray drift, while forecasts for impending rain ensure fertilizers aren't washed away before absorption or pesticides aren't diluted.
    - **Weeding and Intercultural Operations:** Planning these activities during predicted dry periods prevents soil compaction and allows for effective weed removal.
- **Protection Against Immediate Weather Hazards**
    - **Pest and Disease Management:** Many pest and disease outbreaks are highly sensitive to short-term weather conditions (e.g., specific combinations of temperature and high humidity for fungal diseases). Short-range forecasts allow for timely application of preventatives or treatments, minimizing severe crop damage.
    - **Frost Protection:** Warnings for overnight frost enable farmers to deploy immediate protective measures like running irrigation systems, covering sensitive crops, or using smudge pots to raise temperatures.

- **Hailstorm/Thunderstorm Warnings:** Farmers can take rapid protective actions for standing crops or recently harvested produce, move livestock to shelter.
- **Heavy Rain/Wind Warnings:** Helps in securing farm equipment, temporary shelters, and planning for immediate field drainage.

- **Harvesting and Post-Harvest Management**
  - **Harvest Timing:** Foreknowledge of dry windows (e.g., 2-3 days without rain) is vital for harvesting mature crops to prevent spoilage, maintain quality, and ensure machinery access to fields.
  - **Drying and Storage:** Predicting humid conditions or rain immediately after harvest informs decisions on drying harvested produce (e.g., open-air vs. mechanical dryers) and proper storage to prevent mold or rotting.
- **Labor Management**
  - Farmers can effectively plan labor deployment for various field activities based on the short-term weather outlook, maximizing efficiency.
- **Resolution**

Resolution in short-range forecasting refers to the spatial and temporal detail of the forecasts.

- **Spatial Resolution:** For short-range forecasts, particularly for agricultural applications, very high spatial resolution is required to capture localized weather variations. This typically ranges from 1 km to 10 km. This level of detail allows forecasts to differentiate conditions even within small farming areas, considering local topography, elevation, and land use. For example, IMD's operational regional models for short-range forecasting in India often run at resolutions of around 4-12 km.
- **Temporal Resolution:** Forecasts are issued and updated very frequently, often every 3 to 6 hours, with specific hourly outputs available for key parameters like rainfall or temperature for the next 24 hours. For rapidly evolving severe weather, warnings are issued instantly (nowcasts") as conditions develop.

In essence, short-range forecasting, powered by advanced remote sensing, sophisticated NWP, and expert meteorologists, provides the critical, highly accurate, and timely information that empowers farmers to make precise, tactical decisions, mitigating risks and enhancing the productivity and resilience of their operations.

10.11. Bharat Forecasting System

The **Bharat Forecast System (BFS)** is a landmark achievement in India's meteorological capabilities, signifying a major leap towards hyper-local, high-resolution weather prediction. Developed by the Indian Institute of Tropical Meteorology (IITM) in Pune, under the Ministry of Earth Sciences, BFS was officially unveiled and operationalized on May 26, 2025.

- **Key Features and How it Works?**

The BFS is a state-of-the-art weather forecasting platform that integrates cutting-edge technology and scientific advancements:

10.11.1 High-Resolution Numerical Weather Prediction (NWP) Model

- **6-km Resolution:** A defining feature of BFS is its ability to provide forecasts at an unprecedented **6-kilometer spatial resolution** for the tropical region (30^0South to 30^0North latitude, encompassing the entire Indian mainland). This is a significant improvement over the previous 12-km resolution used in India and surpasses the resolution of many global forecast models (which typically range from 9 km to 14 km).
- **Village/Panchayat Level Forecasts:** This enhanced resolution means that forecasts can now be provided for individual villages or panchayats, offering unprecedented granularity and relevance for local communities.

10.11.2 Powered by Supercomputing

- **Arka" Supercomputer:** BFS is powered by the new supercomputer named «Arka» at IITM, Pune. Arka boasts a performance capacity of 11.77 petaflops and a massive storage capability of 33 petabytes. This immense computational power is crucial for processing the vast amounts of meteorological data quickly and running the high-resolution NWP models efficiently.
- **Faster Processing:** The increased processing capacity dramatically reduces the time required to generate forecasts. While earlier models took 12-14 hours, BFS can now produce forecasts in 4-6 hours, enabling more rapid updates.

10.11.3 Extensive Data Integration (Remote Sensing & In-Situ)

- **Doppler Weather Radars (DWRs):** The system heavily leverages data from India's network of 40 operational Doppler Weather Radars positioned across the country. These high-resolution sensors provide real-time tracking of weather patterns, including precipitation, storm intensity, and wind dynamics. There are plans to expand this network to 100 DWRs for nationwide nowcasting coverage (forecasts for the next two hours).

- **Satellite Data:** BFS integrates real-time satellite data from various sources, including India's own INSAT series. This includes visible, infrared, and water vapor imagery, providing continuous information on cloud cover, atmospheric moisture, and storm development.
- **Other Observational Data:** Data from surface weather stations, radiosondes, ocean buoys, and other observational platforms are also assimilated into the system to provide comprehensive initial conditions for the model.

10.11.4 Indigenous Development

- BFS is a testament to India's commitment to Atmanirbhar Bharat" (self-reliant India) and Make in India" initiatives in scientific solutions. It has been indigenously developed by the IITM, Pune, demonstrating India's leadership in meteorological science.
- A notable aspect highlighted during its launch is that the project was spearheaded by four women scientists, reflecting the vision of 'Nari Shakti' (women's power) in Indian science.

10.11.5 Improved Accuracy for Extreme Events

- BFS has demonstrated significant improvements in forecasting extreme weather events. It has shown a 30% improvement in forecasting extreme rainfall events and a 64% enhancement in prediction accuracy within the core monsoon zones. This is crucial for disaster risk reduction.
- It also shows significant progress in accurately tracking cyclone paths.

10.11.6 Significance and Importance

The launch of the Bharat Forecast System marks a paradigm shift in India's weather forecasting capabilities, with far-reaching implications:

- **Enhanced Precision and Localization:** The 6-km resolution allows for highly localized, village-level forecasts, making weather information much more relevant and actionable for specific communities.
- **Improved Disaster Preparedness:** More accurate and timely predictions of extreme rainfall, cyclones, and other severe weather events enable better disaster management planning, early warnings, and potentially saving lives and property.

- **Economic Impact:** By reducing potential losses and facilitating gains in weather-sensitive sectors, BFS is expected to directly contribute to India's economic growth, especially as the nation aims to become a top global economy.
- **Global Leadership:** This indigenous breakthrough positions India among the global leaders in high-resolution weather prediction, especially for the complex tropical regions. The system is also expected to benefit other tropical nations facing similar weather challenges.
- **Scientific Advancement:** It represents a major advancement in India's self-reliance in meteorological sciences and fosters further research and development.

The Bharat Forecast System is not just a technological upgrade; it's a strategic national asset designed to empower citizens, bolster the economy, and enhance India's resilience to the increasing impacts of climate variability.

10.12 Nowcasting

Let's rewrite the explanation of nowcasting, emphasizing the role of radar networks, and then detail their limitations in the context of agricultural applications.

10.12.1 Nowcasting: 0-6 Hours

Nowcasting delivers highly detailed, localized, and rapidly updated weather forecasts for the very short term – from the present up to 6 hours ahead. It's about understanding the immediate evolution of weather. This is the most accurate form of weather prediction, relying heavily on real-time observations and the extrapolation of current weather features.

- **How Nowcasting is Made?**

Nowcasting is fundamentally observation-driven, leveraging high-frequency, high-resolution data streams and rapid processing techniques.

- **Intensive Real-Time Data Collection (Predominantly Remote Sensing):** The cornerstone is the continuous influx of granular data.
  - **Doppler Weather Radar (DWR) Networks:** These are the backbone of precipitation nowcasting and a critical component of any national weather service's infrastructure. DWRs scan the atmosphere frequently (every 5-10 minutes), providing high-resolution 2D and 3D images of precipitation (rain, snow, hail) intensity and movement.
  - **Doppler Capability:** Critically, DWRs use the Doppler effect to measure the *velocity* of precipitation particles, revealing wind patterns

within storms. This is vital for detecting severe phenomena like tornadoes (via mesocyclones), downbursts, and strong wind gusts.

- **Quantitative Precipitation Estimation (QPE):** DWRs provide estimates of rainfall rates, which are essential for flash flood warnings and agricultural water management.
- **Dual-Polarization Radar:** Modern DWRs use dual-polarization technology, which provides even more detailed information about the *type* of precipitation (e.g., distinguishing between rain, hail, and snow), improving the accuracy of QPE and severe weather detection.

- **Weather Satellites (High-Frequency Imagery)**
  - **Geostationary Satellites:** Provide continuous, frequent images (every 10-15 minutes, sometimes even faster) in visible, infrared, and water vapor channels. These are used to track cloud patterns, storm development, and atmospheric flow.
  - **Polar-Orbiting Satellites:** Provide less frequent, but more detailed vertical profiles of atmospheric temperature and humidity.
- **Automated Weather Stations (AWS):** A dense network provides real-time, surface-level measurements of temperature, humidity, wind, and precipitation.
- **Lightning Detection Networks:** Pinpoint lightning strikes, indicating active thunderstorms.
- **Other Sensors:** Wind profilers, ceilometers (cloud base height), disdrometers (rain drop size).
- **Extrapolation and Tracking Methods (Synoptic & AI):** Given the short lead time, nowcasting often relies on simpler, faster techniques than complex numerical models.
  - **Extrapolation (Persistence and Kinematic Methods):** Assumes observed weather features will continue to move and evolve similarly. Algorithms track the movement and deformation of radar echoes or cloud features.
  - **Synoptic Interpretation:** Meteorologists use their knowledge of atmospheric dynamics and **synoptic methods** to interpret radar and satellite data, identifying weather systems and extrapolating their likely behavior.
  - **Convection Permitting Models (CPMs):** For the 3-6 hour range, very high-resolution NWP models (1-3 km grid spacing) are increasingly used to explicitly resolve thunderstorms.

  - **Artificial Intelligence (AI) / Machine Learning (ML):** AI models can learn complex patterns from radar and satellite data to predict the future evolution of weather features.
- **Rapid Processing and Dissemination:** Speed is crucial. Data must be processed, and forecasts disseminated within minutes.
- **Tools Used**
  - **Observation Systems:** High-density DWR networks, rapid-scan geostationary satellites, dense AWS networks, lightning detection systems.
  - **Data Ingestion and Processing Platforms:** High-speed data pipelines.
  - **Nowcasting Software:** Algorithms for echo tracking, and short-term atmospheric modeling.
  - **High-Performance Computing (HPC):** For running CPMs and AI models.
  - **Meteorological Workstations:** Graphical interfaces for real-time analysis of multi-source data.
  - **Automated Alert Systems:** Detect severe weather thresholds and trigger immediate alerts.
  - **Dissemination Channels:** Mobile apps, SMS, radio, TV, social media.

10.12.2 Importance and Applications in Agriculture

Nowcasting enables farmers to make critical, real-time decisions to protect crops, livestock, and machinery from imminent threats.

- **Immediate Protection from Hazards**
  - **Thunderstorm & Hail Warnings:** Allow farmers to move livestock to shelter and secure equipment.
  - **Heavy Rainfall & Flash Flood Warnings:** Help farmers delay operations and initiate drainage.
  - **Sudden Gusty Winds:** Allow farmers to secure temporary structures.
  - **Frost/Heatwave Nowcasts:** Enable last-minute protective actions.
- **Optimizing Time-Sensitive Field Operations**
  - **Spraying:** Farmers can avoid spraying before imminent rain or during high winds.

- **Harvesting:** Farmers can plan harvesting during dry windows.
- **Irrigation:** Farmers can turn off irrigation systems before rain.
- **Tillage/Planting:** Farmers can plan operations around brief dry periods.

• **Resolution:** High spatial (0.5-2 km) and temporal (5-15 minute updates) resolution is essential.

10.12.3 Limitations of Radar Networks for Agricultural Applications

While DWR networks are invaluable for nowcasting, they have limitations specific to agriculture:

• Range Limitations and Earth Curvature

- **Maximum Range:** DWRs have a maximum effective range, typically around 150-200 km for detecting precipitation accurately. Beyond this, the radar beam spreads out, and the sensitivity decreases.
- **Beam Height and Cone of Silence":** Due to the Earth's curvature, the radar beam rises with distance from the radar. At longer ranges, the beam may overshoot low-level precipitation, especially in hilly or mountainous terrain. There's also a cone of silence" directly above the radar, where it cannot observe.
- **Implication for Agriculture:** This means that a single radar cannot provide uniform coverage over a large agricultural region. Areas far from the radar or in complex terrain may have less accurate or incomplete precipitation information.

• Beam Blockage and Clutter

- **Terrain Blockage:** Hills, mountains, and even tall buildings can block the radar beam, creating shadows" where precipitation is not detected.
- **Ground Clutter:** Reflections from the ground, buildings, and other non-weather objects can contaminate the radar signal, especially at low altitudes and near the radar.
- **Biological Clutter:** Insects, birds, and bats can also create radar echoes, which can be difficult to distinguish from light rain.
- **Implication for Agriculture:** This means that radar-derived rainfall estimates may be inaccurate in areas with significant terrain or clutter, affecting decisions about irrigation, spraying, and harvesting.

- **Vertical Resolution and Precipitation Type**
    - DWRs primarily measure precipitation intensity at a certain height above the ground. The type of precipitation (rain vs. snow vs. hail) can affect how it is detected.
    - **Implication for Agriculture:** While dual-polarization radar helps, accurately estimating rainfall reaching the ground, especially in complex terrain or during mixed precipitation events, can be challenging. This is important for irrigation decisions and assessing the risk of crop damage from hail.
- **Cost and Maintenance**
    - DWR systems are expensive to purchase, install, and maintain. They require specialized personnel and regular calibration.
    - **Implication for Agriculture:** While national weather services operate DWR networks, the cost limits the possibility of deploying dense networks specifically for agricultural applications.
- **Data Interpretation and Integration**
    - Raw radar data requires skilled interpretation. While automated algorithms exist, distinguishing between weather and non-weather echoes, correcting for attenuation (signal weakening due to heavy rain), and estimating rainfall accurately can be complex.
    - **Implication for Agriculture:** Farmers typically need readily understandable information, such as rainfall amounts at their specific location. Translating raw radar data into actionable information requires sophisticated processing and integration with other data sources (e.g., surface observations, soil moisture data).
- **Limited Direct Information on Other Agricultural Parameters**
    - DWRs primarily measure precipitation and wind. They do not directly measure other agriculturally important parameters like temperature, humidity, solar radiation, or soil moisture.
    - **Implication for Agriculture:** Radar data needs to be integrated with other weather and environmental information to provide a complete picture for agricultural decision-making.
- **Temporal Resolution vs. Rapidly Changing Conditions**
    - While DWRs provide frequent updates (every 5-10 minutes), extremely rapid changes in precipitation intensity or storm movement, especially

for small-scale convective events, may still be difficult to capture perfectly.

- **Implication for Agriculture:** For very time-sensitive decisions (e.g., stopping spraying just before a sudden downpour), even 5-10 minute updates may not be sufficient.

In conclusion, while radar networks are indispensable for nowcasting and provide crucial information for agriculture, their limitations related to range, blockage, data interpretation, and the lack of direct measurement of other agricultural parameters must be considered. Integrating radar data with other observational data and sophisticated modeling techniques is essential for providing accurate and actionable weather information for agricultural decision-making.

10.13 Forecast Verification

Forecast verification is the systematic and rigorous process of evaluating the quality, accuracy, and skill of weather forecasts by comparing them against observations of the actual weather. This critical step provides invaluable feedback for forecasters and model developers, helps refine Numerical Weather Prediction (NWP) models, and builds trust and understanding with the users of weather information.

10.13.1 Core Principles of Forecast Verification

Before diving into specific methods, it's essential to understand the underlying principles:

- **The Truth":** Forecasts are always compared against observations, which are considered the truth" or ground truth." These observations come from diverse sources:
  - **In-situ:** Surface weather stations, radiosondes (weather balloons), ocean buoys, aircraft sensors.
  - **Remote Sensing:** Doppler Weather Radars (for precipitation), weather satellites (for cloud cover, temperature, water vapor), lightning detection networks.
  - **Analyzed Fields:** Gridded data products that optimally combine observations with model background fields to create the most comprehensive representation of the atmosphere at a given time.
- **Benchmarking (Skill Assessment):** An accurate forecast isn't necessarily a skillful one. Skill is measured by how much better a

forecast performs compared to a simple reference forecast. Common benchmarks include:

- **Persistence Forecast:** The assumption that tomorrow's weather will be exactly the same as today's. This is a surprisingly strong benchmark for very short-range forecasts.
- **Climatology Forecast:** Predicting the average weather for a given date or period based on historical records. This is a robust baseline, especially for long-range and seasonal forecasts.
- **Random Forecast:** A completely arbitrary guess, used to demonstrate that a forecast has *any* predictive ability beyond pure chance.

- **Attributes of Forecast Quality:** Verification aims to quantify various aspects of a forecast's performance:
  - **Accuracy:** How close is the forecast to the observed reality?
  - **Bias:** Does the forecast systematically over-predict or under-predict (e.g., always too warm, always too wet)?
  - **Reliability (for probabilistic forecasts):** Do the forecast probabilities correspond to the observed frequencies? (e.g., if a 70% chance of rain is forecast, does it actually rain 70% of the time that forecast is made?).
  - **Resolution (for probabilistic forecasts):** Can the forecast system distinguish between different outcomes or events?
  - **Sharpness (for probabilistic forecasts):** How often does the forecast system issue confident (high probability) forecasts that are correct?
  - **Skill:** How much better is the forecast than a known reference forecast?

10.13.2 Methods Employed for Different Types of Forecasts

Forecast verification methods are carefully selected to match the nature of the forecast (e.g., whether it predicts a yes/no event, a continuous value, or a probability) and the specific weather element.

- **Verification for Deterministic (Categorical/Yes/No) Forecasts:** These methods apply when the forecast predicts the occurrence or non-occurrence of a discrete event (e.g., rain/no rain,""fog/no fog,""thunderstorm/no thunderstorm").
- **Core Tool - Contingency Table (or Confusion Matrix):** This 2x2 table is the foundation, summarizing the hits, misses, false alarms, and correct negatives:

2x2 Contingency Table		Event Observed	
		Yes	**No**
Event Observed	Yes	A	B
	No	C	D

- **Key Metrics Derived from Contingency Tables**
    - **Probability of Detection (POD) / Hit Rate:** $POD=A/(A+C)$
    - **Purpose:** Measures the fraction of *observed* events that were correctly forecast. High values are good.
  - **Limitation:** Can be artificially high if the forecast over-predicts the event (high false alarms).
  - **False Alarm Ratio (FAR):** $FAR=B/(A+B)$
    - **Purpose:** Measures the fraction of *forecasted* events that did *not* occur. Low values are good.
    - **Limitation:** Can be artificially low if the forecast under-predicts the event (many misses).
  - **Critical Success Index (CSI) / Threat Score (TS):** $CSI=A/(A+B+C)$
    - **Purpose:** Considers hits, false alarms, and misses. It is particularly useful for rare events, as it penalizes both false alarms and misses. Scores range from 0 (no skill) to 1 (perfect).
  - **Bias Score (BIAS):** $BIAS=(A+B)/(A+C)$
    - **Purpose:** Indicates if the forecast has a tendency to over-predict ($BIAS > 1$) or under-predict ($BIAS < 1$) the frequency of an event.
  - **Percentage Correct (PC) / Accuracy:** $PC=(A+D)/N$
    - **Purpose:** Overall proportion of correct forecasts.
    - **Limitation:** Can be misleading for rare events. If an event is very rare, simply forecasting no event" every time will yield a very high PC, even if it completely misses the few occurrences.
  - **Heidke Skill Score (HSS):** $HSS=(PC-Pe)/(1-Pe)$, where Pe is the percentage correct due to chance.
    - **Purpose:** Measures accuracy relative to a random forecast, removing the influence of correct forecasts due to chance. Scores range from $-\infty$ to 1 (perfect), with 0 indicating no skill over random.
  - **Gilbert Skill Score (GSS):** Similar to CSI but designed to remove hits due to random chance.

10.13.3 Verification for Deterministic (Continuous Variable) Forecasts

These methods apply when the forecast predicts a specific numerical value (e.g., temperature in °C, wind speed in km/h, precipitation amount in mm).

- **Core Approach: Point-by-Point Comparison** Each forecast value is directly compared with the corresponding observed value at the same location and time.
- Correlation Coefficient (e.g., Pearson Correlation Coefficient, Anomaly Correlation Coefficient - ACC)
    - **Purpose:** Measures the linear relationship between forecast and observed values. A value of 1 indicates perfect positive correlation, -1 perfect negative, and 0 no linear correlation.
    - **Anomaly Correlation Coefficient (ACC):** Particularly useful for longer-range forecasts. It measures the correlation between forecast anomalies (deviation from climatology) and observed anomalies. High ACC indicates skill in forecasting patterns (e.g., positions of high/low pressure systems).

10.13.4 Verification for Probabilistic Forecasts (Ensemble Forecasts)

These methods are crucial for forecasts that provide a probability of an event (e.g., 60% chance of rain") or a range of possible outcomes (e.g., from an ensemble prediction system).

- **Core Concept: Reliability and Resolution** Probabilistic forecasts are evaluated on two key aspects:
    - **Reliability:** Do the forecast probabilities match the observed frequencies? (e.g., for all instances where a 70% chance of rain was forecast, did it actually rain 70% of the time?).
- **Resolution:** Can the forecast system discriminate between different outcomes? (e.g., are 90% rain forecasts followed by rain more often than 10% rain forecasts?).
    - **Reliability Diagram:** A graphical tool that plots the observed frequency of an event (y-axis) against the forecast probability (x-axis) for various probability bins.
    - **Interpretation:** A perfectly reliable forecast would fall along the 45-degree diagonal. Points above the diagonal indicate under-forecasting (forecasted probabilities were too low), while points below indicate over-forecasting (forecasted probabilities were too high).

- **Ranked Probability Score (RPS) / Continuous Ranked Probability Score (CRPS):**
  - **Purpose:** Generalizations of the Brier Score for multi-category events or continuous variables. They measure the integrated squared difference between the forecast cumulative probability distribution and the observed cumulative distribution. Lower scores are better.
- **Rank Histogram (Talagrand Diagram)**
  - **Purpose:** Specifically for ensemble forecasts. It assesses whether the observed outcome falls within the range of the ensemble members and if the ensemble spread is appropriate. It plots the frequency with which the observation falls into each rank (bin) defined by the sorted ensemble members.
  - **Interpretation:** For a perfectly reliable ensemble, the histogram should be flat (uniform), meaning the observation is equally likely to fall into any rank. A U" shape indicates the ensemble spread is too small (under-dispersive), while a dome" shape indicates it's too large (over-dispersive).

10.13.4 Spatial Verification Methods (for High-Resolution Forecasts)

Traditional point-by-point verification can be too harsh for high-resolution NWP models, as a slight misplacement of a feature (e.g., a thunderstorm 5 km away from its predicted location) is penalized heavily, even if the forecast correctly predicted the *existence* and *intensity* of the storm. Spatial methods address this.

- **Neighborhood Methods (Fuzzy" Verification)**
  - **Principle:** Instead of strict point-to-point comparison, these methods compare the forecast and observation over a local neighborhood" (e.g., a 10km x 10km box). They give credit for forecasts that are close enough" to the observed event.
  - **Example Metrics:** Fractions Skill Score (FSS).
- **Object-Oriented Methods (e.g., MODE - Method for Object-based Diagnostic Evaluation)**
  - **Principle:** These methods first identify objects" (e.g., distinct storm cells, areas of heavy rainfall, frontal boundaries) in both the forecast and the observed fields. Then, they compare the attributes of these objects (e.g., size, shape, intensity, orientation, displacement error).
  - **Purpose:** Provides more diagnostic information than traditional methods, helping model developers understand *why* a forecast might

be good or bad (e.g., did it get the storm's intensity right but misplace its center?).

- **Scale Decomposition Methods (e.g., Wavelet-based methods)**
  - **Principle:** These methods break down the forecast and observed fields into different spatial scales (like different wavelengths in a Fourier analysis). Errors are then analyzed for each scale.
  - **Purpose:** Helps identify at which spatial scales (e.g., synoptic-scale, mesoscale, convective scale) the model performs well or poorly, guiding model improvement efforts.

10.13.5 Current Practices (e.g., by India Meteorological Department - IMD)

IMD, like other leading meteorological organizations, employs a comprehensive suite of these verification methods:

- **Routine Operational Verification:** Daily verification of forecasts across all lead times (short, medium, extended range, seasonal) for various parameters (temperature, rainfall, wind, humidity) at different geographical scales (all-India, regional, sub-divisional, district).
- **Categorical Rainfall Verification:** Widely used for assessing rainfall forecasts in different categories (e.g., dry, isolated, scattered, fairly widespread, widespread) using POD, FAR, CSI, and Bias Score.
- **Probabilistic Verification:** Increasing emphasis on verifying probabilistic forecasts from its **Monsoon Mission Climate Forecasting System (MMCFS)** and other ensemble systems using Brier Score, CRPS, and reliability diagrams.
- **Tropical Cyclone Verification:** Specialized metrics for predicting cyclone tracks, intensity, and landfall points.
- **Verification for Agrometeorological Advisories:** While not a separate methodology, the underlying weather forecasts that feed into agromet advisories are rigorously verified to ensure the reliability of the advice provided to farmers.
- **Model Diagnostics and Research:** Verification results are crucial for internal model diagnostics, identifying systematic errors (biases), and guiding the continuous development and improvement of IMD's NWP models.
- **International Standards:** IMD adheres to verification standards and participates in intercomparison projects organized by the World

Meteorological Organization (WMO) to ensure consistency and compare its performance against global benchmarks.

In essence, forecast verification is an indispensable and continuously evolving discipline that underpins the scientific integrity and practical utility of weather forecasting, driving ongoing improvements in accuracy and reliability.

10.14 Value Added Forecast

In meteorology, a Value-Added Forecast" refers to a weather prediction product that goes beyond the basic raw output of Numerical Weather Prediction (NWP) models. It involves further processing, interpretation, localization, and tailoring of generic weather forecasts to make them directly relevant, actionable, and decision-ready for specific user groups or sectors.

The value" is added by transforming raw meteorological data into actionable intelligence, often incorporating local knowledge, user-specific thresholds, and economic considerations.

10.14.1 How Value-Added Forecasts are Created?

Creating a value-added forecast is a multi-step process that typically involves a combination of human expertise, advanced post-processing techniques, and an understanding of the end-user's needs.

- **Foundation- Raw NWP Model Output**

The process begins with the raw output from sophisticated NWP models (like IMD's Bharat Forecast System or global models like GFS, ECMWF). These models provide predictions for basic meteorological parameters (temperature, humidity, wind, pressure, precipitation) on a defined grid.

- **Bias Correction and Calibration**
    - **Statistical Post-processing (e.g., Model Output Statistics - MOS):** NWP models often have systematic biases (e.g., consistently forecasting temperatures too high or rainfall too low in certain conditions). MOS uses historical data to statistically correct these biases, making the raw model output more accurate.
    - **Ensemble Post-processing:** For probabilistic forecasts from ensembles, advanced techniques like Bayesian Model Averaging (BMA) or Ensemble Model Output Statistics (EMOS) are used to calibrate the ensemble probabilities and provide more reliable probabilistic forecasts.

- **Localization and Downscaling**
  - **High-Resolution Regional Models:** While global models provide large-scale weather patterns, value-added forecasts often rely on downscaled forecasts from high-resolution regional models (e.g., IMD's 6-km BFS) that capture local terrain effects, land-sea breezes, and urban influences.
  - **Microclimate Knowledge:** Meteorologists or agrometeorologists apply their knowledge of local microclimates, topography (e.g., rain shadow areas, frost hollows), and unique geographical features that influence weather at a very fine scale.
  - **Integration with Local Observations:** Incorporating real-time data from local Automated Weather Stations (AWS) and other ground sensors helps fine-tune forecasts for specific locations.
- **Human Expertise and Synoptic Judgment**
  - **Forecaster Intervention:** Experienced meteorologists play a crucial role. They analyze outputs from multiple models, identify inconsistencies, apply their synoptic understanding of weather systems, and integrate the latest observational data (especially radar and satellite imagery). They use their judgment to refine forecasts, particularly for complex or rapidly evolving weather situations. This human touch" adds significant value that purely automated systems cannot provide.
  - **Quality Control:** Meteorologists perform a final quality check on the automated outputs, correcting any obvious errors or anomalies.
- **Tailoring to User-Specific Thresholds and Operations**

This is where the value" truly becomes apparent, as forecasts are translated into actionable advice.

- **Threshold-Based Alerts:** Instead of just wind speed will be 25 km/h," a value-added forecast might trigger an alert: Wind speed will exceed 20 km/h, activate windbreak nets in orchards."
- **Industry-Specific Language:** Using terminology familiar to the end-user (e.g., optimal spraying window,""harvesting conditions").
- **Integration of Non-Meteorological Data:** Combining weather forecasts with data on crop growth stages, pest life cycles, soil moisture, market prices, or operational schedules.

- **Dissemination and Feedback**
  - **Targeted Platforms:** Delivering the customized forecasts through channels most accessible to the user (e.g., mobile apps, SMS alerts, dedicated web portals, radio broadcasts, extension services).
  - **Feedback Loops:** Collecting feedback from users on the utility and accuracy of the value-added forecasts helps refine and improve future products.

10.14.2 Importance and Benefits of Value-Added Forecasts

Value-added forecasts transform raw data into actionable intelligence, leading to significant benefits:

- **Improved Decision-Making:** They provide specific, localized, and timely information that allows users to make optimal decisions, rather than having to interpret complex meteorological data themselves.
- **Reduced Risk and Losses:** By anticipating specific impacts, users can take proactive measures to mitigate risks, protect assets, and prevent losses (e.g., protecting crops from frost, securing infrastructure before a storm).
- **Enhanced Efficiency and Productivity:** Optimizing operations based on precise weather information can lead to significant gains in efficiency, resource utilization, and overall productivity (e.g., precise irrigation scheduling, timely planting).
- **Economic Gains:** Direct and indirect economic benefits accrue from reduced losses, increased yields, optimized resource use, and better planning. This is particularly crucial in weather-sensitive sectors like agriculture.
- **Tailored to Specific Needs:** They address the unique requirements of different sectors, providing information relevant to their operational thresholds and vulnerabilities.
- **Increased User Engagement and Trust:** When forecasts are directly relevant and useful, users are more likely to trust and act upon the information.
- **Support for Policy and Planning:** Value-added forecasts can inform long-term strategic planning for climate resilience, infrastructure development, and resource allocation at regional and national levels.

10.14.3 Examples of Value-Added Forecasts in Agriculture

In agriculture, value-added forecasts (often delivered as Agrometeorological Advisories) are crucial for transforming generic weather information into actionable farm-level advice. IMD's Gramin Krishi Mausam Sewa (GKMS) scheme is a prime example.

- **Sowing/Planting Advisories**
  - **Generic Forecast:**"Light to moderate rain expected in your district over the next 2-3 days."
  - **Value-Added Forecast:**"With the expected rainfall of 20-30 mm over the next 48 hours, soil moisture will be optimal. Farmers in the XYZ block are advised to proceed with sowing of [specific crop, e.g., maize/paddy] varieties by [date]."
- **Irrigation Scheduling**
  - **Generic Forecast:**"No rain expected for the next 5 days. Temperature will be above normal."
  - **Value-Added Forecast:**"A dry spell is forecast for the next 5 days. The Evapotranspiration (ET) demand for [crop type, e.g., wheat] is expected to be high. Schedule light irrigation in [specific field/crop stage] tomorrow morning to prevent moisture stress and optimize water use."
- **Pest and Disease Management**
  - **Generic Forecast:**"High humidity and moderate temperatures expected for the next 3 days."
  - **Value-Added Forecast:**"The forecast of high humidity (above 85%) and moderate temperatures (25-30°C) over the next 72 hours creates favorable conditions for [specific disease, e.g., fungal blight] in [crop, e.g., potato]. Farmers are advised to apply a prophylactic fungicide spray immediately."
- **Fertilizer and Pesticide Application**
  - **Generic Forecast:**"Rain expected on Tuesday. Strong winds on Wednesday."
  - **Value-Added Forecast:**"Avoid fertilizer application on Monday and Tuesday due to anticipated rainfall which could lead to nutrient leaching. Postpone pesticide spraying until Thursday morning, as strong winds (above 15 km/h) are expected on Wednesday, which will cause spray drift."

- **Harvesting and Post-Harvest Management**
    - **Generic Forecast:**"Dry weather expected for the coming week."
    - **Value-Added Forecast:**"A clear, dry window of 5-7 days is forecast. Farmers are advised to commence harvesting of [crop, e.g., paddy/ soybean] once the moisture content is optimal. Ensure proper drying and storage to prevent fungal growth due to recent high humidity."
- **Livestock Management**
    - **Generic Forecast:**"Temperatures may exceed 40^{0}C in the afternoon."
    - **Value-Added Forecast:**"Extreme heat stress is expected for livestock between 1 PM and 4 PM over the next two days. Ensure adequate water availability and provide shade for animals. Avoid grazing during peak afternoon hours."

In essence, value-added forecasts are the bridge between complex scientific data and practical, on-ground decision-making, significantly enhancing the utility and impact of meteorological services for various sectors, especially agriculture.

10.15 Climate Forecasts: Methods, Current Status, Limitations, and Applications

Climate forecasts, distinct from short-term weather predictions, focus on anticipating atmospheric and oceanic conditions averaged over extended periods, ranging from subseasonal (2 weeks to 2 months) to seasonal (3-6 months) and decadal (up to 10 years) timescales. They provide probabilistic outlooks for large-scale climate variables like temperature and precipitation anomalies, offering strategic guidance for various sectors.

- **Methods Used**

Climate forecasting primarily relies on sophisticated coupled global climate models (CGCMs) and increasingly incorporates Representative Concentration Pathways (RCPs) for longer-term projections.

10.15.1 Dynamical Models (Coupled Global Climate Models - CGCMs)

- **Initialization:** Unlike daily weather models, CGCMs for climate forecasts are initialized with the observed state of the slower-evolving components of the Earth system. This includes sea surface temperatures (SSTs), subsurface ocean heat content, soil moisture, and sea ice extent. These components have a longer memory" of past conditions, providing the primary source of predictability over months to years.

- **Ensemble Forecasting:** Climate forecasts are almost universally produced using ensemble methods. Multiple simulations are run from slightly varied initial conditions or by combining different CGCMs from various international modeling centers (Multi-Model Ensembles - MMEs). This approach is crucial for:
  - **Quantifying Uncertainty:** Reflecting the inherent uncertainty in longer-range predictions due to the chaotic nature of the climate system and imperfections in model representation.
  - **Providing Probabilistic Forecasts:** The spread of the ensemble members allows forecasters to express predictions as probabilities (e.g., a 60% chance of above-normal rainfall), which is more realistic for these timescales.
- **Representation of Climate Drivers:** CGCMs aim to accurately simulate the dynamics and interactions of major global and regional climate oscillations that significantly influence climate patterns:
  - **El Niño-Southern Oscillation (ENSO):** The irregular warming (El Niño) or cooling (La Niña) of SSTs in the equatorial Pacific, which is a dominant driver of global climate variability.
  - **Indian Ocean Dipole (IOD):** A seesaw pattern of SST anomalies in the tropical Indian Ocean, a key modulator of the Indian monsoon.
  - **Madden-Julian Oscillation (MJO):** An eastward-moving pulse of clouds and rainfall that affects tropical weather on weekly to monthly timescales, influencing subseasonal predictability.
  - **Other Teleconnections:** Including phenomena like the North Atlantic Oscillation (NAO) and Pacific Decadal Oscillation (PDO).
- **Representative Concentration Pathways (RCPs):** For projections spanning several decades into the future (e.g., decadal forecasts or long-term climate change assessments), CGCMs are forced with **Representative Concentration Pathways (RCPs)**. RCPs are scenarios that describe different possible future trajectories of greenhouse gas (GHG) concentrations and land-use change, based on varying assumptions about population growth, economic development, technological advancements, and climate policies. They are named after the radiative forcing (energy imbalance) level they are expected to reach by 2100 (e.g., RCP2.6, RCP4.5, RCP6.0, RCP8.5), representing futures with strong mitigation (low emissions) to business-as-usual" (high emissions).

10.15.2 Statistical Models

- **Principle:** These models identify and leverage historical statistical relationships (correlations) between predictors (e.g., past SST anomalies in key regions, global atmospheric circulation patterns) and predictands (e.g., seasonal rainfall, regional temperatures).
- **Methodology:** Common techniques include regression analysis, principal component analysis, and various machine learning algorithms trained on long observational datasets.
- **Role:** While dynamical models are now at the forefront, statistical models continue to serve as valuable tools for benchmarking, providing complementary insights, and sometimes offering comparable skill for specific regional phenomena. They are also vital for bias correcting and downscaling dynamical model outputs.

10.15.3 Hybrid Approaches

- Most operational climate forecast systems now blend dynamical model outputs with statistical post-processing techniques. This involves correcting systematic biases in model output and calibrating probabilistic forecasts to ensure their reliability.
- **Downscaling:** Since CGCMs operate at coarse resolutions, downscaling techniques (both statistical and dynamical) are used to translate large-scale climate information to finer regional and local scales, making it more relevant for specific applications.

10.15.4 Current Status

Climate forecasting capabilities have advanced significantly, with continuous improvements driven by increased computational power, better understanding of climate processes, and enhanced observational networks.

- **Improved Skill:** The skill of seasonal forecasts, particularly for major climate drivers like ENSO, IOD, and the aggregate Indian monsoon rainfall, has shown noticeable improvement over the last two decades. Multi-model ensembles consistently outperform individual models.
- **Operational Systems Globally:** Many national and international centers routinely issue climate forecasts across various timescales.
- **Indian Context (IMD and Monsoon Mission)**
    - The **India Meteorological Department (IMD)**, under the **Monsoon Mission** program, has made substantial progress. It employs a Multi-Model Ensemble (MME) system, integrating its own Monsoon

Mission Climate Forecasting System (MMCFS) and other global CGCMs for operational monthly and seasonal forecasts.

- This has led to a demonstrable reduction in the average absolute forecast error for All India Seasonal Rainfall and an improved correlation with observed rainfall.
- IMD issues two-stage forecasts for the Southwest Monsoon (April and June updates), alongside forecasts for the Northeast Monsoon, summer maximum temperatures/heat waves, and winter minimum temperatures/cold waves.
- All these forecasts are fundamentally probabilistic, reflecting the inherent uncertainty at longer lead times.
- Research and development continue to enhance skill for extreme events and provide finer spatial resolution.

10.15.5 Limitations

Despite advancements, inherent challenges persist in climate forecasting:

- **Inherent Predictability Limits**
  - **Chaotic Nature:** The Earth's climate system is chaotic. While initial conditions of slow-varying components offer predictability for months, the exact evolution of the atmospheric component remains challenging to predict beyond about two weeks.
  - **Spring Predictability Barrier" for ENSO:** Forecast skill for ENSO often drops significantly during the Northern Hemisphere spring, making predictions beyond this period more uncertain.
  - **Internal Variability:** Natural, unforced internal variability of the climate system poses a fundamental limit to predictability, especially at decadal timescales where anthropogenic forcing may still be relatively weak.
- **Model Limitations**
  - **Resolution:** Even with supercomputers, CGCMs operate at coarse spatial resolutions (tens to hundreds of kilometers), making it difficult to explicitly resolve smaller-scale processes (e.g., individual clouds, specific regional weather phenomena, complex topography). These processes must be approximated through parameterizations, which introduce uncertainties.
  - **Incomplete Understanding/Parameterization:** Processes like cloud formation, aerosol interactions, ocean-atmosphere coupling in complex

regions, and land surface dynamics are not fully understood or perfectly represented in models, leading to biases and uncertainties. Cloud feedbacks are a major challenge.

- **Systematic Biases:** Models often exhibit persistent errors (biases) that require statistical correction.
- **Computational Cost:** The sheer computational demand of running high-resolution, long-duration ensemble simulations limits the number of ensemble members and achievable resolution.

- **Data Limitations**
  - **Observational Gaps:** Despite global efforts, gaps exist in observation coverage, especially over remote oceans and the deep ocean, affecting the accuracy of initial conditions for models.
  - **Historical Data:** Long, consistent, and high-quality historical observational data are essential for training and validating models, but these are not always readily available for all regions and variables.

- **Scenario Uncertainty (Related to RCPs)**
  - For decadal and longer-term climate projections, the biggest source of uncertainty is not the climate model itself, but **human behavior and societal choices**. We cannot know which **Representative Concentration Pathway (RCP)** the world will actually follow (e.g., will global emissions drastically reduce, stabilize, or continue to rise?).
  - The chosen RCP directly dictates the projected future greenhouse gas concentrations, which in turn drive the projected warming and climate impacts. This means climate projections are if-then" statements (if society follows RCPX, then the climate will change by Y amount") rather than definitive forecasts.

10.15.6 Applications (Used For)

Despite the inherent uncertainties and limitations, climate forecasts provide invaluable information for strategic planning, risk management, and adaptation across various sectors, especially in a climate-vulnerable nation like India.

- **Agriculture and Food Security**
  - **Crop Planning:** Guiding decisions on suitable crop varieties, sowing dates, and cropping patterns for a coming season based on expected rainfall and temperature probabilities.
  - **Water Management:** Strategic planning for reservoir levels, long-term irrigation strategies, and drought/flood mitigation measures.

- **Pest and Disease Projections:** Anticipating seasonal conditions favorable for widespread pest or disease outbreaks, enabling proactive management.
- **Resource Allocation:** Farmers and policymakers can make informed decisions about resource acquisition (seeds, fertilizers) and allocation.
- **Food Policy:** Informing government policies on food grain procurement, storage, and import/export.

- **Water Resources Management**
  - **Reservoir Operations:** Optimizing water allocation for irrigation, hydropower generation, and drinking water supply over the long term.
  - **Infrastructure Planning:** Informing the design and construction of new dams, canals, and flood control systems based on projected changes in water availability and extreme events under different RCP scenarios.
- **Energy Sector**
  - **Hydropower Planning:** Estimating seasonal hydropower potential.
  - **Demand Forecasting:** Projecting heating/cooling energy demand based on anticipated seasonal temperature anomalies.
  - **Renewable Energy Integration:** Planning for solar and wind energy output variability over seasons.
- **Disaster Risk Reduction and Management**
  - **Early Preparedness:** Providing lead time for preparing for large-scale climate hazards like prolonged droughts, widespread heatwaves, or shifts in monsoon patterns.
  - **Vulnerability Assessment:** Using different RCP scenarios to assess future risks and vulnerabilities, informing the development of climate-resilient policies and infrastructure.
- **Public Health**
  - **Disease Surveillance:** Anticipating conditions favorable for vector-borne disease outbreaks (e.g., malaria, dengue) and heat-related illnesses based on seasonal temperature and rainfall patterns.
  - **Health System Preparedness:** Informing public health campaigns and resource allocation for potential climate-sensitive health challenges.
- **Economy and Policy**
  - **Macroeconomic Planning:** Governments use climate forecasts for budgeting, assessing inflation risks, and estimating sectoral growth.

- **Sectoral Adjustments:** Industries sensitive to climate (e.g., tourism, textiles, construction) can adjust their operational strategies.
- **Climate Change Adaptation Policies:** RCPs are fundamental for assessing future climate impacts, guiding national and international climate policies, and informing investments in adaptation and mitigation strategies. They help policymakers understand the consequences of different emission pathways.

In essence, climate forecasts, especially when informed by RCPs, provide the critical scientific foundation for proactive, strategic decision-making in a world grappling with climate variability and change.

10.16 Conclusion

In conclusion, understanding the diverse principles and methodologies of weather forecasting is paramount, whether we're talking about the immediate future or long-term climate trends. From the rapid, observation-driven techniques of nowcasting (0-6 hours), which leverage dense radar networks and satellite imagery for critical, real-time decisions, to the short-range (1-3 days) and medium-range (3-10 days) forecasts heavily reliant on advanced Numerical Weather Prediction (NWP) models and ensemble forecasting for daily and weekly planning, each type serves a distinct purpose. Beyond this, subseasonal (2 weeks-2 months) and long-range (seasonal to multi-year) forecasts shift their focus to broader climate drivers like ENSO, La Niña, the Indian Ocean Dipole, and the Madden-Julian Oscillation, providing probabilistic outlooks crucial for strategic adaptation.

The ultimate goal of all these forecasting efforts, exemplified by advancements like India's Bharat Forecast System (BFS), is to move beyond mere prediction towards Value-Added Forecasts." These customized advisories translate complex meteorological data into actionable intelligence, directly benefiting critical sectors like agriculture. While inherent limitations exist—from the chaotic nature of the atmosphere and resolution constraints to the practical challenges of radar coverage and data interpretation—continuous advancements in modeling, observation technologies (especially remote sensing), and verification methods are steadily enhancing forecast accuracy and utility. By embracing these sophisticated tools and tailoring them to specific user needs, weather forecasting will continue to play an increasingly vital role in building resilience, optimizing resource use, and safeguarding livelihoods in the face of our dynamic climate.

Multiple Choice Quesitons (Bold means correct answer)

1. Which type of forecast primarily focuses on very localized, rapidly evolving weather events for a timeframe of 0-6 hours?
   a) Short-range forecast
   b) Medium-range forecast
   c) Nowcasting
   d) Seasonal forecast
2. What is the core technique used in nowcasting that involves projecting the observed movement and evolution of current weather features forward in time?
   a) Numerical Weather Prediction
   b) Ensemble forecasting
   c) Extrapolation
   d) Climate modeling
3. Which observational tool is considered paramount for nowcasting due to its ability to track precipitation type, intensity, and movement in near real-time?
   a) Weather balloons
   b) Surface weather stations
   c) Doppler Weather Radar
   d) Ocean buoys
4. For short-range forecasts (1-3 days), what is the primary scientific principle involved?
   a) Statistical correlation of climate indices
   b) Extrapolation of current conditions
   c) Treating the atmosphere as an initial value problem
   d) Analyzing historical climate trends
5. What is the dominant tool used to make short-range weather forecasts?
   a) Manual synoptic analysis
   b) Numerical Weather Prediction models
   c) Simple persistence forecasts
   d) Satellite picture interpretation only
6. In short-range forecasting, what does data assimilation" refer to?
   a) The process of collecting data from various sources.
   b) Combining observations with a previous forecast to define the atmosphere's initial state for a model.
   c) Analyzing historical weather patterns.
   d) Disseminating forecast information to the public.

7. Why is ensemble forecasting" increasingly used in medium-range forecasts?
    a) To save computational power.
    b) To produce a single, most likely deterministic forecast.
    c) To quantify forecast uncertainty due to the chaotic nature of the atmosphere.
    d) To directly measure atmospheric conditions.
8. For medium-range forecasts (3-10 days), which aspect of the forecast becomes more significant compared to short-range?
    a) Localized thunderstorm initiation b) Daily maximum temperature
    c) Probabilistic forecasting d) Precise time of rainfall onset
9. What is the primary source of predictability for subseasonal forecasts (2 weeks - 2 months)?
    a) Exact initial atmospheric conditions
    b) Slowly varying boundary conditions and large-scale oscillations
    c) Daily weather patterns from 10 days ago
    d) The timing of the next solar flare
10. Which major tropical oscillation is a key source of predictability for subseasonal forecasts, particularly for the Indian monsoon?
    a) El Niño-Southern Oscillation (ENSO)
    b) Atlantic Multi-decadal Oscillation (AMO)
    c) Madden-Julian Oscillation (MJO)
    d) Pacific Decadal Oscillation (PDO)
11. What kind of models are often used for subseasonal forecasts to capture interactions between different Earth system components?
    a) Simple statistical models
    b) Very high-resolution short-range models
    c) Coupled atmosphere-ocean-land models
    d) Extrapolation models
12. Seasonal forecasts (3-6 months) primarily treat the atmosphere as what kind of problem?
    a) An initial value problem
    b) A chaotic system with no predictability
    c) A boundary value problem
    d) A pure extrapolation problem

13. What is the most critical oceanic phenomenon that drives predictability for seasonal forecasts globally, including the Indian monsoon?
    a) North Atlantic Oscillation (NAO)
    b) El Niño-Southern Oscillation (ENSO)
    c) Arctic Oscillation (AO)
    d) Southern Annular Mode (SAM)
14. How are seasonal forecasts typically expressed due to high uncertainty over long lead times?
    a) Precise daily weather events
    b) Exact temperature and rainfall amounts
    c) Probabilities of conditions being above/below/near normal
    d) Fixed weather for the entire season
15. What is the main focus of long-range forecasts (beyond 6 months to years)?
    a) Predicting daily weather patterns in the distant future.
    b) Analyzing climate trends, decadal variability, and climate change projections.
    c) Providing precise cyclone tracks for next year.
    d) Detailing specific rainfall events for a future season.
16. What kind of models are used for long-range climate projections, integrating various Earth system components?
    a) Simple statistical regression models
    b) Very high-resolution NWP models
    c) Earth System Models (ESMs)
    d) Persistence models
17. Which of the following is not a direct application of Nowcasting in agriculture?
    a) Deciding the optimal time for harvesting fragile crops.
    b) Planning irrigation for the next week.
    c) Moving livestock to shelter before a sudden storm.
    d) Deciding whether to spray pesticides in the next hour.
18. What is a key limitation of long-range forecasts of monsoon rainfall for direct agricultural planning at the local level?
    a) They are too accurate.
    b) They provide too much daily detail.

c) They lack fine spatial and temporal granularity.

d) They are too cheap to produce.

19. Why is the Long Range Forecast of Southwest Monsoon rainfall of immense significance for India?

a) It only affects the tourism industry.

b) It primarily impacts industrial production.

c) It is crucial for agriculture, water resources, and the economy.

d) It predicts the stock market performance.

20. Which agency is primarily responsible for issuing operational LRFs for the Southwest Monsoon in India?

a) Ministry of Finance

b) National Disaster Response Force (NDRF)

c) Indian Institute of Tropical Meteorology (IITM)

d) India Meteorological Department (IMD)

21. What is the general correlation between El Niño and Indian Southwest Monsoon rainfall?

a) Positive correlation (El Niño leads to more rain)

b) Negative correlation (El Niño leads to less rain)

c) No correlation

d) Varies randomly each year

22. How does a positive Indian Ocean Dipole (IOD)" typically influence the Indian monsoon?

a) It suppresses monsoon rainfall.

b) It generally favors good monsoon rainfall.

c) It has no influence on the monsoon.

d) It causes heavy snowfall in the Himalayas.

23. What is the typical relationship between extensive Eurasian snow cover in winter/spring and the subsequent Indian Summer Monsoon Rainfall?

a) Positive correlation (more snow, more rain)

b) Negative correlation (more snow, less rain)

c) No consistent relationship

d) Snow cover is a consequence, not a cause.

24. Which major oscillation provides significant predictability for the within-season variability (active and break spells) of the Indian monsoon?
    a) El Niño
    b) La Niña
    c) Madden-Julian Oscillation (MJO)
    d) Indian Ocean Dipole (IOD)
25. What is the typical lead time for Medium-Range Weather Forecasting?
    a) 0-6 hours
    b) 1-3 days
    c) 3-10 days
    d) 3-6 months
26. Which of the following is not a primary observation source for Medium-Range Weather Forecasting?
    a) Weather balloons
    b) Commercial aircraft data
    c) Real-time, minute-by-minute radar updates for local storms
    d) Global satellite data
27. What is the purpose of parameterization schemes" in NWP models?
    a) To make the models run faster without affecting accuracy.
    b) To represent the effects of atmospheric processes too small to be explicitly resolved.
    c) To adjust the model output based on human judgment.
    d) To input data from ground stations directly into the model.
28. Why is ensemble forecasting" crucial for medium-range forecasts?
    a) It simplifies the model equations.
    b) It allows for purely deterministic outputs.
    c) It quantifies forecast uncertainty due to atmospheric chaos.
    d) It reduces the need for observations.
29. What type of model output is typically used for medium-range forecasts to express uncertainty?
    a) Daily specific rainfall amounts.
    b) Hourly temperature predictions for 10 days.
    c) Probabilities of events or ranges of possible outcomes.
    d) A single, fixed forecast for the entire period.

30. Which of the following is a key limitation of medium-range forecasts?
    a) They are always perfectly accurate.
    b) Their accuracy decreases with increasing lead time.
    c) They can perfectly resolve every individual thunderstorm.
    d) They don't require any observations.
31. In agriculture, what is a key application of medium-range forecasts regarding inputs?
    a) Deciding what crops to plant next year.
    b) Optimizing the timing of fertilizer and pesticide application.
    c) Planning for long-term water storage in reservoirs.
    d) Predicting global food prices.
32. What is the typical lead time for Short-Range Weather Forecasting?
    a) 0-6 hours **b) 1-3 days**
    c) 3-10 days d) 2 weeks to 2 months
33. Which observational tool is considered a backbone" of precipitation nowcasting and short-range forecasting due to its high temporal and spatial resolution?
    a) Weather balloons b) Automated Weather Stations
    c) Doppler Weather Radar d) Ocean buoys
34. How do geostationary satellites contribute to short-range forecasting?
    a) By providing detailed vertical profiles of atmospheric gases.
    b) By offering continuous, frequent imagery of cloud cover and atmospheric moisture.
    c) By measuring ocean currents directly.
    d) By detecting earthquakes.
35. What does data assimilation" aim to create for NWP models in short-range forecasting?
    a) A complete historical record of past weather.
    b) The most accurate and consistent initial conditions for the model.
    c) The final forecast product ready for dissemination.
    d) A map of global climate zones.

36. Why are high-resolution regional models often preferred for short-range forecasting over global models?
    a) They are cheaper to run.
    b) They can resolve smaller-scale atmospheric features more accurately.
    c) They don't require any observational data.
    d) They are specifically designed for seasonal prediction.
37. What role do rapid update cycles" (RUCs) play in short-range and nowcasting?
    a) They reduce the amount of data needed.
    b) They allow models to quickly ingest the latest observations for fresh forecasts.
    c) They extend the forecast lead time to months.
    d) They are used for long-range climate projections only.
38. Which of the following is a key application of short-range forecasts in agriculture related to field operations?
    a) Deciding whether to invest in new farm machinery for next year.
    b) Precisely timing sowing, irrigation, and pesticide application.
    c) Planning for crop diversification over multiple seasons.
    d) Predicting the average monsoon rainfall for the entire country.
39. What is a typical spatial resolution for short-range weather forecasts, especially for agricultural applications?
    a) Hundreds of kilometers
    b) 1 km to 10 km
    c) Global averages
    d) Entire continents
40. What is the typical temporal resolution for updates in nowcasting products?
    a) Daily
    b) Weekly
    c) Every 5-15 minutes
    d) Monthly
41. What is the name of the new high-resolution forecasting system operationalized by IITM, Pune, for India?
    a) Monsoon Mission
    b) Global Forecast System
    c) Bharat Forecast System (BFS)
    d) Agromet Advisory System

42. What is the key spatial resolution achieved by the Bharat Forecast System (BFS) for the tropical region?

    a) 12-kilometer
    b) 25-kilometer
    c) 6-kilometer
    d) 100-kilometer

43. Which supercomputer powers the Bharat Forecast System (BFS) at IITM, Pune?

    a) Param Siddhi
    b) Mihir
    c) Arka
    d) Pratyush

44. A key benefit of BFS for India is its expected improvement in forecasting which type of events?

    a) Geological events
    b) Extreme rainfall events and cyclone paths
    c) Economic recessions
    d) Solar flares

45. The indigenous development of BFS aligns with which Indian initiative?

    a) Digital India
    b) Swachh Bharat Abhiyan
    c) Atmanirbhar Bharat
    d) Startup India

46. What does Forecast Verification" primarily involve?

    a) Disseminating forecasts to the public.
    b) Evaluating the quality and accuracy of forecasts against observations.
    c) Collecting new weather data.
    d) Developing new weather models.

47. When verifying a forecast, what is the term used for correctly predicting the occurrence of an event?

    a) Miss
    b) False Alarm
    c) Hit
    d) Correct Negative

48. If a forecast predicts rain, but it does not rain, what is this called in verification terms?

    a) Hit
    b) Miss
    c) Falsc Alarm
    d) Correct Negative

49. Which score is commonly used for probabilistic forecasts of binary events and measures the mean squared difference between forecast probabilities and observed outcomes?
    a) Mean Absolute Error (MAE)
    b) Critical Success Index (CSI)
    c) Brier Score (BS)
    d) Anomaly Correlation Coefficient (ACC)
50. What does a Reliability Diagram" visually represent in probabilistic forecast verification?
    a) The relationship between temperature and pressure.
    b) The observed frequency of an event versus its forecast probability.
    c) The track of a cyclone.
    d) The amount of rainfall over a region.
51. What is the purpose of comparing a forecast against a persistence forecast" in verification?
    a) To see if the forecast is always wrong.
    b) To measure the forecast's skill relative to simply assuming conditions won't change.
    c) To evaluate long-range climate trends.
    d) To check for model biases.
52. For high-resolution models, why are Spatial Verification Methods" often preferred over traditional point-by-point comparisons?
    a) They are easier to compute.
    b) They give credit for forecasts that are close" to the observation, even if slightly misplaced.
    c) They only work for global models.
    d) They don't require any observations.
53. What does a Value-Added Forecast" primarily do?
    a) Provides raw model output directly to users.
    b) Transforms generic weather forecasts into actionable, decision-ready information for specific users.
    c) Focuses solely on long-range climate projections.
    d) Is only applicable to aviation.

54. Which of the following is not a characteristic of how value is added" to a forecast?
    a) Incorporating local knowledge and user-specific thresholds.
    b) Translating forecasts into industry-specific language.
    c) Simply displaying raw model output without further processing.
    d) Integrating non-meteorological data like crop growth stages.
55. What is a key benefit of Value-Added Forecasts for farmers?
    a) They tell farmers exactly what to do every hour.
    b) They provide generic weather information for the entire country.
    c) They enable precise, tactical decisions to optimize operations and mitigate risks.
    d) They help in predicting global commodity prices.
56. What is a limitation of Doppler Weather Radar networks for agricultural applications?
    a) They can measure temperature very accurately.
    b) They have range limitations due to Earth's curvature, potentially missing low-level precipitation far away.
    c) They are very cheap to deploy nationwide.
    d) They can directly measure soil moisture.
57. When is a persistence forecast" most likely to be a strong benchmark for forecast skill?
    a) For seasonal forecasts.
    b) For very short-range forecasts (nowcasting).
    c) For long-range climate projections.
    d) For predicting cyclone intensity.
58. Which verification metric is specifically designed to assess whether an ensemble forecast's spread is appropriate for representing uncertainty?
    a) Mean Absolute Error (MAE)
    b) Critical Success Index (CSI)
    c) Rank Histogram (Talagrand Diagram)
    d) Anomaly Correlation Coefficient (ACC)
59. What is the term for a systematic tendency of a forecast to consistently over-predict or under-predict a particular value or event?
    a) Accuracy
    b) Skill
    c) Bias
    d) Resolution

60. Why is human interpretation and synoptic judgment still important in short-range and nowcasting, despite advanced models?
    a) To replace all model output.
    b) Models are never accurate.
    c) To refine model outputs, integrate local effects, and issue warnings based on expert analysis.
    d) To collect raw observational data manually.

11

Indigenous Technical Knowledge for Weather Forecasts and Its Validity

11.1 Introduction

Following our comprehensive exploration of scientific crop modeling and the rigorous processes of model assessment, we now pivot to an equally vital, yet distinct, realm of understanding weather in agriculture: Indigenous Technical Knowledge (ITK). This chapter, Indigenous Technical Knowledge for Weather Forecasts and Its Validity," delves into the profound and often overlooked wisdom embedded within local communities worldwide, knowledge systems accumulated over generations that offer unique insights into anticipating future weather patterns.

In the upcoming sections, we will embark on an in-depth examination of the diverse forms that ITK related to weather forecasting takes. This includes detailed discussions on:

- **Observational Indicators:** How traditional communities keenly observe subtle changes in the natural world, such as specific patterns in animal behavior (e.g., bird migration patterns, insect activity, animal calls), plant phenology (e.g., early or late flowering, unusual fruiting patterns of wild plants), and the behavior of flora and fauna as harbingers of upcoming weather shifts.
- **Atmospheric and Astronomical Signs:** The interpretation of sky conditions (e.g., cloud formations, specific colors of sunrise/sunset, halos around the sun or moon), wind patterns, atmospheric pressure changes (often perceived through bodily sensations or changes in local sound travel), and celestial phenomena (e.g., star brightness, moon phases) as indicators for short-term or seasonal forecasts.
- **Traditional Ecological Calendars and Proverbs:** The role of traditional calendars that map out expected seasonal weather based on historical observations, and folk proverbs or sayings that encapsulate generations of empirical weather wisdom in concise, memorable forms, guiding agricultural practices.

- **Hydro-meteorological Observations:** Localized interpretations of water body behavior (e.g., river levels, spring flows, dew formation) as predictors.

Furthermore, a significant portion of this chapter will be dedicated to a critical examination of the validity and reliability of these traditional forecasts. We will explore the methodologies employed to assess the accuracy of ITK, discussing its strengths, particularly its high relevance and accuracy within specific local contexts and microclimates. We will also acknowledge its inherent limitations, such as its localized nature, often qualitative rather than quantitative predictions, and its potential vulnerability to rapid, unprecedented climate change. Finally, the chapter will explore the immense potential and inherent challenges involved in integrating ITK with modern scientific meteorological predictions, recognizing that a synergistic approach can significantly enhance agricultural resilience, especially for vulnerable smallholder farmers who rely heavily on both traditional wisdom and contemporary scientific information for their livelihoods.

11.2 Observational Indicators in Indigenous Technical Knowledge for Weather Forecasting

Indigenous Technical Knowledge (ITK) for weather forecasting is deeply rooted in generations of acute observation and intimate understanding of the local environment. Unlike modern meteorology, which relies on quantitative data from instruments and complex atmospheric models, ITK leverages qualitative indicators from the natural world. These observational indicators are signs and patterns noticed in the behavior of animals, the phenology of plants, and the characteristics of the atmosphere and land, which have, over centuries, been empirically linked to impending weather changes. This section explores these rich observational categories in detail.

11.2.1 Animal Behavior (Zoological Indicators)

Traditional communities, through prolonged cohabitation with nature, have developed an intricate understanding of how various animal species respond to subtle shifts in atmospheric pressure, humidity, temperature, and other precursors to weather changes. These observations are often highly localized and specific to regional fauna.

- **Birds**
    - **Low Flying Birds:** Many traditional beliefs suggest that birds flying low, especially before a storm, are indicative of falling atmospheric pressure, which makes it harder for them to gain altitude (Fig.11.1).

Fig. 11.1: Low flying birds – indicator of incoming stormy conditions

- **Increased Chirping/Activity:** Some observations link increased or frantic bird activity, or a sudden quietness, to impending bad weather. For example, excessive chirping of sparrows or restlessness of chickens might precede a storm.
- **Migration Patterns:** Unusual timing or routes of migratory birds can sometimes be interpreted as signs of long-term seasonal shifts or more severe weather in their usual destinations.

- **Insects**
  - **Ants:** Ants building higher mounds or sealing their entrances can be seen as a sign of impending heavy rain or floods, as they try to protect their colonies.
  - **Bees:** Bees staying close to their hives or returning early may indicate an approaching storm, as they avoid being caught in bad weather.
  - **Mosquitoes/Flies:** Increased biting or swarming activity of certain insects might be linked to changes in humidity or atmospheric pressure before rain.

- **Reptiles and Amphibians**
  - **Frogs and Toads:** Increased croaking or a change in their usual vocalizations can sometimes be associated with impending rain, as they are highly sensitive to humidity.

- **Snakes:** Snakes moving to higher ground or exhibiting unusual restlessness may be seen as a sign of impending floods.

- **Livestock and Domestic Animals**
  - **Cattle/Livestock Behavior:** Cattle huddling together, lying down more frequently, or being unusually restless might be interpreted as a sign of approaching bad weather. Dogs and cats may become more agitated or seek shelter.
  - **Sheep:** Sheep gathering in a corner of a field can be seen as a sign of impending strong winds or rain.

- **Fish**
  - Fish jumping out of water or behaving unusually can sometimes be linked to changes in water temperature or atmospheric pressure preceding a storm.
  - **Explanation of Underlying Principles:** Many of these animal behaviors are rooted in genuine physiological responses. Falling barometric pressure, for example, can affect insects‘ flight mechanisms or the air sacs in birds, making low-altitude flying more comfortable. Increased humidity often triggers reproductive cycles or activity in amphibians and certain insects. While not always perfectly reliable, these observations represent empirical correlations built over vast periods of time.

11.2.2 Plant Phenology (Botanical Indicators)

Plants, being stationary, cannot escape adverse weather. Instead, their growth, flowering, fruiting, and leaf behavior are highly attuned to environmental cues, making them sensitive natural barometers. These phenological indicators are often used for seasonal or medium-range forecasts.

- **Early/Late Flowering or Fruiting**
  - **Unusual Timing:** If a particular wild plant or tree flowers significantly earlier or later than its historical average, or if it produces an unusually abundant or sparse fruit set, it might be interpreted as a sign of the character of the upcoming season (e.g., a mild winter, a long dry summer).
  - **Example:** In some regions, abundant flowering of certain trees might be linked to a forecast of good monsoon rains, while sparse flowering might indicate a drier season.

- **Leaf Behavior**
  - **Leaf Curling/Drooping:** Certain plants' leaves might curl, droop, or show increased turgidity (stiffness) in response to changes in humidity or atmospheric pressure, signaling impending rain.
  - **Leaf Turning:** Some traditional beliefs suggest that if leaves on trees show their undersides (due to wind or humidity changes), rain is coming.
- **Seed Dispersal**

The behavior of seed pods or cones (e.g., fir cones opening or closing) can be linked to changes in humidity, providing an indicator of dry or wet conditions.

- **Mushroom Growth**
  - An unusual flush of mushrooms, particularly certain types, might be seen as a strong indicator of prolonged wet conditions or high humidity.
  - **Explanation of Underlying Principles:** Plant phenology is strongly driven by accumulated thermal time (GDD), photoperiod, and water availability. Anomalies in these environmental factors can trigger unusual phenological responses. For example, a warmer-than-average winter might lead to earlier bud break and flowering. Plants' stomata and cell turgor are directly affected by humidity and water potential, leading to visible changes in leaf posture.

11.2.3 Atmospheric and Terrestrial Observations

These are direct observations of the sky, wind, and ground, often used for short-term (hours to a day) predictions.

- **Sky Conditions**
  - **Cloud Formations:** Specific types of clouds (e.g., dark, ominous clouds; high, wispy cirrus clouds; low-hanging fog) are universally recognized as precursors to certain weather events (rain, clear skies, cold).
  - **Color of Sky (Sunrise/Sunset):**Red sky at night, sailors' delight; red sky in morning, sailors take warning. This proverb holds some scientific basis, as red sunrises can indicate high-pressure systems moving east, bringing stormy weather.
  - **Halos around Sun/Moon:** Often associated with high-altitude ice crystals (cirrus clouds), which can precede warm fronts and rain (Fig.11.2).

- **Visibility:** Exceptionally clear visibility can sometimes precede a frontal system, while hazy conditions might indicate stable air or an approaching change.

Fig. 11.2: Halo around moon – indicator of high humidity/rainfall

- **Wind Patterns**
  - **Direction and Strength:** Consistent shifts in wind direction (e.g., from west to east, or sudden changes) are often linked to approaching weather fronts. Increased wind strength, especially after a period of calm, can signal a storm.
  - **Local Wind Signs:** Traditional knowledge often includes very specific local wind patterns that are reliable indicators for a microclimate.
- **Dew and Fog**
  - **Heavy Dew:** Often indicates clear, calm nights and no immediate rain, as conditions are conducive to condensation. Absence of dew can imply cloud cover or strong winds, suggesting a change.
  - **Morning Fog:** Can sometimes be associated with stable atmospheric conditions that will burn off to a clear day.
- **Smoke Behavior**
  - Smoke rising straight up implies high pressure and stable conditions, while smoke hanging low or falling indicates low pressure and potential rain.
- **Sounds**
  - Distant sounds being heard more clearly can indicate a high humidity or a specific atmospheric layering associated with impending rain.

- **Explanation of Underlying Principles:** These observations are directly linked to meteorological phenomena. Cloud types reflect atmospheric stability and moisture content. Sky colors are due to light scattering by particles, which vary with atmospheric conditions. Wind shifts indicate pressure gradients and frontal movements. Dew formation requires specific temperature and humidity conditions.

11.2.4 Other Environmental Indicators

- **Soil and Rock Behavior**
    - **Salt Dampness:** In some arid regions, if naturally occurring salts on rocks or ground become damp to the touch, it indicates increased humidity and potential for rain.
    - **Soil Smell:** A distinct earthy smell just before rain is due to geosmin, a chemical released by soil bacteria activated by humidity.
- **Human Body Aches:** Some individuals claim to feel joint aches or headaches before changes in atmospheric pressure or humidity, predicting rain.

Indigenous Technical Knowledge represents a rich tapestry of localized empirical observations, deeply intertwined with the specific ecosystems in which communities reside. While not always easily translatable into universal, quantitative predictions, these observational indicators reflect a profound and invaluable understanding of local environmental cues that have historically provided reliable guidance for agricultural practices and daily life. The challenge and opportunity lie in systematically evaluating their validity and exploring how they can complement modern scientific forecasting tools.

11.3 Atmospheric and Astronomical Signs for Weather Forecasting (ITK)

Beyond observations of the living world, Indigenous Technical Knowledge (ITK) places significant emphasis on interpreting signs from the non-living elements of the environment, particularly the atmosphere and celestial bodies. These observations, honed over centuries of close interaction with local climates, serve as powerful, often immediate, indicators of impending weather changes. While modern meteorology relies on instruments to measure atmospheric pressure, humidity, and cloud height, traditional communities have developed sophisticated qualitative methods to perceive these changes through direct sensory experience and patterned observation.

11.3.1 Atmospheric Signs

These indicators are derived from direct observation of the sky and ambient air conditions, often providing short to medium-range forecasts.

- **Cloud Formations**
  - **Specific Cloud Types:** Traditional knowledge often categorizes clouds not just by their shape but by their texture, movement, and color, associating them with particular weather outcomes. For example:
    - **Dark, Low-Hanging, or Fast-Moving Clouds:** Almost universally indicate impending rain or stormy weather. The darker the clouds, the heavier the expected precipitation.
    - **High, Wispy (Cirrus) Clouds:** Often seen as precursors to a change in weather, sometimes signaling an approaching warm front and rain, particularly if they thicken and lower.
    - **"Mackerel Sky" (Altocumulus/Cirrocumulus):** Small, rippled, high-level clouds are often linked to unsettled weather or rain within a day or two ("Mackerel sky, not long dry").
    - **Flat, Featureless Clouds:** May indicate stable air and fair weather, while rapidly building, towering clouds (cumulonimbus) are clear indicators of thunderstorms.
  - **Cloud Movement:** Observing the direction and speed of clouds at different altitudes can give clues about wind patterns at various atmospheric layers, hinting at approaching weather systems. Clouds moving in different directions at different heights might signal changing wind shear.
- **Color of the Sky (Sunrise/Sunset)**
  - **Red sky at night, sailor's delight; Red sky in morning, sailor's warning."** This famous proverb has a scientific basis.
    - **Red Sunset:** Often occurs when dust and small particles are trapped in the atmosphere by a high-pressure system, scattering blue light and leaving red hues. In mid-latitude regions where weather systems generally move from west to east, a clear, red sky in the west (where the sun sets) means that fair weather (associated with the high-pressure system) is moving towards the observer.
    - **Red Sunrise:** A red sky in the east (where the sun rises) implies that the good weather system has already passed to the east, and a low-pressure system (often bringing wet and windy conditions) is approaching from the west.

- **Yellow/Pale Sky:** A bright yellow sunset might indicate strong winds; a pale yellow, wet conditions. Dull, leaden skies often foretell rain.

- **Wind Patterns**
  - **Directional Shifts:** Changes in prevailing wind direction are critical indicators. For instance, a persistent shift from a dry, continental wind to a moist, oceanic wind direction is a strong signal for impending rain.
  - **Wind Strength and Sound:** A sudden calm before a storm, or a specific "moaning" or "whistling" sound of the wind, can be traditional cues. Wind picking up strength in a particular direction might indicate a developing pressure gradient.
  - **Local Wind Behavior:** Indigenous communities often have very specific knowledge about how local topographical features (mountains, valleys, water bodies) influence wind patterns, and how these localized winds relate to upcoming weather. For example, specific valley breezes might indicate fair weather, while a sudden gust from an unusual direction could mean a storm.

- **Atmospheric Pressure Changes (Indirect Perception)**
  - Traditional people don't measure barometric pressure with instruments, but they observe its effects:
    - **Smoke Behavior:** Smoke from fires rising straight up into the air often indicates high pressure and stable, fair weather. Conversely, smoke hanging low to the ground or swirling downwards suggests low pressure and likely rain.
    - **Aches and Pains:** Some individuals report feeling joint aches or headaches before significant drops in atmospheric pressure, preceding stormy weather.
    - **Sounds Traveling:** Sounds carrying further than usual can indicate a stable, dense atmosphere (often associated with high pressure or specific temperature inversions) which might precede a change in weather.

11.3.2 Astronomical Signs

Observations of celestial bodies, particularly the Moon and stars, are also integrated into ITK for weather forecasting, often providing clues about humidity, atmospheric stability, and long-range seasonal patterns.

Moon Halos and Rings

- **Appearance:** A hazy ring or halo appearing around the Moon (or sometimes the Sun, though harder to observe) is a common traditional indicator of approaching bad weather, usually rain or snow.
- **Scientific Basis:** These halos are caused by the refraction of moonlight (or sunlight) through hexagonal ice crystals suspended in high-altitude cirrus clouds. These cirrus clouds often precede a warm front or a low-pressure system, which typically brings precipitation within 12-48 hours. The saying When halo rings the moon or sun, rain‘s approaching on the run is widely recognized and has strong scientific validity.
- **Distance of Halo:** Some lore suggests that the further the halo appears from the moon, the nearer the coming rain.

- **Star Observation (Scintillation/Twinkling)**
  - **Increased Twinkling:** When stars appear to twinkle rapidly or irregularly, it is often interpreted as a sign of turbulent, unstable air in the upper atmosphere, which can precede stormy or unsettled weather (e.g., strong winds, rain).
  - **Clear, Steady Stars:** Conversely, stars that appear clear and steady, with minimal twinkling, often indicate calm, stable atmospheric conditions and fair weather.
  - **Scientific Basis:** The twinkling of stars (scintillation) is caused by variations in the refractive index of different layers of air as starlight passes through the Earth‘s atmosphere. Turbulent or humid air causes more refraction and thus more twinkling.
  - **Visibility of Specific Stars/Constellations:** In some cultures, the visibility or position of certain stars or constellations at particular times of the year is linked to seasonal weather patterns (e.g., the onset of a rainy season or a cold spell).

- **Moon Phases**
  - While less scientifically robust for short-term prediction, some traditional beliefs link specific moon phases (e.g., new moon, full moon) to tendencies for certain weather conditions or a higher likelihood of weather changes. For instance, wet moon (horns pointing up, holding water) or dry moon (horns pointing down, spilling water) are common folklore.

Indigenous communities have, for millennia, built sophisticated empirical weather forecasting systems based on these meticulous observations of atmospheric and astronomical phenomena. While qualitative and local, this ITK represents an invaluable repository of environmental wisdom. Understanding these indicators, and the scientific principles that often underpin them, offers significant potential for complementing modern meteorological forecasts, especially in localized agricultural contexts where traditional knowledge remains a primary guide for decision-making.

11.4 Traditional Ecological Calendars and Proverbs (Indigenous Technical Knowledge)

Beyond direct observations of biotic and atmospheric signs, Indigenous Technical Knowledge (ITK) for weather forecasting is also deeply embedded in long-standing cultural frameworks: Traditional Ecological Calendars and Weather Proverbs. These represent highly distilled, generations-long accumulations of environmental wisdom, transforming complex patterns into accessible and actionable knowledge for communities, particularly in guiding agricultural and livelihood activities.

11.4.1 Traditional Ecological Calendars (Ethno-Calendars)

Traditional ecological calendars are dynamic systems of time-keeping that organize the year not solely by astronomical or Gregorian dates, but by ecological and meteorological events that are significant to a local community's way of life. They are fundamentally different from conventional calendars because their divisions and markers are often tied to recurring natural phenomena, which inherently carry weather-related information.

- **Concept and Structure**
  - Unlike fixed-date calendars, these calendars mark periods based on events like:
    - **Phenological Events:** The flowering of specific plants, the fruiting of particular trees, the emergence of certain insects, or the migration of birds.
    - **Meteorological Patterns:** The expected onset of dry seasons, rainy seasons, specific wind patterns (e.g., monsoon winds), or periods of intense heat or cold.
    - **Hydrological Cycles:** The swelling of rivers, the drying of wetlands, or changes in spring flows.

  - **Agricultural Cycles:** The time for planting specific crops, weeding, harvesting, or preparing land for the next season.
- These calendars are often cyclical, reflecting the predictable rhythm of the local ecosystem. They might have fewer, longer "months" or "seasons defined by clusters of these events.

- **Relationship to Weather Forecasting**
  - **Predictive Framework:** The ecological events themselves serve as reliable indicators. For example, the ripening of a particular wild fruit might signal that the long dry season is about to begin, or the arrival of a specific migratory bird might indicate the imminent onset of the rains.
  - **Seasonal Outlook:** By knowing which ecological events define each period, communities gain a nuanced understanding of the *expected* weather conditions for that time of year, including typical temperatures, rainfall, and wind patterns.
  - **Adaptive Scheduling:** Farmers use these calendars to make crucial decisions about planting, harvesting, and other agricultural operations. If the ecological indicators suggest an early or delayed monsoon, planting schedules can be adjusted accordingly.
  - **Intergenerational Knowledge Transfer:** These calendars are powerful tools for transmitting complex environmental knowledge across generations, often through oral traditions, songs, stories, and community rituals.

- **Examples**
  - **African Contexts:** Many African communities have calendars marked by the blooming of specific trees or the arrival of certain animals, which correspond directly to the expected start or end of rainy seasons or periods of drought. For instance, the hungry season might be defined by the period before the harvest when food stores are low, and the calendar cues help prepare for this.
  - **Indigenous North American Calendars:** Often track the Moon of the Buds, Moon of the Berries, or Snow Moon, each intrinsically linked to expected weather and resource availability.
  - **Southeast Asian Rice Calendars:** Highly detailed calendars exist that correlate specific lunar phases or constellations with the optimal timing for different rice cultivation tasks, given the expected monsoon rains.

Strengths

- **Highly Localized and Relevant:** Perfectly adapted to the specific microclimate and ecosystem of the community.
- **Holistic Integration:** Connects weather with ecology, agriculture, and cultural practices, providing a comprehensive worldview.
- **Resilience Building:** Enables communities to adapt their practices to the natural variability of their climate.

- **Limitations**

  - **Localized:** May not be easily transferable to different geographical areas.
  - **Qualitative:** Often provides qualitative predictions (wet season,cold spell) rather than precise quantitative data (e.g., 50 mm of rain).
  - **Vulnerability to Climate Change:** Rapid and unprecedented shifts in climate can disrupt the reliability of traditional ecological cues, as the expected timing of events might become erratic.

11.4.2 Proverbs and Sayings (Weather Lore)

Weather proverbs, also known as weather lore, are concise, memorable phrases or rhymes that encapsulate generations of empirical observations about weather patterns. They are often passed down orally and serve as practical, easily recallable forecasting rules.

- **Concept and Characteristics**

  - **Empirical Generalizations:** Each proverb is a distillation of repeated observations linking a specific natural sign to a subsequent weather event.
  - **Rhyme and Rhythm:** Often employ rhyme, alliteration, or rhythm to make them easier to remember and transmit.
  - **Localized but Some Universal:** While many are local, some (like the red sky proverb) have widespread variations due to universal atmospheric principles.
  - **Actionable Advice (Implicit or Explicit):** They provide implicit or explicit advice for action (e.g., Mackerel sky and mares' tails make tall ships carry low sails implies a storm is coming, so reduce sail).

- **Relationship to Weather Forecasting**
  - **Short-Term Prediction:** Many proverbs relate to short-term weather changes (within hours or a day or two), such as signs of approaching rain or clear weather.
  - **Seasonal Prediction:** Some proverbs offer long-range, seasonal outlooks, often linked to observations in specific months or about the abundance of natural resources (e.g., A green Christmas makes a fat graveyard - a mild winter leads to disease).
  - **Rule of Thumb:** They serve as readily accessible rules of thumb that farmers and others can use to make immediate decisions based on simple observations.
- **Examples (beyond those in previous sections)**
  - When the dew is on the grass, rain will never come to pass. When the grass is dry at morning light, look for rain before the night." (Relates to humidity and clear skies.)
  - No dew in the morning, look for a shower or storm." (Absence of dew indicates cloud cover or wind, disrupting condensation.)
  - If March comes in like a lion, it will go out like a lamb." (Suggests an inverse relationship between early and late March weather).
  - When smoke from a chimney goes straight up, fair weather it will be. When smoke descends to the ground, foul weather it will be found." (Relates to atmospheric pressure and stability).
  - Swallows high, a clear sky. Swallows low, rain we shall know." (Relates to insects, which swallows hunt, flying lower in high humidity/low pressure before rain).
- **Strengths**
  - **Memorability and Accessibility:** Easy to learn and apply without instruments.
  - **Practicality:** Directly link observations to actionable weather information.
  - **Cultural Significance:** Deeply woven into the cultural fabric, reinforcing traditional ways of life.
- **Limitations**
  - **Variable Reliability:** Not all proverbs hold universal or scientific validity; some are more accurate than others, and some are highly localized.

- **Qualitative and Imprecise:** Lack quantitative detail (e.g., rain vs. 10 mm of rain).
- **Static Nature:** May become less reliable in the face of rapid climate change, as the underlying environmental patterns shift.
- **Causation vs. Correlation:** Like all empirical observations, they represent correlations; the underlying mechanisms might not be fully understood by the users.

In summary, traditional ecological calendars and weather proverbs are powerful examples of how ITK translates complex environmental observations into actionable knowledge. They provide invaluable frameworks for anticipating weather, guiding agricultural practices, and fostering resilience within communities that have long depended on their intimate connection with the natural world. While facing challenges in an era of rapid climate change, their local relevance and cultural depth underscore their continued importance.

11.5 Hydrometeorological Observations (Indigenous Technical Knowledge)

In addition to interpreting signs from biotic life, the atmosphere, and celestial bodies, Indigenous Technical Knowledge (ITK) for weather forecasting draws significant insights from hydrometeorological observations. These refer to the careful monitoring and interpretation of the behavior of water in its various forms and locations – be it dew, springs, rivers, or the very ground itself. Traditional communities, often living in close harmony with their local water sources, have developed nuanced understandings of how these elements reflect or foretell changes in precipitation, humidity, and temperature.

11.5.1 Dew and Fog Formation

Dew and fog are visible manifestations of atmospheric moisture and temperature dynamics. Their presence or absence, and their characteristics, provide immediate clues about upcoming weather.

- **Heavy Dew at Dawn:** A strong indicator of a clear, calm night and often precedes a fair, sunny day. Dew forms when the ground surface cools rapidly (due to outgoing longwave radiation under clear skies) to below the dew point temperature, causing water vapor in the air to condense on surfaces. The absence of wind also allows the moisture to settle.
  - ***Proverbial Link:*** When the dew is on the grass, rain will never come to pass. When the grass is dry at morning light, look for rain before the night.

- **Absence of Dew:** Conversely, if there is no dew on a morning when it would typically be expected, it can signal either significant cloud cover (which prevents the ground from cooling sufficiently for dew formation) or strong winds (which mix the air and prevent saturation near the surface). Both conditions can be precursors to a change in weather, often indicating an approaching frontal system or increased likelihood of rain.
- **Morning Fog:** Depending on its characteristics and dissipation, morning fog can be interpreted. If fog burns off quickly and rises, it often portends a clear and sunny day. If it lingers or moves in specific directions, it might indicate persistent moisture or an approaching weather front.
- **Underlying Principle:** Dew and fog formation are directly tied to atmospheric humidity, temperature gradients, and radiative cooling, all of which are fundamental meteorological processes.

11.5.2 River and Stream Levels/Flows

Communities living near rivers and streams often observe their behavior as an indicator of broader rainfall patterns, especially for seasonal predictions.

- **Unusual River Levels**
  - **Lower-than-expected levels:** If rivers or streams are unusually low for a particular time of year, it can signal an impending drought or a drier-than-average rainy season in the upstream catchment areas.
  - **Higher-than-expected levels:** Conversely, unusually high levels outside of normal flood seasons can sometimes be a sign of persistent heavy rainfall upstream, potentially indicating an extended wet spell or a more intense monsoon.
- **Changes in Flow Velocity/Clarity**
  - Subtle changes in the speed of water flow or the clarity of the water might be interpreted as precursors to changes in precipitation or land runoff. For example, sudden murkiness could indicate recent heavy rains in distant catchment areas.
  - ***Underlying Principle:*** River flow is a direct integration of precipitation over a large catchment area, reflecting the cumulative hydrological conditions that directly correlate with regional rainfall amounts.

11.5.3 Spring and Well Water Levels

Changes in groundwater levels, as reflected in springs and wells, are often slower indicators but can provide valuable long-term insights into water availability and seasonal rainfall.

- **Fluctuations in Spring Flow:** Indigenous communities often monitor the volume and consistency of flow from natural springs. A significant reduction or complete drying of springs during periods when they should be active can be a strong indicator of prolonged drought conditions or a poor recharge season.
- **Well Water Depths:** Traditional wells provide a direct window into the local water table. A noticeable drop in well water levels earlier than usual or reaching historically low depths can signal an impending dry spell or a severe long-term water deficit. Conversely, a rapid rise might indicate recent or expected heavy rainfall.
    - *Underlying Principle:* Spring and well levels are directly linked to the groundwater table, which is recharged by precipitation. Their fluctuations reflect the balance between rainfall infiltration and water abstraction, offering insights into hydrological conditions over longer periods.

11.5.4 Soil Moisture and Ground Dampness

Direct observations of the ground itself can also serve as local weather indicators, particularly for short-range rainfall predictions.

- **Surface Dampness/Smell:** A sudden increase in ground dampness, or a distinct earthy smell (geosmin, produced by soil bacteria) even before rain begins to fall, often signals a rapid increase in atmospheric humidity and an approaching rain event.
- **Rocks Sweating":** In some regions, rocks or cave walls appearing sweaty or damp to the touch can indicate high humidity and condensation, suggesting rain is near.
- **Ants Moving Eggs/Food to Higher Ground:** While an animal behavior, this is often a response to perceived rising groundwater or impending heavy rains, signaling changes in soil moisture.
- **Underlying Principle:** The direct interaction of the soil surface with atmospheric moisture and temperature gradients provides immediate cues about humidity and imminent precipitation.

Hydrometeorological observations, as part of ITK, demonstrate an acute awareness of the water cycle's subtle manifestations. These indicators, developed through generations of empirical learning, offer practical and often reliable local forecasts, particularly valuable for communities whose livelihoods are intrinsically tied to water availability and predictable hydrological patterns. Integrating these deep local insights with modern scientific models holds immense potential for enhancing regional weather forecasting and supporting adaptive agricultural practices.

11.6 Conclusions

This chapter has embarked on a journey into the rich and profound world of Indigenous Technical Knowledge (ITK) for weather forecasting, revealing it as a vital, centuries-old repository of environmental wisdom. We have seen how communities, through generations of keen observation and intimate connection with their local ecosystems, have developed sophisticated systems for anticipating weather patterns. The diverse forms of ITK explored—ranging from the subtle cues in animal behavior and the intricate seasonal rhythms of plant phenology, to the direct interpretations of atmospheric and astronomical signs, and the nuanced readings of hydrometeorological observations—all underscore a holistic understanding of nature's interconnectedness.

These traditional forecasting methods, while largely qualitative and inherently localized, possess undeniable strengths. Their deep contextual relevance within specific microclimates, their low-cost accessibility, and their inherent integration with agricultural calendars and cultural practices make them invaluable, particularly for smallholder farmers who often lack access to modern scientific forecasts. However, we also critically acknowledged their limitations, including their localized applicability, their often qualitative nature, and a growing vulnerability to the unprecedented and rapid shifts brought about by contemporary climate change, which can disrupt the very natural indicators upon which ITK relies.

Ultimately, the exploration of ITK for weather forecasts reveals a compelling opportunity: not to replace modern scientific meteorology, but to foster a powerful synergy. By systematically assessing the validity of ITK, understanding its empirical underpinnings, and exploring effective mechanisms for its integration with cutting-edge scientific predictions, we can forge more comprehensive, resilient, and locally relevant weather information systems. Such an integrated approach holds immense promise for empowering farmers, enhancing adaptive capacity, and ultimately strengthening agricultural productivity and food security in the face of an increasingly unpredictable global climate.

Multiple Choice Questions (Bold means correct answer)

1. Indigenous Technical Knowledge (ITK) for weather forecasting is primarily based on:
   a) Quantitative data from advanced meteorological instruments.
   b) Generations of acute observation and intimate understanding of the local environment.
   c) Complex mathematical models run on supercomputers.
   d) Global satellite imagery analysis.
2. Which of the following is NOT typically an observational indicator used in ITK for weather forecasting?
   a) Animal behavior. b) Plant phenology.
   c) Economic market trends. d) Atmospheric signs.
3. According to ITK, low-flying birds before a storm can indicate:
   a) Rising atmospheric pressure.
   b) Falling atmospheric pressure, making it harder to gain altitude.
   c) Abundant food sources.
   d) A clear, sunny day ahead.
4. If ants are observed building unusually high mounds or sealing their entrances, ITK might interpret this as a sign of:
   a) Impending drought.
   b) A very mild winter.
   c) Impending heavy rain or floods.
   d) A stable high-pressure system.
5. Unusual timing or routes of migratory birds in ITK might suggest:
   a) A change in wind direction for the next day.
   b) Long-term seasonal shifts or more severe weather in their usual destinations.
   c) A specific lunar phase.
   d) Increased local insect populations.
6. Plant phenology, as an ITK indicator, typically provides insights for:
   a) Immediate, hourly weather changes.
   b) Short-term (daily) forecasts.
   c) Seasonal or medium-range forecasts.
   d) Predicting volcanic eruptions.

7. If a wild plant flowers significantly earlier than its historical average, ITK might interpret this as a sign of:
   a) A very cold, harsh upcoming season.
   b) An unusually mild or warmer upcoming season.
   c) Impending strong winds.
   d) Increased soil salinity.
8. The proverb Red sky at night, sailor's delight; Red sky in morning, sailor's warning has a basis in:
   a) Pure superstition with no scientific correlation.
   b) The scattering of light by particles in the atmosphere and typical weather system movement.
   c) Astronomical alignments that dictate ocean tides.
   d) The behavior of nocturnal animals.
9. A halo appearing around the Moon or Sun is often interpreted in ITK as a sign of:
   a) Extremely clear and stable weather.
   b) Approaching precipitation (rain or snow).
   c) Decreased atmospheric humidity.
   d) A solar flare.
10. The scientific basis for a halo around the Moon or Sun involves:
   a) Lunar magnetism.
   b) Refraction of light through ice crystals in high-altitude clouds.
   c) Absorption of light by water droplets in the lower atmosphere.
   d) The Moon's gravitational pull on clouds.
11. What does increased or rapid twinkling of stars typically indicate in ITK for weather forecasting?
   a) Very stable and calm atmospheric conditions.
   b) Turbulent, unstable air in the upper atmosphere, potentially preceding unsettled weather.
   c) Decreased atmospheric moisture.
   d) A sign of good harvest.
12. Traditional ecological calendars are distinct from conventional calendars because they:
   a) Only track lunar cycles.
   b) Organize the year based on astronomical or Gregorian dates only.

c) Divide the year by significant ecological and meteorological events.

d) Are used exclusively by meteorologists.

13. The Mackerel sky (small, rippled, high-level clouds) in ITK is often linked to:

a) Immediate clear skies.

b) Prolonged periods of drought.

c) Unsettled weather or rain within a short period.

d) A strong, consistent high-pressure system.

14. Weather proverbs are best described as:

a) Complex mathematical equations.

b) Concise, memorable phrases encapsulating empirical weather observations.

c) Detailed scientific reports.

d) Instrumental readings.

15. The proverb When smoke from a chimney goes straight up, fair weather it will be. When smoke descends to the ground, foul weather it will be found relates to ITK's understanding of:

a) Wind direction only.

b) Atmospheric pressure and stability.

c) The amount of fuel in the chimney.

d) Cloud temperature.

16. Which observation is an example of a hydrometeorological indicator in ITK?

a) Bird migration patterns.

b) Cloud formations.

c) Dew formation.

d) Star twinkling.

17. Heavy dew at dawn is typically interpreted in ITK as a sign of:

a) Impending heavy rainfall.

b) A clear, calm night and fair weather ahead.

c) Very strong winds.

d) High humidity leading to immediate fog.

18. If springs are unusually low or dry for a particular time of year, ITK might interpret this as an indicator of:

a) An upcoming period of abundant rainfall.

b) Prolonged drought conditions or poor seasonal recharge.

c) Increased groundwater availability.

d) A shift in tidal patterns.

19 The absence of dew on a morning when it would typically be expected can signal:

a) Perfect sunny weather for the entire day.

b) Significant cloud cover or strong winds preventing condensation.

c) Decreased atmospheric moisture.

d) An imminent rise in air temperature.

20. What is a key strength of ITK weather forecasts?

a) Their ability to provide precise quantitative predictions globally.

b) Their perfect reliability under all conditions, including rapid climate change.

c) Their deep contextual relevance within specific local microclimates.

d) Their independence from any human observation.

21. A major limitation of ITK weather forecasts, particularly in the face of rapid climate change, is their:

a) High cost of implementation.

b) Inability to be passed down through generations.

c) Potential disruption of traditional indicators due to unprecedented shifts.

d) Exclusive focus on short-term predictions.

22. What does an integrated approach (combining ITK and modern science) to weather forecasting aim to achieve?

a) To replace ITK with scientific methods entirely.

b) To solely rely on ITK for all weather predictions.

c) To enhance agricultural resilience through comprehensive and locally relevant information.

d) To make weather forecasting more complex and less accessible.

23. The observation that specific plants flower earlier or later than usual due to weather changes is an example of which ITK indicator?

a) Zoological indicator.

b) Astronomical indicator.

c) Botanical/Phenological indicator.

d) Hydro-meteorological indicator.

24. A distinct earthy smell just before rain is attributed in ITK to:
    a) Increased wind speed.
    b) Geosmin, a chemical released by soil bacteria activated by humidity.
    c) The presence of a rainbow.
    d) A sudden drop in air pressure.
25. While ITK is primarily qualitative, what benefit does it offer to smallholder farmers?
    a) Access to global long-range climate models.
    b) High-precision numerical forecasts.
    c) Low-cost, actionable insights integrated with local practices.
    d) Advanced statistical analysis of weather data.
26. The traditional belief that When the grass is dry at morning light, look for rain before the night relates to the absence of:
    a) Wind.
    b) Clouds.
    c) Dew.
    d) Sunlight.
27. The concept of hungry season in some traditional ecological calendars is often defined by:
    a) A specific astronomical alignment.
    b) The period before harvest when food stores are low.
    c) The blooming of a rare flower.
    d) A particular wind direction.
28. Which of the following is a direct observation of **terrestrial signs** in ITK for weather forecasting?
    a) The color of the sunset.
    b) Fish jumping out of water.
    c) Soil dampness or sweating rocks.
    d) The timing of bird migration.
29. The scientific basis of many ITK observations suggests that:
    a) They are purely coincidental.
    b) They reflect genuine physiological or atmospheric responses to precursor meteorological changes.
    c) They are only valid in the past.
    d) They contradict all modern scientific findings.

30. The primary challenge in integrating ITK with modern scientific weather predictions is often:

    a) The high cost of ITK.

    b) The lack of any data from ITK.

    c) Bridging the gap between qualitative, localized traditional knowledge and quantitative, large-scale scientific data.

    d) Farmers' unwillingness to accept new information.

12

Crop Weather Calendars

12.1 Introduction

Building upon our foundational understanding of how weather intrinsically shapes crop development, from the micro-scale physiological responses to the broader patterns revealed by Indigenous Technical Knowledge, we now transition to a chapter dedicated to a highly practical and actionable synthesis of this critical interrelationship: Crop Weather Calendars. These calendars are far more than mere chronological charts; they represent systematically organized and dynamic compilations of meteorological information meticulously integrated with precise crop-specific data, designed to serve as an indispensable guide for optimizing agricultural practices.

In this chapter, we will embark on a comprehensive exploration of Crop Weather Calendars, moving beyond the often informal or culturally embedded ecological calendars previously discussed. We will delve into how these powerful tools systematically integrate historical weather patterns, which involves analyzing long-term meteorological data to establish reliable averages and identify probabilities of specific weather events, such as the typical onset of monsoons, average frost dates, or periods of peak solar radiation. They also incorporate real-time meteorological data, bringing in current weather observations to track actual conditions against historical norms and initial forecasts. Furthermore, Crop Weather Calendars utilize future weather forecasts, drawing on short-, medium-, and long-range meteorological predictions to anticipate upcoming conditions (Fig.12.1).

This integrated weather information is then critically aligned with the precise phenological stages of a given crop cultivar, recognizing that a crop's sensitivity and requirements vary significantly throughout its life cycle. Our exploration will detail the essential components that constitute a robust crop weather calendar, including the identification of crucial growth periods, pinpointing the most sensitive and influential developmental stages of the crop like germination, tillering, flowering, and grain filling, where weather impacts are most pronounced. It also involves the establishment of weather thresholds, which means defining the optimal, critical, and damaging ranges of key weather parameters such as temperature, rainfall, solar radiation, and humidity for each specific growth stage, a process crucial for achieving both high yield and quality. Moreover, these calendars facilitate the strategic alignment of various farm operations with the most favorable climatic conditions to maximize efficiency and minimize risk. This includes determining the optimal timing for land preparation based on soil moisture and expected rainfall, aligning planting and sowing dates with ideal temperature and moisture for germination and establishment, and optimizing nutrient application to coincide with periods of high nutrient demand and favorable weather for uptake. The calendars also aid in pest and disease management by forecasting ideal conditions for outbreaks or development to enable timely interventions, precisely planning irrigation scheduling based on actual crop water demand and anticipated rainfall, and aligning harvesting with conditions that ensure optimal yield, quality, and drying.

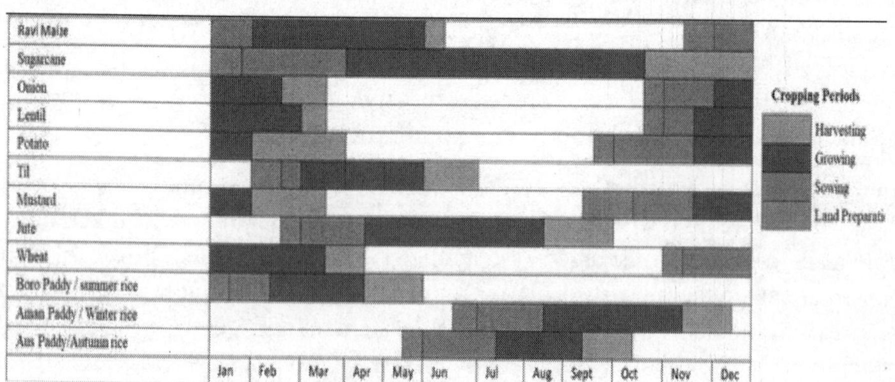

Fig. 12.1: Basic crop calander

Ultimately, a deep understanding and effective utilization of Crop Weather Calendars are paramount for modern agriculture. These tools empower farmers and agricultural managers to enhance farm productivity, strategically mitigate weather-related risks, improve resource use efficiency, and fundamentally foster greater resilience in agricultural systems that increasingly confront the challenges posed by escalating climate variability and change.

12.2 Objectives and Purpose of Crop Weather Calendars

Crop Weather Calendars are sophisticated tools meticulously designed with a clear set of objectives and a singular overarching purpose: to optimize agricultural productivity and enhance resilience by systematically integrating meteorological information with crop-specific requirements throughout the growing season. Their fundamental aim is to empower farmers, extension workers, and agricultural planners with actionable knowledge that facilitates informed decision-making, moving beyond reactive responses to proactive management in the face of climatic variability. A Crop Weather Calendar (CWC) is a valuable agrometeorological tool used to align the crop growth stages with prevailing weather conditions to optimize agricultural practices. It helps farmers, extension workers, and planners make informed decisions by integrating weather information with the crop's physiological needs (Fig.12.2).

The crop-weather calendars are used in agromet-advisory services. With the week as the basic time unit, the compiled information can assist forecasters in framing weather warnings and forecasts directed at farmers. Crop-weather calendars provide information on crop growth stages, normal weather for crop growth, warnings to be issued based on prevailing weather condition, water requirement of crops during their various phenophases and weather conditions favourable for development of crop pests and diseases.

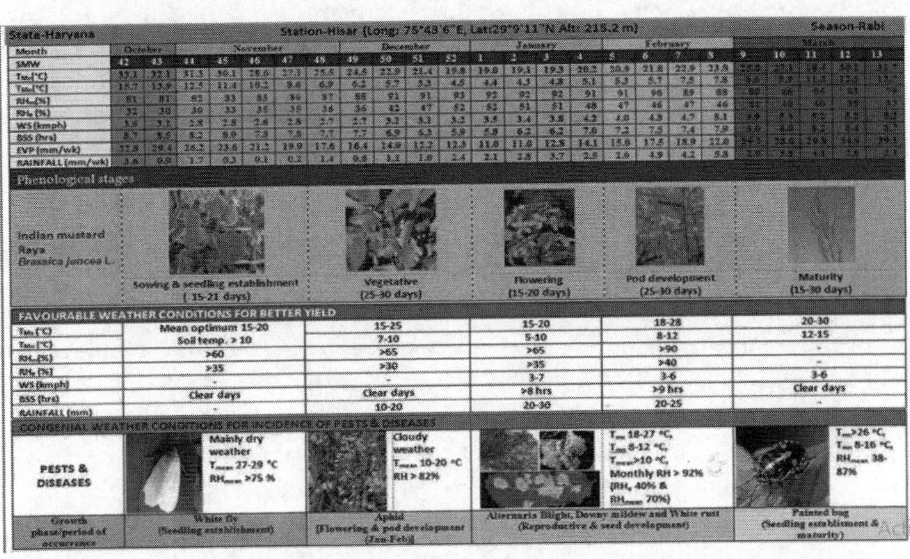

Fig. 12.1: Crop weather calendar for mustard

These calendars are useful for crop planning, irrigation scheduling and plant protection measures. Information contained in the calendars gives broad indications of the direction of development which may prove useful to the planners, agricultural administrators, plant breeders and the farmers in formulating policy matters regarding plant breeding, crop adoption, drought, supplemental irrigation, maximizing the yield. Crop weather calendars can act as guiding tool while issuing crop weather advisories for the farming community.

The primary objectives and purposes of developing and utilizing Crop Weather Calendars include:

Firstly, a core objective is to optimize resource use efficiency, particularly concerning water and nutrients. By precisely aligning irrigation schedules with actual crop water demand, determined by prevailing weather conditions and the crop's physiological stage, these calendars help minimize water wastage while preventing moisture stress that can significantly curb yield. Similarly, they guide the timing of fertilizer applications to coincide with peak nutrient uptake periods, ensuring that valuable inputs are utilized effectively, reducing leaching and runoff, and maximizing plant response. This precise timing ensures that resources are applied when the crop can best utilize them, enhancing return on investment and promoting environmental sustainability.

Secondly, Crop Weather Calendars are instrumental in mitigating weather-related risks and minimizing potential losses. By identifying critical weather thresholds for each growth stage – such as heat limits for flowering or frost susceptibility during emergence – the calendars enable early warnings for adverse conditions. This foresight allows farmers to implement preventive measures or adopt adaptive strategies like adjusting planting dates, applying stress-alleviating treatments, or considering alternative crops. By foreseeing periods of high rainfall, they can also guide decisions on drainage or disease management, thereby reducing vulnerability to extreme events and protecting potential yield.

Thirdly, they serve to enhance overall farm productivity and stabilize yields. By providing a clear roadmap of optimal conditions for each developmental phase, the calendars help ensure that crops experience favorable environments during their most sensitive stages. This proactive management, guided by weather insights, directly contributes to achieving the crop's full genetic potential, maximizing biomass accumulation and harvestable yield. Moreover, by adapting practices to seasonal climate predictions, they help reduce year-to-year yield variability, fostering greater stability in agricultural output.

Fourthly, Crop Weather Calendars act as a vital support system for informed farm-level decision-making. They distill complex meteorological and agronomic data into easily understandable formats, allowing farmers to plan crucial operations with precision. This includes determining the most opportune time for land preparation, selecting suitable planting dates based on expected temperature and moisture, scheduling pest and disease surveillance, and identifying the ideal window for harvesting to maximize both quantity and quality. The calendars translate scientific knowledge into practical, timely advice that directly impacts daily farm operations.

Fifthly, they play a crucial role in facilitating adaptation to climate change. As climatic patterns become increasingly unpredictable, with more frequent extreme events and shifts in seasonal norms, historical agricultural practices may no longer suffice. Crop Weather Calendars, by integrating long-term climate projections with real-time data, offer a dynamic tool to adjust practices, evaluate the suitability of new stress-tolerant cultivars, and develop resilient farming systems capable of thriving under future climatic realities. They help identify emerging risks and inform strategies for long-term sustainability.

Finally, while primarily focused on yield quantity, these calendars also aim to improve crop quality. By ensuring optimal growing conditions during critical periods, especially during grain or fruit filling, they contribute to better nutritional content, physical attributes, and market value of the harvested produce.

In essence, the objectives and purpose of Crop Weather Calendars converge on creating a more intelligent, responsive, and robust agricultural system. They transform raw weather data into strategic insights, empowering stakeholders to harness favorable conditions, confront adverse ones, and ultimately ensure food security and sustainable livelihoods in a changing world.

12.3 Inputs Required for Crop Weather Calendars

The effectiveness and accuracy of a Crop Weather Calendar hinge entirely on the quality, comprehensiveness, and timeliness of the input data it utilizes. Building such a calendar is an exercise in integrating diverse datasets that span meteorological, agronomic, and edaphic (soil) domains. These inputs allow the calendar to accurately define optimal conditions, identify critical thresholds, and provide relevant advisories for specific crops in specific locations.

Here are the key inputs required for developing a robust Crop Weather Calendar:

12.3.1 Meteorological Data (Weather Data):

This is the cornerstone of any crop weather calendar, providing the environmental context for crop growth. It needs to be comprehensive and cover different temporal scales.

- **Historical Weather Data (Long-Term Averages and Variability)**
    - **Purpose:** To establish climatic norms, identify typical seasonal patterns, calculate long-term averages for various parameters, and understand historical variability and the probability of extreme events (e.g., probability of first/last frost, onset of monsoon, dry spells). This helps define the *expected* weather conditions for different periods of the year.

- **Specifics:** Daily or weekly records (ideally for 20-30+ years) of:
  - **Temperature:** Daily maximum (Tmax), minimum (Tmin), and mean temperatures. Crucial for calculating growing degree days (GDD) and identifying heat/cold stress periods.
  - **Precipitation:** Daily rainfall amount, frequency, and intensity. Essential for water balance calculations and assessing drought/flood risk.
  - **Solar Radiation (Sunshine Hours/Incoming Radiation):** Total daily solar radiation or sunshine hours. Important for photosynthesis and biomass accumulation.
  - **Relative Humidity:** Daily mean, max, and min humidity. Affects transpiration, dew formation, and disease development.
  - **Wind Speed and Direction:** Influences evapotranspiration, lodging risk, and pesticide drift.
  - **Evapotranspiration (Reference ET):** Can be measured or calculated from other weather parameters, providing a measure of atmospheric water demand.

- **Real-Time (Current) Weather Data**
  - **Purpose:** To track the actual weather conditions as they unfold during the current growing season. This allows for comparison with historical averages and forecasts, enabling dynamic adjustments to advisories.
  - **Specifics:** Daily or even hourly measurements of the same parameters as historical data, often from local weather stations, automatic weather stations (AWS), or remote sensing platforms.

- **Weather Forecasts (Short, Medium, and Long-Range)**
  - **Purpose:** To anticipate future weather conditions and provide proactive guidance for upcoming farm operations.
  - **Specifics**
    - **Short-Range (1-3 days):** Detailed forecasts for daily temperature, precipitation, wind, etc. Crucial for immediate operational decisions (e.g., spraying, irrigation, harvest).
    - **Medium-Range (3-10 days):** Less precise but still valuable for planning weekly operations.
    - **Long-Range/Seasonal Forecasts (Weeks to Months):** Provide general outlooks for temperature and rainfall anomalies (e.g., above-average monsoon,""warmer-than-normal summer"). Essential for

strategic planning, cultivar selection, and risk assessment for the entire growing season.

12.3.2 Crop-Specific Data

This category defines the biological responses and requirements of the target crop, often varying significantly by cultivar.

- **Phenological Information**
  - **Purpose:** To link specific crop growth stages to accumulated thermal time (GDD) or calendar days. This defines the critical windows of sensitivity.
  - **Specifics**
    - **GDD Requirements:** For each major phenological stage (e.g., planting to emergence, emergence to flowering, flowering to physiological maturity) for the specific cultivar.
    - **Typical Duration of Stages:** In days, if GDD data is not available or as a supplementary guide.
- **Crop Water Requirements**
  - **Purpose:** To determine the amount of water needed by the crop at different growth stages and its sensitivity to water stress.
  - **Specifics**
    - **Crop Coefficients (Kc):** Factors that relate reference evapotranspiration (ET0) to actual crop water use (ETc) at different developmental stages.
    - **Sensitivity to Water Stress (**Ky **factors):** How much yield is reduced for a given level of water deficit at different stages.
    - **Rooting Depth:** Maximum effective rooting depth at different stages, influencing water extraction capacity.
- **Critical Temperature Thresholds**
  - **Purpose:** To identify temperatures that are optimal, sub-optimal, or damaging to the crop at various stages.
  - **Specifics**
    - **Base Temperature** (Tbase)**, Optimal Temperature** (Topt)**, Upper Thresholds** (Tupper) **for growth and development.**
    - **Specific thresholds for heat stress** (e.g., during pollen viability, grain filling).

- **Specific thresholds for chilling or frost injury** (e.g., at seedling, flowering, or maturity stages).

- **Solar Radiation Requirements/Efficiency**
  - **Purpose:** To link light availability to biomass accumulation.
  - **Specifics**
    - **Radiation Use Efficiency (ϵ):** For the specific crop/cultivar.
    - **Light Interception Characteristics (e.g., light extinction coefficient):** Relates LAI to intercepted radiation.
    - **Nutrient Response Curves (Optional but useful for advanced calendars):**
    - **Purpose:** To relate nutrient demand to specific growth stages and environmental conditions.
    - **Specifics:** Information on peak nutrient uptake periods and how water stress or temperature extremes can affect nutrient availability and uptake.

12.3.3 Soil Data

Soil properties directly influence water availability and nutrient cycling.

- **Soil Physical Properties**
  - **Purpose:** To determine the soil's capacity to store and supply water and to assess drainage characteristics.
  - **Specifics**
    - **Soil Texture:** Proportion of sand, silt, and clay (influences water holding capacity, drainage).
    - **Bulk Density:** Affects root penetration and water infiltration.
    - **Water Holding Capacity:** Field capacity (FC) and permanent wilting point (PWP) – the amount of water the soil can hold and the amount available to plants.
    - **Infiltration Rate:** How quickly water enters the soil.
    - **Drainage Characteristics:** Important for flood/waterlogging risk assessment.
- **Soil Chemical Properties (Optional)**
  - **Purpose:** To understand nutrient availability and pH, which interact with weather.

- **Specifics:** pH, organic matter content, nutrient levels (N, P, K, micronutrients).

12.3.4 Management Data (Local Practices)

The specific farming practices employed influence how a crop interacts with weather.

- **Planting Information**
    - **Specifics:** Typical planting dates, row spacing, planting density for the region and crop.
- **Fertilizer Application**
    - **Specifics:** Typical timing, type, and amount of fertilizer applied.
- **Irrigation Practices**
    - **Specifics:** Existing irrigation infrastructure, methods used, and typical scheduling.
- **Pest and Disease Management**
    - **Specifics:** Common pest/disease cycles in the region and typical control measures.

By carefully collecting, organizing, and integrating these diverse data inputs, Crop Weather Calendars transform raw information into a powerful, dynamic tool that supports precise agricultural management and adaptation strategies.

12.3.4 Structure of a Crop Weather Calendar

The design of a Crop Weather Calendar is crucial for its effectiveness as an agricultural planning tool. It is fundamentally structured to integrate complex meteorological and agronomic information into an intuitive format, typically resembling a matrix or timeline. This organization allows users to easily visualize the interplay between a crop's growth stages and the prevailing or anticipated weather conditions, leading to informed decision-making.

The calendar provides information along two main axes: horizontally for the temporal progression and crop development, and vertically for specific environmental parameters and recommended actions.

- **Horizontal Information: The Temporal and Developmental Flow:** The horizontal axis of a Crop Weather Calendar maps out the progression of the growing season, connecting time with the crop's biological journey.

- **Time Scale:** This defines the chronological divisions used. Common units include:
  - **Calendar Weeks or Months:** These are the most universally understood temporal units (e.g., Week 1, Week 2, or specific months like June, July).
  - **Days After Sowing (DAS):** Provides a direct count from the planting date, offering precision especially when phenological development is primarily tied to accumulated heat.
- **Crop Developmental Stages:** More biologically precise, this categorizes time based on the crop's physiological progression (e.g., Vegetative Phase, Reproductive Phase, Grain Filling). This often requires conversion from calendar dates using tools like Growing Degree Days (GDD).
- **Crop Phenological Stages:** For each defined time unit, the calendar explicitly indicates the predominant developmental stage the crop is expected to be in. This is a vital section, as a crop's sensitivities, resource requirements, and vulnerabilities vary dramatically from one stage to another. Examples include seedling emergence, tillering or branching, panicle or flower initiation, flowering (anthesis), grain/pod/fruit development, physiological maturity, and harvest maturity. This section effectively provides the what" of the crop's life at any given point in the season.
- **Vertical Information: Parameters, Requirements, and Actions:** The vertical axis details the specific types of information provided for each segment of the timeline or developmental stage. This allows for a comprehensive breakdown of relevant weather conditions, the crop's specific needs and vulnerabilities, and the practical management responses.
- **Meteorological Parameters (Historical, Current, Forecast):** This section presents the environmental context.
  - **Historical Norms and Probabilities:** Includes long-term averages for key weather variables such as average daily maximum and minimum temperatures (^{0}C), average total rainfall (mm) and number of rainy days, and average sunshine hours or solar radiation (MJ/m^2). Crucially, it may also include probabilities of specific events like the first/last frost, typical monsoon onset/withdrawal dates, or the likelihood of consecutive dry days. This helps in understanding the typical climatic conditions and inherent risks.

    - **Critical Thresholds:** This defines the optimal ranges and the damaging limits of weather parameters for the crop at each stage. Examples include minimum/maximum temperatures for growth, specific heat stress thresholds (e.g., above 35°C during flowering), frost damage thresholds (e.g., below 0^{0}C at emergence), and water stress thresholds (e.g., critical soil moisture deficit).
    - **Current/Forecasted Conditions:** In more dynamic or digital calendars, a section may integrate real-time weather observations and short- to long-range forecasts, allowing the calendar's advisories to adapt throughout the season based on unfolding and anticipated weather.
- **Crop Requirements and Vulnerability:** This segment clearly outlines what the crop needs and what it is sensitive to during the indicated stage. For example, it might state High water demand" during flowering, Highly sensitive to heat stress" during anthesis, High nutrient uptake" during early vegetative growth, or Vulnerable to lodging" during grain filling if combined with strong winds.
- **Recommended Agricultural Operations and Advisories:** This is the most actionable component, offering concrete guidance for farm management. This can cover a wide array of activities:
    - **Land Preparation:** Advising on optimal timing based on soil moisture and expected rainfall.
    - **Sowing/Planting:** Suggesting ideal planting windows and cultivar choices based on forecasted conditions.
    - **Nutrient Application:** Recommending the type, amount, and precise timing of fertilizers to coincide with peak nutrient demand.
    - **Irrigation Scheduling:** Providing guidance on when and how much water to apply, often based on real-time water balance or predicted dry spells.
    - **Pest and Disease Management:** Alerting to conditions conducive to specific pest outbreaks or disease development, with recommendations for scouting and control.
    - **Weed Management:** Highlighting optimal periods for effective weed control.
    - **Harvesting:** Advising on the ideal timing for harvest to ensure optimal yield, quality, and proper post-harvest handling.
    - **Contingency Planning:** Offering proactive advice for coping with abnormal weather events, such as suggesting supplemental irrigation during a prolonged dry spell or considering drought-tolerant varieties.

- **Example of a Conceptual Calendar Layout**

Below is a simplified conceptual representation of how this information might be structured in a Crop Weather Calendar:

Time Frame (e.g., Week/DAS)	Crop Phenological Stage	Avg. Max/ Min Temp (^{0}C)	Avg. Rainfall (mm)	Critical Weather Risk	Crop Needs / Vulnerabilities	Recommended Farm Operations / Advisories
Week 1 (DAS 1-7)	Sowing/Germination	28/18	25	Early drought / Heavy rain	Adequate moisture, Temp > 15°C for emergence	Ensure good seedbed. Sow with onset of monsoon.
Week 4 (DAS 22-28)	Vegetative / Tillering	32/20	15	Mid-season dry spell	High nutrient uptake, sufficient moisture	Apply 1st Nitrogen dose. Monitor for moisture stress.
Week 8 (DAS 50-56)	Flowering / Anthesis	36/25	10	Heat stress	Highly sensitive to heat, water deficit	Ensure sufficient irrigation. Monitor for pollen viability.
Week 12 (DAS 78-84)	Grain Filling	30/22	50	Heavy rain / Wind (Lodging)	High assimilate demand, susceptible to lodging	Ensure good drainage. Monitor for lodging, apply K.
Week 16 (DAS 106-112)	Harvest Maturity	25/15	5	Early frost	Low moisture content for harvest	Harvest when grain moisture content is optimal. Ensure proper post-harvest drying.

This systematic framework, encompassing location and cultivar-specific details, empowers farmers with actionable insights, translating complex scientific data into practical guidance for enhancing productivity and building resilience against the vagaries of weather.

12.4 Dynamic Crop Weather Calendars

Building upon the concept of a structured Crop Weather Calendar, Dynamic Crop Weather Calendars represent a significant evolution. Unlike their static predecessors, which are primarily based on historical climatic normals and fixed phenological timings, dynamic calendars continuously adapt and update their recommendations throughout the growing season. They achieve this by integrating real-time weather observations and up-to-date weather forecasts, providing highly relevant and actionable advisories that respond to the unfolding environmental conditions.

The shift from static to dynamic calendars is a direct response to the increasing unpredictability of weather patterns driven by climate change. In an era where historical averages alone are insufficient for reliable planning, dynamic calendars offer the agility and responsiveness needed to manage modern agriculture effectively.

12.4.1 What Makes Them Dynamic"?

The dynamic" nature of these calendars stems from several key features:

- **Integration of Real-Time Weather Data**
    - Instead of relying solely on historical averages, dynamic calendars continuously ingest live weather data from local automatic weather stations (AWS), remote sensing platforms (satellites, drones), and ground sensors.
    - This allows the system to track the actual accumulated Growing Degree Days (GDD), current soil moisture deficits, prevailing temperature extremes, and recent rainfall patterns as they occur.
    - For example, if a week is unusually warm, the calendar can adjust the predicted onset of the next phenological stage (e.g., flowering) to be earlier than a static calendar would suggest.
- **Incorporation of Weather Forecasts (Multi-Range)**
    - Dynamic calendars actively utilize short-range (1-3 days), medium-range (3-10 days), and even updated long-range (seasonal) weather forecasts.

- This forward-looking component enables proactive advice. For instance, if a heatwave is forecast for the flowering stage, the calendar can issue an immediate alert and recommend preventive measures *before* the stress occurs. If heavy rain is expected, it can advise delaying fertilizer application to prevent leaching.

- **Automated Updates and Adjustments of Advisories**
  - As real-time data and forecasts change, the calendar's advisories and recommendations are automatically re-evaluated and updated.
  - This means that advice for irrigation, pest management, or nutrient application is not fixed but changes based on actual and anticipated conditions. For example, if unexpected rainfall occurs, the irrigation recommendation will be dynamically reduced or postponed.

- **Phenological Tracking and Recalculation**
  - Instead of assuming fixed dates for phenological stages, dynamic calendars often employ simple crop models (like GDD models) that run in real-time. Based on actual accumulated GDD, the model predicts the *actual*timing of the next growth stage, which might deviate from historical averages.
  - This ensures that advisories are aligned with the crop's true physiological state, not just a calendar date.

- **Feedback Loops (Implicit in Elementary Models)**
  - While more complex models incorporate explicit feedback, in elementary dynamic calendars, the feedback loop is more about the system responding to observed deviations. If the crop is experiencing stress due to current weather, the calendar's advisories adjust to mitigate that stress or recommend compensatory actions.

- **Technologies Enabling Dynamism**

The shift to dynamic calendars is largely facilitated by advancements in:

- **Automatic Weather Stations (AWS):** Providing continuous, real-time local weather data.
- **Remote Sensing (Satellites and Drones):** Offering broad-scale and localized data on vegetation indices (e.g., NDVI), land surface temperature, and soil moisture, which can be fed into the calendar system.

- **Numerical Weather Prediction (NWP) Models:** The scientific backbone for generating accurate and updated short-, medium-, and long-range forecasts.
- **Geographic Information Systems (GIS):** For integrating spatial weather, soil, and crop data.
- **Data Platforms and Cloud Computing:** For processing vast amounts of incoming data efficiently.
- **Mobile Applications and SMS Services:** For disseminating dynamic advisories directly to farmers in a timely and accessible manner.

- **Benefits of Dynamic Crop Weather Calendars**

The enhanced responsiveness of dynamic calendars offers significant advantages:

- **Increased Accuracy and Relevance:** Advisories are tailored to the precise conditions of the current season and location, making them far more accurate and directly applicable than static historical calendars.
- **Improved Responsiveness to Unforeseen Events:** Farmers can react quickly to sudden weather shifts (e.g., unseasonal rains, unexpected heatwaves, or early frost warnings), implementing timely preventive or mitigating measures.
- **Enhanced Decision-Making Under Uncertainty:** By providing updated probabilities and risk assessments based on evolving forecasts, dynamic calendars help farmers make more confident decisions in an uncertain climate.
- **Better Risk Management:** Proactive warnings and adaptive strategies reduce crop losses from adverse weather, contributing to more stable yields and farm incomes.
- **Optimized Resource Use:** More precise guidance on irrigation and nutrient application prevents over-application and under-application, leading to greater efficiency, cost savings, and reduced environmental impact.
- **Facilitating Climate Change Adaptation:** Dynamic calendars are crucial tools for adapting to changing climate patterns, as they allow for flexible responses to altered phenologies, new stress periods, and shifted rainfall regimes.
- **Increased Farm Productivity:** By minimizing losses and optimizing conditions, dynamic calendars contribute to maximizing the crop's potential yield.

- **Challenges in Implementation**

Despite their benefits, dynamic crop weather calendars face challenges:

- **Data Availability and Quality:** Requires a dense network of reliable weather stations and robust remote sensing data, which may not be available in all regions.
- **Technological Infrastructure:** Needs reliable internet connectivity and access to computing resources for data processing and dissemination.
- **Model Complexity and Maintenance:** Even with elementary models as a base, integrating real-time data and forecasts requires sophisticated systems and ongoing maintenance.
- **User Adoption and Training:** Farmers need training and trust in the system to adopt dynamic advisories, which might differ significantly from traditional practices.
- **Cost:** Initial setup and ongoing operation can be expensive.

In conclusion, Dynamic Crop Weather Calendars represent the forefront of applying agro-meteorological knowledge to practical agriculture. By continuously integrating the latest weather information with crop-specific needs, they empower farmers to navigate the complexities of a changing climate with greater precision and resilience, ultimately contributing to more sustainable and productive farming systems.

12.5 Crop Weather Calendars of India Meteorological Department

The India Meteorological Department (IMD), under the Ministry of Earth Sciences, plays a pivotal role in providing agro-meteorological services to the farming community in India. A cornerstone of these services has been the development and dissemination of Crop Weather Calendars (CWCs). These calendars are designed to bridge the gap between complex meteorological information and the practical needs of farmers, helping them make informed decisions to enhance agricultural productivity and resilience.

12.5.1 Historical Context and Evolution

IMD's engagement with agricultural meteorology dates back to the establishment of its Agricultural Meteorology Division in Pune in 1932. The concept of Crop Weather Calendars was developed to systematically compile and present weather-related information relevant to specific crops and their growth stages. Initially, these calendars were largely static, based on long-term historical climatic normals (averages) for various districts and crops across the country.

Over the years, with advancements in meteorological forecasting, computing power, and communication technologies, IMD's approach to CWCs has evolved significantly, moving towards more dynamic and integrated advisory services. This evolution is largely driven by the increasing need to adapt to climate variability and change, where historical averages alone are no longer sufficient.

12.5.2 Structure and Information Provided (Traditional IMD CWCs)

Traditionally, IMD's Crop Weather Calendars were often presented in a pictorial or tabular format, typically divided into three main parts:

- **Crop Husbandry / Life History (Bottom Portion)**
  - This section diagrammatically illustrates the typical life cycle of the major crops in a given district, from the date of sowing to harvest and threshing.
  - It indicates important phenological phases, such as sowing, germination, transplantation (for rice), tillering, elongation, flowering, grain formation, and harvest.
  - These phases are typically aligned with Standard Meteorological Weeks (SMW) and corresponding calendar months, providing a clear temporal progression.
  - It may also include information on common crop varieties, their duration, and suitable soil types.
- **Normal Weather Requirements (Middle Portion)**
  - This central part provides long-term weekly or monthly climatic normals for the specific location.
  - **Average Meteorological Parameters:** Includes normal maximum and minimum temperatures, total rainfall, number of rainy days, sunshine hours, relative humidity, and evaporation.
  - **Normal Phase-wise Water Requirement:** Indicates the typical water needs of the crop during its various phenological stages, often derived from potential evapotranspiration multiplied by crop coefficients.
  - This section helps farmers understand the *expected* weather conditions that are generally favorable for crop growth in that region during different periods.

- **Weather Warnings and Pest/Disease Conditions (Upper Portion)**
  - This crucial section provides actionable insights by linking weather conditions to potential risks.
  - **Favorable Conditions for Pests and Diseases:** It outlines the specific meteorological conditions (e.g., high humidity and moderate temperature for blast disease in rice) that are conducive to the incidence, multiplication, and spread of major pests and diseases relevant to the crop.
  - **Nature of Weather Warnings:** It specifies the type of weather warnings that may be issued by the forecasting office for different crop periods and weather situations (e.g., warnings for heavy rainfall, drought, heat waves, cold waves, hailstorms, strong winds/cyclones). This helps forecasters and extension workers to tailor their advisories.

12.5.3 How IMD's Crop Weather Calendars are Used by Farmers

While originally designed for agricultural meteorologists and planners, the information from these calendars is disseminated to farmers, primarily through the Gramin Krishi Mausam Sewa (GKMS - Rural Agricultural Weather Service) program. This involves:

- **Agromet-Advisory Services (AAS):** The district-level Agro-Meteorological Field Units (AMFUs) and Krishi Vigyan Kendras (KVKs) interpret the information from CWCs, combine it with short- and medium-range weather forecasts, and issue weekly or bi-weekly **Agromet Advisories** in local languages.
- **Decision Support:** Farmers use these advisories for various farm operations:
  - **Sowing/Planting Dates:** Deciding the optimal time to sow based on expected monsoon onset or favorable temperatures.
  - **Irrigation Scheduling:** Knowing when to irrigate based on actual rainfall and crop water demand.
  - **Nutrient Management:** Timing fertilizer application to coincide with active growth phases and favorable weather for uptake.
  - **Pest and Disease Control:** Preparing for or implementing control measures when weather conditions are conducive to outbreaks.
  - **Harvesting:** Planning harvest to avoid adverse weather (e.g., heavy rains or strong winds at maturity).

- **Contingency Planning:** The calendars provide a basis for preparing for and mitigating the impact of adverse weather events like droughts or floods.

12.5.4 Evolution Towards Dynamism and Integration

Recognizing the limitations of static calendars in a changing climate, IMD, often in collaboration with ICAR (Indian Council of Agricultural Research) and State Agricultural Universities (SAUs), has been moving towards Dynamic Crop Weather Calendars and more sophisticated Agromet Advisory Systems.

- **Real-time Data Integration:** Efforts are underway to integrate real-time weather observations from a denser network of Automatic Weather Stations (AWS) and remote sensing products (like NDVI, soil moisture maps) into the advisory generation process.
- **Forecast-Based Advisories:** Greater emphasis is placed on generating advisories based on daily updated short- and medium-range forecasts, making the recommendations more precise and immediate.
- **Mobile Apps and Digital Dissemination:** Platforms like Meghdoot" (a joint initiative of IMD, IITM, and ICAR) deliver location-specific, stage-wise agro-advisories directly to farmers' mobile phones via SMS and mobile apps, providing dynamic updates throughout the season.
- **Refinement for Climate Change:** IMD continuously updates its climatic normals and conducts research to understand and integrate the impacts of climate change (e.g., shifts in monsoon patterns, increased frequency of extreme events) into its CWC framework and advisory services, to ensure continued relevance and accuracy for future agricultural planning.

In essence, IMD's Crop Weather Calendars, evolving from static guides to dynamic decision-support tools, form a critical backbone of agricultural extension services in India, empowering millions of farmers to respond more effectively to weather challenges and opportunities.

12.6 Conclusion

This chapter has underscored the pivotal role of Crop Weather Calendars as indispensable tools in modern agriculture, serving as a powerful bridge between complex meteorological science and actionable farm-level decision-making. We have moved beyond the historical and qualitative insights of traditional ecological calendars to explore formalized systems that systematically integrate historical climate data, real-time weather observations, and future forecasts with the precise phenological requirements of specific crops.

The fundamental objectives of Crop Weather Calendars — namely, to optimize resource use, mitigate weather-related risks, enhance productivity, and facilitate informed decision-making — highlight their critical importance in the agricultural landscape. By meticulously structuring information horizontally across time and crop stages, and vertically for detailed meteorological parameters, crop vulnerabilities, and recommended operations, these calendars transform abstract data into practical guidance. The evolution towards dynamic Crop Weather Calendars, exemplified by the ongoing efforts of the India Meteorological Department (IMD), further amplifies their value by continuously adapting advisories to the unfolding environmental reality.

Ultimately, Crop Weather Calendars represent a proactive approach to managing the inherent uncertainties of agriculture. They empower farmers to optimize critical farm operations, from planting to harvesting, based on scientific understanding and timely weather intelligence. In an era defined by increasing climate variability and the growing imperative for sustainable food security, the widespread adoption and continuous refinement of robust Crop Weather Calendars are not merely beneficial, but essential for building resilient agricultural systems capable of thriving in a changing world.

Multiple Choice Questions (Bold means correct answer)

1. Crop Weather Calendars primarily synthesize understanding from:
   a) Economic market analysis and global trade routes.
   b) Meteorological information and crop-specific data.
   c) Soil microbial activity and plant genetics only.
   d) Historical rainfall patterns and pest outbreaks exclusively.
2. A key distinction of Crop Weather Calendars from traditional ecological calendars is their:
   a) Reliance solely on animal behavior.
   b) More formalized and systematically organized compilation of data.
   c) Lack of connection to agricultural practices.
   d) Exclusive use of qualitative observations.
3. Which type of weather data is *not* typically integrated into Crop Weather Calendars?
   a) Historical weather patterns.
   b) Real-time meteorological data.
   c) Future weather forecasts.
   d) Paleoclimatic data from ice cores.

4. One of the core objectives of Crop Weather Calendars is to:
   a) Promote a single, uniform cropping system globally.
   b) Optimize resource use efficiency, particularly for water and nutrients.
   c) Predict global food prices with high accuracy.
   d) Replace all traditional farming knowledge.
5. By identifying critical weather thresholds for each growth stage, Crop Weather Calendars primarily help in:
   a) Determining the exact market value of the crop.
   b) Enabling early warnings and implementing preventive measures for adverse conditions.
   c) Calculating the total biomass produced irrespective of stress.
   d) Designing new agricultural machinery.
6. Crop Weather Calendars contribute to enhancing farm productivity by:
   a) Promoting random planting dates.
   b) Ensuring crops experience favorable environments during sensitive stages.
   c) Reducing the need for any kind of fertilizer.
   d) Ignoring pest and disease risks.
7. For which crucial farm operation do Crop Weather Calendars provide guidance, ensuring proper soil moisture and expected rainfall?
   a) Post-harvest storage. **b) Land preparation.**
   c) Marketing of produce. d) Financial accounting.
8. The input data for Crop Weather Calendars is expected to be comprehensive, covering which domains?
   a) Social, political, and cultural.
   b) Economic, legal, and regulatory.
   c) Meteorological, agronomic, and edaphic (soil).
   d) Geological, astronomical, and marine.
9. For historical weather data, what is the ideal duration of daily or weekly records required for a robust Crop Weather Calendar?
   a) 1-2 years. b) 5-10 years.
   c) 20-30+ years. d) Less than 1 year.

10. Which specific meteorological parameter is crucial for calculating Growing Degree Days (GDD) and identifying heat/cold stress periods in a Crop Weather Calendar?
    a) Wind speed.
    b) Relative humidity.
    c) Daily maximum and minimum temperatures.
    d) Atmospheric pressure.
11. Crop Coefficients (Kc) are an important crop-specific input to Crop Weather Calendars, as they help determine:
    a) The crop's sensitivity to pest attacks.
    b) The crop's water requirements at different growth stages.
    c) The optimal planting density.
    d) The amount of solar radiation intercepted.
12. The maximum effective rooting depth of a crop is a soil/crop input that influences:
    a) Its sensitivity to heat stress.
    b) Its susceptibility to lodging.
    c) Its water extraction capacity from the soil.
    d) Its nutrient application rate.
13. On a Crop Weather Calendar, the horizontal axis typically represents:
    a) Specific weather parameters like temperature or rainfall.
    b) The progression of time and crop developmental stages.
    c) The various recommended farm operations.
    d) Different types of soil.
14. If a calendar uses Vegetative,""Reproductive," and Grain Filling" as its primary horizontal divisions, it is organized by:
    a) Calendar weeks.
    b) Days After Sowing (DAS).
    c) Crop developmental stages.
    d) Standard Meteorological Weeks (SMW).
15. The vertical axis of a Crop Weather Calendar typically provides:
    a) A singular historical yield average.
    b) Detailed meteorological parameters, crop requirements, and recommended actions.

c) A general overview of global weather patterns.

d) Economic forecasts for crop prices.

16. Which type of information is found in the Recommended Agricultural Operations / Advisories" section of a Crop Weather Calendar?

a) Details on the genetic makeup of the crop.

b) Precise timing and type of fertilizer application.

c) The history of agricultural policy.

d) Global trade agreements for commodities.

17. What distinguishes **Dynamic Crop Weather Calendars** from static ones?

a) They only use historical averages.

b) They continuously adapt using real-time weather observations and updated forecasts.

c) They are always printed on paper.

d) They are exclusively used for subsistence farming.

18. The dynamic" nature of these calendars is largely enabled by technologies such as:

a) Hand tools and manual irrigation.

b) Traditional rain gauges and physical notebooks.

c) Automatic Weather Stations (AWS) and remote sensing.

d) Manual observation of animal behavior.

19. A key benefit of Dynamic Crop Weather Calendars is their:

a) Reduced need for any kind of weather data.

b) Increased accuracy and relevance of advisories tailored to current conditions.

c) Fixed and unchanging recommendations throughout the season.

d) Sole reliance on expert intuition.

20. When an unexpected rainfall event occurs, a dynamic crop weather calendar would typically:

a) Ignore the event and stick to the original plan.

b) Automatically recalculate and potentially postpone or reduce irrigation recommendations.

c) Issue a warning about heat stress.

d) Advise increasing fertilizer application.

21. The India Meteorological Department (IMD) initially developed Crop Weather Calendars based on:
    a) Daily satellite imagery.
    b) Long-term historical climatic normals for districts and crops.
    c) Real-time sensor networks in every field.
    d) Farmer feedback only.
22. What information is found in the Weather Warnings and Pest/Disease Conditions" section of traditional IMD CWCs?
    a) Global market prices for pesticides.
    b) Favorable meteorological conditions for pest/disease incidence and types of warnings to be issued.
    c) Detailed chemical composition of fertilizers.
    d) Historical crop yield statistics only.
23. How are IMD's Crop Weather Calendar insights primarily disseminated to farmers in India?
    a) Through international news channels.
    b) Via the Gramin Krishi Mausam Sewa (GKMS) program and Agromet Advisories.
    c) Directly by supercomputer models.
    d) Through academic journals.
24. A challenge in implementing dynamic crop weather calendars is:
    a) The simplicity of their underlying models.
    b) The widespread availability of high-quality, real-time data in all regions.
    c) The need for robust technological infrastructure and data quality.
    d) Farmers' immediate and universal adoption of new technologies without training.
25. The ultimate goal of Crop Weather Calendars in modern agriculture is to:
    a) Eliminate all risk from farming.
    b) Increase reliance on a single crop variety.
    c) Foster greater resilience and enhance productivity in a changing climate.
    d) Centralize all agricultural decision-making away from farmers.

13

Agroadvisory Based on Weather Forecast

13.1 Introduction

Having meticulously explored the foundations of crop-weather interactions, the development and validation of crop growth models, the wisdom embedded in Indigenous Technical Knowledge, and the systematic organization of Crop Weather Calendars, we now arrive at the culmination of this knowledge synthesis: the Agroadvisory Based on Weather Forecast. This chapter delves into the critical process of transforming raw meteorological data and scientific understanding into actionable, timely, and location-specific advice for the farming community.

An agroadvisory, at its core, is a targeted recommendation provided to farmers, guiding them on various agricultural operations by integrating current weather conditions, short- to medium-range weather forecasts, and the specific needs of their crops at different growth stages. It represents the final, practical output of complex agro-meteorological research and forecasting efforts, designed to empower farmers to make adaptive decisions that can significantly impact their yield, resource efficiency, and resilience against climatic shocks.

In the upcoming sections, we will comprehensively examine the multi-faceted nature of weather-based agroadvisories. Our exploration will detail:

- **The Conceptual Framework:** Understanding the underlying principles and objectives of these advisories, emphasizing their shift from reactive responses to proactive planning.
- **Essential Components of an Agroadvisory:** Breaking down the key pieces of information typically included, such as weather forecasts (temperature, rainfall, wind, humidity), crop-specific phenological stages, potential risks (pests, diseases, extreme weather), and precise recommendations for farm operations (sowing, irrigation, fertilization, pest control, harvesting).

- **The Data Flow and Processing Pipeline:** Tracing how raw meteorological data from diverse sources (satellites, radars, ground stations, numerical weather models) are collected, processed, analyzed, and integrated with agronomic insights to generate the advisories. This will include a discussion on the role of specialized agrometeorological centers and expert systems.
- **Dissemination Mechanisms:** Investigating the various channels through which these advisories reach the farmers, from traditional methods like community meetings and radio broadcasts to modern approaches like SMS alerts, mobile applications, and digital platforms, highlighting the challenges of effective last-mile delivery.
- **Impact Assessment and Feedback Mechanisms:** Examining how the effectiveness of these advisories is evaluated and how feedback from farmers is incorporated to continuously refine and improve the content and delivery of future recommendations.
- **Challenges and Opportunities:** Discussing the hurdles in developing and delivering robust agroadvisories, such as data gaps, technological infrastructure limitations, and the need for greater farmer literacy, alongside the immense opportunities for enhancing climate resilience, food security, and sustainable agricultural development.

This chapter underscores that effective agroadvisory services are not merely about predicting weather; they are about interpreting that prediction in the context of specific agricultural systems, translating uncertainty into actionable strategies, and ultimately fostering a more informed and adaptive farming community capable of navigating the complexities of a changing climate.

13.2 Conceptual Framework of Agroadvisory Based on Weather Forecast

The conceptual framework of agroadvisory services based on weather forecasts is a sophisticated, multi-layered system built on the fundamental principle that timely, relevant, and actionable weather information can significantly transform agricultural outcomes. It represents a paradigm shift from traditional, often reactive, farming practices to a proactive, science-driven, and adaptive approach (Fig.13.1).

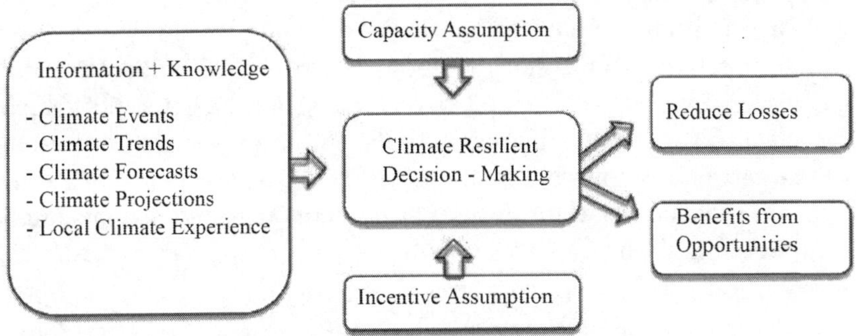

Fig. 13.1: Conceptual framework of climatic risk management/farm advisories

This framework systematically links cutting-edge meteorological science with profound agronomic expertise, all channeled through effective communication mechanisms to empower farmers with the insights necessary to optimize their daily operations.

At its core, this framework functions as a dynamic decision-support system. It is designed to furnish farmers with precise guidance, recognizing that weather is the single most influential abiotic factor in agriculture. Its impact, however, is not monolithic; it varies intricately depending on the specific crop, its precise developmental stage, the local soil conditions, and the prevailing farm management practices. The system aims to translate complex scientific data into comprehensible instructions that farmers can immediately apply on their fields.

- **Key Pillars of the Conceptual Framework (Detailed)**

The entire framework operates through the seamless, iterative interaction of four interconnected pillars, each contributing essential components to the final agroadvisory.

13.2.1 High-Quality Weather Information (Forecasting & Monitoring)

This pillar forms the indispensable foundation, ensuring that all subsequent advice is rooted in the most accurate and up-to-date understanding of the atmospheric conditions.

- **Continuous Monitoring:** It begins with the relentless collection of real-time weather parameters. A dense network of ground-based observatories, including Automatic Weather Stations (AWS), provides continuous measurements of temperature (maximum, minimum, mean), rainfall (amount and intensity), relative humidity, wind speed and direction, and solar radiation. These ground observations are complemented by advanced remote sensing technologies, such as weather radar for tracking precipitation and storm systems, and satellites (like those providing Normalized Difference Vegetation Index (NDVI) or soil moisture data) for broader spatial coverage and insights into land surface conditions.

- **Multi-Temporal Scale Prediction:** Beyond current observations, the crucial element is the generation of highly accurate weather forecasts across various time horizons, each serving distinct decision-making needs:
  - **Nowcasting (0-6 hours):** Provides ultra-short-range, highly localized predictions critical for immediate, urgent decisions. For instance, advising farmers to immediately begin harvesting a ripe crop before an imminent thunderstorm or to delay pesticide application if heavy rainfall is expected within the next two hours.
  - **Short-Range Forecasts (1-3 days):** Offers daily operational planning guidance. This informs decisions such as initiating or postponing irrigation for the next day, scheduling pest scouting activities when conditions are favorable for their emergence, or planning minor field operations like weeding or inter-cultivation.
  - **Medium-Range Forecasts (3-10 days):** Supports weekly tactical planning. Farmers can use this to time the top-dressing of fertilizers with expected light rainfall to enhance nutrient uptake, plan major weeding operations for a dry spell, or initiate preventive disease management if prolonged humid conditions are forecast.
  - **Extended Range Forecasts (2-4 weeks):** Provides a more general outlook for strategic tactical adjustments. This could include recommending short-duration crop varieties if a prolonged dry spell is predicted after initial sowing, issuing early warnings for potential widespread pest or disease outbreaks, or advising on adjustments to irrigation water budgeting.
  - **Long-Range/Seasonal Forecasts (Months):** Offers broad strategic guidance for major decisions. This informs choices such as *selecting* appropriate crop cultivars (e.g., drought-tolerant vs. high-yielding), adjusting overall cropping patterns for the season (e.g., reducing area under water-intensive crops during anticipated weak monsoons), and planning for water resource allocation at a regional level.

- **Data Quality and Reliability:** Throughout this process, stringent quality control measures are paramount. Inaccurate or untimely weather data can lead to erroneous advisories, eroding farmer trust and potentially causing significant losses.

13.2.2 Agronomic Expertise and Crop-Weather Relationships

This pillar is where raw meteorological data gains agricultural meaning. It involves the critical translation of weather conditions into their likely impact on specific crops within local contexts.

- **Interpreting Weather for Crop Impact:** This is not merely reporting temperature numbers but explaining what those numbers *mean* for the crop. For example, a forecast of 38°C during the flowering stage of maize isn't just a number; agronomic expertise translates it to a high risk of pollen sterility and reduced grain set due to heat stress.
- **Crop Phenology and Stage-Specific Vulnerability:** A deep understanding of the crop's developmental stages is crucial. Each stage has unique requirements and vulnerabilities:
  - **Germination & Emergence:** Sensitive to soil moisture and temperature; too cold or too dry, and germination fails.
  - **Vegetative Growth (Tillering/Branching):** High demand for nutrients and water for biomass accumulation. Sensitivity to waterlogging (e.g., paddy at transplanting) or prolonged drought.
  - **Reproductive Stage (Flowering/Anthesis/Pollination):** Often the most critical and sensitive phase. Heat stress can cause pollen infertility (e.g., in wheat, maize, rice); water stress can lead to flower abortion; heavy rain can wash away pollen or hinder insect pollinators.
  - **Grain/Fruit Filling:** Requires sustained photosynthetic activity and assimilate translocation. Heat stress can shorten the filling period, leading to shriveled, lighter grains. Excessive rain can lead to spoilage or disease.
  - **Maturity/Harvest:** Requires dry conditions to prevent lodging, mold, or reduced grain quality.
  - Crop Weather Calendars, as discussed, systemize these stage-specific needs and vulnerabilities, often using Growing Degree Days (GDD) to predict the *actual* progression of these stages based on real-time temperatures.

- **Local Soil Conditions as Modifiers:** Soil characteristics profoundly influence how a crop experiences weather. Agroadvisories consider:
  - **Soil Texture and Water Holding Capacity:** A given amount of rainfall will be held differently by sandy soils (low capacity, quick drainage) versus clay soils (high capacity, slow drainage), impacting water availability for the crop.
  - **Drainage:** Heavy soils in low-lying areas pose a higher risk of waterlogging during heavy rainfall.
  - **Nutrient Status:** Soil nutrient availability can interact with water and temperature to affect plant health and stress tolerance.
- **Integration with Local Management Practices:** The advice must be practical and fit into existing local farming systems. This includes considering typical planting methods, irrigation infrastructure, and common fertilization regimes.
- **Pest and Disease Ecology:** Agro-advisories often predict pest and disease outbreaks based on weather conditions. For example, a period of continuous high humidity and moderate temperatures might indicate a high risk of fungal diseases like blast in rice or late blight in potato. Understanding these specific links allows for early warning and preventive action.

13.2.3 Advisory Generation and Customization

This pillar is the nexus where scientific understanding is transformed into actionable guidance.

- **Multi-disciplinary Synthesis:** Teams of agrometeorologists, agronomists, soil scientists, plant pathologists, and entomologists collaborate to craft the advisories. They synthesize the weather forecasts, crop models, soil data, and pest/disease information.
- **High Specificity:** Advisories are typically generated at a highly localized level – district, block, or even village level – ensuring relevance. They are also tailored to specific crops and, ideally, to predominant cultivars within those regions, as their responses to weather can vary significantly.
- **Actionable and Timely Recommendations:** The advice is designed to be clear, concise, and prescriptive. It tells farmers *what to do*, *when to do it*, and *why* it is important. Examples include:
  - **For Sowing/Planting:**Soil moisture is currently optimal. Sow short-duration maize cultivar ‚X‘ within the next 3 days, as a dry spell is forecast towards the end of the week, which could otherwise delay germination.

- **For Irrigation:**Due to anticipated above-normal temperatures and no significant rainfall forecast for the next 5 days, apply 25-30mm of irrigation to your wheat crop, particularly at the tillering stage, to prevent moisture stress. or Postpone irrigation for the next 48 hours as heavy rainfall (30-50mm) is expected.
- **For Nutrient Management:**Top-dress the second dose of nitrogen fertilizer tomorrow, as light rainfall is predicted for the evening, which will facilitate nutrient dissolution and uptake and minimize volatilization losses.
- **For Weed Management:**Undertake weeding operations within the next 2-3 days while the soil is still moist from recent rainfall, as drier conditions are expected to make manual weeding more difficult.
- **For Pest & Disease Control:**Monitor paddy fields for early symptoms of leaf blast disease this week, as current high humidity (above 85%) and moderate temperatures (25-28°C) are highly favorable for its development. If symptoms appear, apply fungicide ‚Y' as a preventive measure.
- **For Harvesting:**Harvest your groundnut crop in the next 2-3 days under clear skies, before the forecast of unseasonal heavy rainfall, to prevent pod rotting and maintain quality.
- **For Contingency Planning:**In case of prolonged waterlogging due to heavy rains, ensure drainage channels are cleared immediately to prevent root damage and yellowing in non-paddy crops.

- **Dynamic Updates:** Advisories are not static documents; they are regularly updated (e.g., twice a week or as needed) to reflect changes in real-time weather, evolving forecasts, and the crop's actual progression.

13.3 Conceptual Framework of Agroadvisory Based on Weather Forecast

The conceptual framework of agroadvisory services based on weather forecasts is a sophisticated, multi-layered system built on the fundamental principle that timely, relevant, and actionable weather information can significantly transform agricultural outcomes. It represents a paradigm shift from traditional, often reactive, farming practices to a proactive, science-driven, and adaptive approach. This framework systematically links cutting-edge meteorological science with profound agronomic expertise, all channeled through effective communication mechanisms to empower farmers with the insights necessary to optimize their daily operations.

At its core, this framework functions as a dynamic decision-support system. It is designed to furnish farmers with precise guidance, recognizing that weather is the single most influential abiotic factor in agriculture. Its impact, however, is not monolithic; it varies intricately depending on the specific crop, its precise developmental stage, the local soil conditions, and the prevailing farm management practices. The system aims to translate complex scientific data into comprehensible instructions that farmers can immediately apply on their fields.

- **Key Pillars of the Conceptual Framework**

The entire framework operates through the seamless, iterative interaction of four interconnected pillars, each contributing essential components to the final agroadvisory.

13.3.1 High-Quality Weather Information (Forecasting & Monitoring)

This pillar forms the indispensable foundation, ensuring that all subsequent advice is rooted in the most accurate and up-to-date understanding of the atmospheric conditions.

- **Continuous Monitoring:** It begins with the relentless collection of real-time weather parameters. A dense network of ground-based observatories, including Automatic Weather Stations (AWS), provides continuous measurements of temperature (maximum, minimum, mean), rainfall (amount and intensity), relative humidity, wind speed and direction, and solar radiation. These ground observations are complemented by advanced remote sensing technologies, such as weather radar for tracking precipitation and storm systems, and satellites (like those providing Normalized Difference Vegetation Index (NDVI) or soil moisture data) for broader spatial coverage and insights into land surface conditions.
- **Multi-Temporal Scale Prediction:** Beyond current observations, the crucial element is the generation of highly accurate weather forecasts across various time horizons, each serving distinct decision-making needs:
    - **Nowcasting (0-6 hours):** Provides ultra-short-range, highly localized predictions critical for immediate, urgent decisions. For instance, advising farmers to *immediately begin harvesting a ripe crop before an imminent thunderstorm* or to *delay pesticide application if heavy rainfall is expected within the next two hours*.

- **Short-Range Forecasts (1-3 days):** Offers daily operational planning guidance. This informs decisions such as initiating or postponing irrigation for the next day, scheduling pest scouting activities when conditions are favorable for their emergence, or planning minor field operations like weeding or inter-cultivation.
- **Medium-Range Forecasts (3-10 days):** Supports weekly tactical planning. Farmers can use this to time the top-dressing of fertilizers with expected light rainfall to enhance nutrient uptake, plan major weeding operations for a dry spell, or initiate preventive disease management if prolonged humid conditions are forecast.
- **Extended Range Forecasts (2-4 weeks):** Provides a more general outlook for strategic tactical adjustments. This could include recommending short-duration crop varieties if a prolonged dry spell is predicted after initial sowing, issuing early warnings for potential widespread pest or disease outbreaks, or advising on adjustments to irrigation water budgeting.
- **Long-Range/Seasonal Forecasts (Months):** Offers broad strategic guidance for major decisions. This informs choices such as selecting appropriate crop cultivars (e.g., drought-tolerant vs. high-yielding), adjusting overall cropping patterns for the season (e.g., reducing area under water-intensive crops during anticipated weak monsoons), and planning for water resource allocation at a regional level.

- **Data Quality and Reliability:** Throughout this process, stringent quality control measures are paramount. Inaccurate or untimely weather data can lead to erroneous advisories, eroding farmer trust and potentially causing significant losses.

13.3.2 Agronomic Expertise and Crop-Weather Relationships

This pillar is where raw meteorological data gains agricultural meaning. It involves the critical translation of weather conditions into their likely impact on specific crops within local contexts.

- **Interpreting Weather for Crop Impact:** This is not merely reporting temperature numbers but explaining what those numbers *mean* for the crop. For example, a forecast of 38°C during the flowering stage of maize isn‘t just a number; agronomic expertise translates it to a high risk of pollen sterility and reduced grain set due to heat stress.
- **Crop Phenology and Stage-Specific Vulnerability:** A deep understanding of the crop‘s developmental stages is crucial. Each stage has unique requirements and vulnerabilities:

- **Germination & Emergence:** Sensitive to soil moisture and temperature; too cold or too dry, and germination fails.
- **Vegetative Growth (Tillering/Branching):** High demand for nutrients and water for biomass accumulation. Sensitivity to waterlogging (e.g., paddy at transplanting) or prolonged drought.
- **Reproductive Stage (Flowering/Anthesis/Pollination):** Often the most critical and sensitive phase. Heat stress can cause pollen infertility (e.g., in wheat, maize, rice); water stress can lead to flower abortion; heavy rain can wash away pollen or hinder insect pollinators.
- **Grain/Fruit Filling:** Requires sustained photosynthetic activity and assimilate translocation. Heat stress can shorten the filling period, leading to shriveled, lighter grains. Excessive rain can lead to spoilage or disease.
- **Maturity/Harvest:** Requires dry conditions to prevent lodging, mold, or reduced grain quality.
- Crop Weather Calendars, as discussed, systemize these stage-specific needs and vulnerabilities, often using Growing Degree Days (GDD) to predict the *actual* progression of these stages based on real-time temperatures.

- **Local Soil Conditions as Modifiers:** Soil characteristics profoundly influence how a crop experiences weather. Agroadvisories consider:
  - **Soil Texture and Water Holding Capacity:** A given amount of rainfall will be held differently by sandy soils (low capacity, quick drainage) versus clay soils (high capacity, slow drainage), impacting water availability for the crop.
  - **Drainage:** Heavy soils in low-lying areas pose a higher risk of waterlogging during heavy rainfall.
  - **Nutrient Status:** Soil nutrient availability can interact with water and temperature to affect plant health and stress tolerance.
- **Integration with Local Management Practices:** The advice must be practical and fit into existing local farming systems. This includes considering typical planting methods, irrigation infrastructure, and common fertilization regimes.
- **Pest and Disease Ecology:** Agro-advisories often predict pest and disease outbreaks based on weather conditions. For example, a period of continuous high humidity and moderate temperatures might indicate a high risk of fungal diseases like blast in rice or late blight in potato. Understanding these specific links allows for early warning and preventive action.

13.3.3 Advisory Generation and Customization

This pillar is the nexus where scientific understanding is transformed into actionable guidance.

- **Multi-disciplinary Synthesis:** Teams of agrometeorologists, agronomists, soil scientists, plant pathologists, and entomologists collaborate to craft the advisories. They synthesize the weather forecasts, crop models, soil data, and pest/disease information.
- **High Specificity:** Advisories are typically generated at a highly localized level – district, block, or even village level – ensuring relevance. They are also tailored to specific crops and, ideally, to predominant cultivars within those regions, as their responses to weather can vary significantly.
- **Actionable and Timely Recommendations:** The advice is designed to be clear, concise, and prescriptive. It tells farmers *what to do*, *when to do it*, and *why* it is important. Examples include:
    - **For Sowing/Planting:**Soil moisture is currently optimal. Sow short-duration maize cultivar ‚X' within the next 3 days, as a dry spell is forecast towards the end of the week, which could otherwise delay germination.
    - **For Irrigation:**Due to anticipated above-normal temperatures and no significant rainfall forecast for the next 5 days, apply 25-30 mm of irrigation to your wheat crop, particularly at the tillering stage, to prevent moisture stress. or Postpone irrigation for the next 48 hours as heavy rainfall (30-50 mm) is expected.
    - **For Nutrient Management:**Top-dress the second dose of nitrogen fertilizer tomorrow, as light rainfall is predicted for the evening, which will facilitate nutrient dissolution and uptake and minimize volatilization losses.
    - **For Weed Management:**Undertake weeding operations within the next 2-3 days while the soil is still moist from recent rainfall, as drier conditions are expected to make manual weeding more difficult.
    - **For Pest & Disease Control:**Monitor paddy fields for early symptoms of leaf blast disease this week, as current high humidity (above 85%) and moderate temperatures (25-28°C) are highly favorable for its development. If symptoms appear, apply fungicide 'Y' as a preventive measure.
    - **For Harvesting:**Harvest your groundnut crop in the next 2-3 days under clear skies, before the forecast of unseasonal heavy rainfall, to prevent pod rotting and maintain quality.

- **For Contingency Planning:**In case of prolonged waterlogging due to heavy rains, ensure drainage channels are cleared immediately to prevent root damage and yellowing in non-paddy crops.

- **Dynamic Updates:** Advisories are not static documents; they are regularly updated (e.g., twice a week or as needed) to reflect changes in real-time weather, evolving forecasts, and the crop‘s actual progression.

13.4 Data Flow and Processing Pipeline for Agroadvisories

The journey of an agroadvisory, from raw atmospheric measurements to actionable advice for a farmer, involves a sophisticated data flow and processing pipeline. This pipeline is a multi-stage, integrated system that collects, processes, analyzes, and synthesizes vast amounts of diverse data. Its efficiency and accuracy are paramount, as any bottleneck or error at an earlier stage can compromise the utility of the final advisory. This process ensures that farmers receive timely, relevant, and reliable guidance.

The pipeline typically consists of several interconnected stages:

13.4.1 Data Acquisition (The Foundation)

This initial stage focuses on gathering comprehensive meteorological and related environmental data from various sources. It's about capturing the most accurate picture of both current and future atmospheric conditions.

- **Ground-Based Observations:** This includes data from a widespread network of:
  - **Manual Weather Stations:** Providing traditional readings of temperature, rainfall, humidity, wind, and sunshine hours.
  - **Automatic Weather Stations (AWS):** Offering continuous, real-time, high-resolution data for the same parameters, transmitting them automatically to central databases. A denser AWS network provides more localized insights.
  - **Agrometeorological Observatories:** Specialized stations that also monitor soil temperature, soil moisture, and sometimes phenological observations directly from fields.
- **Remote Sensing Data:** This involves capturing large-scale spatial data from above:
  - **Satellites:** Provide broad coverage for parameters like cloud cover, rainfall estimation (e.g., from geostationary and polar-orbiting satellites), land surface temperature, and crucial vegetation indices

  (like NDVI - Normalized Difference Vegetation Index), which indicate crop health and growth stages over vast areas.

  - **Weather Radars:** Detect precipitation, its intensity, and movement, crucial for nowcasting and short-range rainfall warnings.
  - **Lightning Detection Networks:** Identify lightning strikes, indicating convective storm activity.

- **Numerical Weather Prediction (NWP) Model Outputs:** These are the backbone of modern forecasting. Global and regional atmospheric models (e.g., Global Forecast System - GFS, Weather Research and Forecasting - WRF model) run on supercomputers, generating forecasts for various meteorological parameters across different time ranges (short, medium, extended, long). These outputs are crucial raw forecast data that need further interpretation.

13.4.2 Data Pre-processing and Quality Control

Raw data, especially from diverse sources, can be noisy, incomplete, or contain errors. This stage is critical for ensuring data integrity before analysis.

- **Collection and Aggregation:** Data from all sources are collected in central servers or cloud platforms.
- **Quality Checks:** Automated and manual checks are performed to identify and correct errors, outliers, and missing values. This might involve:
  - **Range Checks:** Ensuring values fall within plausible meteorological limits.
  - **Consistency Checks:** Comparing data from adjacent stations or different instruments.
  - **Gap Filling:** Estimating missing data using statistical methods or interpolation from surrounding stations.
- **Standardization and Formatting:** Data are converted into a uniform format and units to ensure compatibility for subsequent processing.

13.4.3 Meteorological Forecasting and Analysis

This is where the future weather is predicted and characterized, moving beyond simple observations.

- **NWP Model Integration:** The pre-processed historical and real-time data are fed into sophisticated NWP models. These models use complex physics and atmospheric equations to simulate future atmospheric states.

- **Multi-Range Forecast Generation:** Outputs from NWP models are used to generate forecasts across different time horizons:
  - **Nowcasting:** Often uses radar data and high-resolution models for immediate, very localized predictions.
  - **Short-Range:** Relies on detailed NWP runs for daily forecasts.
  - **Medium-Range:** Uses ensemble forecasting (running models multiple times with slight variations) to provide probabilities and ranges for weather parameters over several days.
  - **Extended and Long-Range:** Utilizes seasonal climate models and statistical techniques to provide broader outlooks on temperature and rainfall anomalies over weeks to months.
- **Forecaster Intervention:** Human meteorologists play a crucial role in interpreting model outputs, especially for complex weather situations or local nuances that models might miss. They apply their expertise to refine forecasts.

13.4.4 Agronomic Integration and Impact Assessment:

This is the bridge where meteorological insights are translated into agricultural implications. This stage draws heavily on the concepts of Crop Weather Calendars and crop modeling.

- **Crop-Specific Overlay:** The weather forecasts (both current and future) are overlaid with detailed **crop phenological information**. This involves:
  - **GDD Calculation:** Using forecasted temperatures to project the progression of crop developmental stages (e.g., when flowering is expected).
  - **Water Balance Models:** Calculating soil moisture status and crop water deficit based on rainfall forecasts, actual evapotranspiration, and soil characteristics.
  - **Identification of Critical Stages:** Recognizing when the crop is in its most sensitive periods (e.g., flowering, grain filling) where specific weather conditions can have maximal impact.
- **Risk Assessment and Vulnerability Analysis**
  - **Weather Threshold Application:** Comparing forecasted weather parameters against established critical thresholds for the specific crop at its current or upcoming stage (e.g., is the temperature forecast above the heat stress threshold for pollen viability?).

  - **Pest and Disease Forecasting Models:** Meteorological data (temperature, humidity, rainfall duration) are fed into specific pest and disease forewarning models to predict the likelihood of outbreaks (e.g., conditions favorable for blast disease in paddy).
  - **Extreme Event Impact:** Assessing the likely impact of forecasted extreme events (e.g., hail, frost, strong winds) on physical crop damage or yield components.
- **Agronomic Data Integration:** Information on local soil types, typical management practices (sowing dates, irrigation methods), and common cultivars is integrated to ensure context-specific interpretations. This forms the basis of Crop Weather Calendars themselves, which systematize expected interactions.

13.4.5 Advisory Generation (The Actionable Output)

This is the stage where the synthesized information is transformed into clear, actionable advice for farmers.

- **Expert Interpretation:** A team of interdisciplinary experts (agrometeorologists, agronomists, plant pathologists, entomologists, soil scientists) convenes to review the integrated data.
- **Formulation of Recommendations:** Based on the assessed impacts and risks, precise recommendations are formulated. These are:
  - **Specific:** Detail *what* action to take (e.g., Apply irrigation,Scout for pests,Harvest).
  - **Timely:** Indicate *when* to take the action (e.g., Within the next 24 hours,Next week,Before the forecast rain).
  - **Location/Crop-Specific:** Tailored to the particular district, block, crop, and even cultivar.
  - **Justified:** Often include a brief explanation of *why* the action is recommended, linking it to the weather forecast and crop status (e.g., to avoid moisture stress, due to high risk of fungal infection).
- **Contingency Planning:** Advice for coping with anticipated adverse weather conditions or unexpected events.
- **Formatting:** The advisory is structured in a user-friendly format, often in local languages, with clear headings and concise bullet points or short paragraphs.

13.5 Impact Assessment and Feedback Mechanism for Agroadvisories

The provision of agroadvisory services based on weather forecasts is not a one-way street. To ensure their continued relevance, effectiveness, and the optimal allocation of resources, it is crucial to implement robust impact assessment and feedback mechanisms. These processes create a continuous learning loop, allowing service providers to understand how advisories are being used, what benefits they deliver, and where improvements are needed.

13.5.1 Importance of Impact Assessment and Feedback

- **Verifying Effectiveness:** Quantifies whether the advisories are actually leading to positive changes at the farm level (e.g., increased yields, reduced losses, better resource efficiency).
- **Demonstrating Value:** Provides evidence of the socio-economic benefits of agrometeorological services, justifying continued investment and expansion.
- **Identifying Gaps and Weaknesses:** Pinpoints areas where the advisories might be unclear, irrelevant, or not reaching the intended audience effectively.
- **Improving Accuracy and Content:** Allows for the refinement of forecasting models, the content of advisories, and the timing of recommendations.
- **Enhancing User Trust and Adoption:** When farmers see that their input is valued and leads to better services, their trust in and adoption of advisories increase.
- **Policy and Planning:** Provides crucial data for agricultural policymakers and planners to formulate strategies for climate resilience and food security.
- **Adaptation to Climate Change:** Helps identify if current advisories are still relevant under changing climatic conditions and if new types of advice are needed.

13.5.2 Impact Assessment Methods and Key Metrics

Impact assessment typically involves systematic studies to quantify the changes attributable to the agroadvisory services. This often involves comparing outcomes for farmers who received and adopted the advisories versus those who did not.

- **Methods**
  - **Farmer Surveys and Questionnaires:** Direct interaction with farmers (beneficiaries and non-beneficiaries) to gather perceptions on utility, adoption rates, clarity, timeliness, and perceived benefits. These can be conducted at regular intervals (e.g., annually, post-harvest).
  - **Field-Level Data Collection:** Collecting quantitative data directly from farms (e.g., actual yield, input use – fertilizer, water, pesticides, labor) from both advisory users and non-users.
  - **Economic Analysis (Cost-Benefit Analysis):** Quantifying the financial benefits (e.g., increased net income, reduced cost of cultivation) derived from following the advisories compared to the costs of providing the service. Studies often compare the economic returns of farmers who adopted advisories versus those who did not.
  - **Yield Gap Analysis:** Comparing the actual yields achieved by farmers using advisories against the potential yield for a given region and crop, aiming to reduce the gap.
  - **Randomized Controlled Trials (RCTs):** In some research settings, a robust method involves providing advisories to a randomly selected group of farmers while a control group does not, allowing for a clearer attribution of impact.
  - **Case Studies:** In-depth qualitative studies of specific farming communities or individual farmers to understand the nuanced ways advisories influence decisions and outcomes.
- **Key Metrics for Assessment**
  - **Adoption Rate:** The percentage of farmers receiving the advisories who actually use and act upon the recommendations.
  - **Yield Increase:** Quantifiable increase in crop yield (e.g., kg/hectare) attributed to advisory adoption.
  - **Loss Reduction:** Quantifiable reduction in crop losses due to adverse weather events or pest/disease outbreaks.
  - **Input Use Efficiency:** Reduction in water use (e.g., liters/kg of produce), fertilizer use, or pesticide application without compromising yield.
  - **Net Returns/Income:** Increase in farmers' net income and improvement in benefit-cost ratio.
  - **Timeliness of Operations:** Whether farmers are conducting operations (sowing, irrigation, harvest) at more optimal times.

- **Awareness and Knowledge Levels:** Increase in farmers' understanding of weather impacts and adaptive strategies.
- **Farmers' Satisfaction:** Perceived usefulness and satisfaction with the advisory service.

13.5.3 Feedback Mechanisms: Structured Channels for Farmer Input

Feedback mechanisms are the formal and informal pathways through which the service providers gather insights from the end-users (farmers). This is a continuous process, not just a post-season assessment.

- **Channels for Collecting Feedback**
  - **Direct Surveys and Questionnaires:** As part of impact assessment, but also more frequently to gather real-time feedback on specific advisories. These can be administered by extension workers or through digital means.
  - **Mobile-Based Feedback**
    - **SMS/IVR Surveys:** Simple questions sent via SMS or automated calls to elicit quick responses on advisory utility.
    - **Mobile App Features:** Dedicated feedback sections, rating systems, or direct messaging within agroadvisory apps (e.g., IMD's Meghdoot often incorporates user feedback features).
  - **Community Meetings and Farmer Field Schools (FFS):** Regular face-to-face interactions allow for qualitative feedback, discussions, and a deeper understanding of farmers' challenges and needs. Extension agents can collect anecdotal evidence and specific queries (Fig.13.2).

Fig. 13.2: Farmer field school

- **Local Agricultural Offices/Krishi Vigyan Kendras (KVKs):** Farmers can visit these centers to provide feedback directly to subject matter specialists.
- **Dedicated Helplines/Call Centers:** Farmers can call in with questions, report issues, or provide feedback.
- **Social Media/WhatsApp Groups:** Increasingly used platforms where farmers can share observations, ask questions, and provide informal feedback on advisories.
- **Pilot/Demonstration Farms:** Closely monitored farms that implement advisories, providing detailed feedback on their practical applicability and outcomes.

13.5.4 Utilization of Feedback -The Learning Loop

The collected feedback is invaluable only if it is systematically analyzed and used to improve the agroadvisory system.

- **Refinement of Weather Forecasts:** Feedback on forecast accuracy (e.g., rain was predicted but didn‘t come, or temperature was much higher than forecast) helps meteorologists fine-tune their models and local adjustments.
- **Improvement of Advisory Content**
    - **Clarity and Simplicity:** Feedback can reveal if advisories are too technical, too long, or use unfamiliar terminology, leading to simplification and better phrasing.
    - **Relevance:** If farmers report that advice is not relevant to their specific crop variety, local soil, or current farming practices, the advisory content can be made more granular and customized.
- **Actionability:** Ensuring that the advice is practical and feasible for farmers to implement.
- **Enhancement of Dissemination Channels:** Feedback on preferred channels, timeliness of delivery, and accessibility issues (e.g., network connectivity problems) can guide improvements in dissemination strategies.
- **Training and Capacity Building:** Identifying common misconceptions or areas where farmers lack understanding can inform the development of targeted training programs for both farmers and extension workers.
- **Service Provider Accountability:** Regular feedback mechanisms foster a sense of accountability for service providers, pushing for continuous improvement.

- **Research and Development:** Feedback often highlights new challenges or emerging needs (e.g., new pest outbreaks linked to changing weather, novel crop varieties) that can inform future research priorities in agro-meteorology and agricultural science.

In summary, robust impact assessment and effective feedback mechanisms are not optional extras; they are integral to the lifecycle of agroadvisory services. They close the loop between service providers and end-users, ensuring that the advisories remain relevant, accurate, and truly beneficial for enhancing agricultural productivity and resilience in the face of evolving climatic challenges.

13.6 Use of Crop Simulation Models and DSS in Agroadvisory Formulation

The formulation of sophisticated, precise, and dynamic agroadvisories relies heavily on advanced analytical tools, notably Crop Simulation Models (CSMs) and Decision Support Systems (DSS). These technologies transform raw meteorological data and general agronomic knowledge into highly specific, actionable recommendations tailored to local conditions and crop requirements. They represent a significant leap from traditional, experience-based advisories to science-driven, data-intensive approaches.

13.6.1 Crop Simulation Models (CSMs)

- **What are CSMs?** Crop Simulation Models are mathematical representations that mimic the growth, development, and yield of a crop as influenced by environmental factors (weather, soil), genetic characteristics of the cultivar, and management practices (sowing date, fertilizer, irrigation). They use fundamental physiological and physical processes (e.g., photosynthesis, respiration, transpiration, nutrient uptake) to simulate daily or hourly changes in plant biomass, leaf area, root growth, and ultimately, final yield.

- **How CSMs are used in Agroadvisory Formulation**

- **Predicting Phenological Stages (Growth Progression)**

  - CSMs accurately calculate **Growing Degree Days (GDD)** or other thermal time units based on real-time and forecasted temperatures. This allows them to predict the precise timing of critical phenological stages (e.g., emergence, flowering, maturity) for a given crop and cultivar, which may deviate from historical averages due to current season‘s weather.

- **Advisory Application:** If a CSM predicts flowering will occur five days earlier than normal due to warmer temperatures, the agroadvisory can then warn farmers about potential heat stress during this critical window and recommend proactive measures.

- **Yield Forecasting under Different Scenarios**
  - CSMs can simulate potential yield under optimal conditions and compare it to yield under various stress scenarios (e.g., drought, heat stress, nutrient deficiency). By feeding current and forecasted weather into the model, they can provide updated yield estimations throughout the season.
  - **Advisory Application:** If a prolonged dry spell is forecast, the model can estimate potential yield loss, prompting advisories on immediate irrigation if water is available, or recommending drought-mitigation strategies.

- **Optimizing Resource Use (Water and Nutrients)**
  - CSMs integrate soil water balance modules that track soil moisture based on rainfall, irrigation, and evapotranspiration. They can predict when the crop will experience water stress.
  - Similarly, nitrogen balance modules can track nutrient availability and uptake.
  - **Advisory Application:** The model can trigger advisories for irrigation when soil moisture drops below critical thresholds, specifying the exact amount of water needed to bring the soil back to field capacity. For nutrients, it can advise on precise top-dressing timings aligned with periods of high uptake and favorable weather conditions.

- **Forecasting Pest and Disease Risk**
  - Many CSMs are coupled with or integrated into specialized pest and disease models. These models use weather parameters (temperature, humidity, rainfall duration, leaf wetness) to predict the likelihood of specific pest outbreaks or disease epidemics.
  - **Advisory Application:** If a model predicts high risk of a particular fungal disease due to prolonged humidity and moderate temperatures, the agroadvisory can issue early warnings for scouting and recommend preventive or curative sprays, thus enabling targeted and efficient pesticide use.

- **Evaluating Management Options and Contingency Planning**
  - CSMs allow for what-if scenario analysis. Farmers or advisors can test different planting dates, irrigation schedules, fertilizer rates, or cultivar choices under various climate scenarios (e.g., normal, dry, wet years, or climate change projections) without conducting costly field experiments.
  - **Advisory Application:** This helps in formulating contingency plans for adverse weather. For instance, if early monsoon failure occurs, the model can compare outcomes for different short-duration alternative crops, leading to specific recommendations for replanting.

13.6.2 Decision Support Systems (DSS)

- **What are DSS?** Decision Support Systems (DSS) are interactive computer-based systems that help users make decisions by organizing and analyzing data from various sources, processing them through models (including CSMs), and presenting information in an understandable format. In agriculture, a DSS acts as an intelligent intermediary, translating complex scientific outputs into practical recommendations.

- **How DSS Integrate CSMs for Agroadvisory Formulation**
  - **Data Integration Hub:** A DSS serves as the central platform that seamlessly integrates all necessary inputs:
  - **Weather Data:** Real-time observations, historical data, and multi-range forecasts (from meteorological departments).
  - **Crop Model Outputs:** Predictions from CSMs on phenology, yield potential, stress levels, and resource requirements.
  - **Soil Data:** Information on soil type, water holding capacity, and nutrient status.
  - **Geographic Information Systems (GIS):** For spatial analysis and mapping of relevant data (e.g., soil maps, rainfall distribution, pest hotspots).
  - **Farmer/Farm Specific Data:** Input from individual farms or local communities regarding specific cultivars, management practices, and observed conditions.
  - **Rule-Based and Expert Systems:** Many DSS incorporate rule-based systems or expert systems that codify the knowledge of agronomists and other specialists. These rules interpret the combined data to trigger specific advisories.

  - **Example Rule:** IF (crop_stage == flowering AND forecast_max_temp > 35^{0}C AND soil_moisture_status == deficit) THEN (RECOMMEND immediate light irrigation AND advise morning irrigation to reduce heat stress AND warn of potential pollen sterility).
- **Scenario Analysis and Optimization:** DSS can present multiple management scenarios and their probable outcomes (e.g., yield difference if irrigated vs. not irrigated). Some advanced DSS can even suggest optimal management strategies based on user-defined objectives (e.g., maximize profit, minimize water use).
- **User-Friendly Interface:** A crucial aspect of DSS is their ability to present complex information and recommendations in an intuitive, accessible format (e.g., dashboards, tables, maps, mobile app interfaces) that can be easily understood by extension workers and farmers, often in local languages.
- **Feedback Loop Integration:** DSS platforms often include modules for collecting farmer feedback on the advisories received, which then helps in refining the models and rules for future iterations.

13.6.3 Benefits of Using CSMs and DSS in Agroadvisory Formulation

- **Precision and Accuracy:** Advisories become highly precise, accounting for local weather, soil, and crop-specific conditions, leading to more accurate recommendations.
- **Proactive Risk Management:** By simulating future scenarios, these tools enable early warnings for potential threats (e.g., drought, heat stress, disease outbreaks), allowing farmers to implement preventive measures rather than reactive ones.
- **Resource Use Efficiency:** Optimizing irrigation and nutrient application based on real-time crop needs reduces wastage, lowers input costs, and minimizes environmental impact.
- **Enhanced Productivity and Yield Stability:** Informed decision-making at critical stages helps maximize crop potential and reduces yield variability, contributing to food security.
- **Climate Change Adaptation:** These models are invaluable for understanding and adapting to changing climate patterns, allowing for testing of new varieties or management strategies under future climate scenarios.

- **Knowledge Transfer and Capacity Building:** They distill complex scientific knowledge into actionable insights, making advanced agronomy accessible to a wider farming community.
- **Data-Driven Decision Making:** Moves farming from intuition to evidence-based strategies, fostering more resilient and profitable agricultural practices.

13.6.4 Challenges

Despite their immense potential, the effective implementation of CSMs and DSS for agroadvisories faces challenges, including the need for extensive data (weather, soil, crop genetic coefficients), computational resources, skilled personnel for model calibration and interpretation, and robust dissemination infrastructure to reach all farmers. However, ongoing advancements continue to make these powerful tools more accessible and user-friendly.

In essence, Crop Simulation Models provide the scientific engine for understanding crop responses to weather, while Decision Support Systems provide the interface and intelligence to integrate these insights with other data to generate timely, specific, and actionable agroadvisories that are indispensable for modern, climate-smart agriculture.

13.7 Customized Agroadvisories

In the realm of agricultural meteorology, the progression from general weather forecasts to generic agroadvisories, and then to highly specific, dynamic advisories, culminates in the concept of **Customized Agromet Advisories**. This represents the highest level of precision and relevance in agricultural guidance, moving beyond district or block-level recommendations to advice tailored to the unique conditions and specific needs of individual farmers, or even individual fields within a farm.

While a general agroadvisory provides broad guidance for a region (e.g., a district or an agro-climatic zone), a customized agroadvisory recognizes that one size does not fit all in agriculture. Factors like micro-climate variations, specific soil types within a village, the exact planting date of a particular farmer, the chosen crop variety, irrigation access, and local pest pressure all influence how a farmer should respond to a weather forecast.

13.7.1 What Defines Customized"?

Customized agroadvisories are characterized by their extreme specificity, taking into account a granular level of detail about the farm and farmer. The customization typically revolves around:

- **Farm/Field-Level Specificity**
  - **Geographic Precision:** Instead of district-wide averages, advisories consider the precise latitude and longitude of a farmer's field, allowing for the application of micro-climatic variations (e.g., differences in rainfall, temperature, or frost pockets within a small area).
  - **Soil Type and Health:** Taking into account the specific soil texture, water holding capacity, nutrient status, and even pH of *that particular field.* This directly influences irrigation needs and fertilizer recommendations.
- **Crop and Cultivar Specificity**
  - Advisories are tailored not just for rice but for long-duration irrigated Basmati rice variety Pusa 1121 or short-duration rainfed maize hybrid JK-MV. Different varieties have distinct phenologies, water requirements, and stress tolerances.
- **Phenological Stage (Real-time Tracking)**
  - While general advisories might state for crops at vegetative stage, customized advisories track the *actual*phenological stage of the farmer's specific crop, accounting for actual accumulated Growing Degree Days (GDD) since sowing. This ensures advice is aligned with the crop's true biological needs.
- **Farmer's Management Practices**
  - Understanding the farmer's specific practices:
    - **Sowing Date:** Crucial for tracking GDD and phenology.
    - **Irrigation Infrastructure:** Whether the farmer has access to irrigation, and what type (drip, sprinkler, flood), influences water management advice.
    - **Fertilization History:** What nutrients have already been applied.
    - **Pest & Disease History:** Previous issues on the farm or in the immediate vicinity.
- **Farmer-Specific Information/Preferences (in advanced systems)**
  - This can include the farmer's risk tolerance, economic constraints, access to labor, or even preferred communication channels. While highly advanced, some systems try to incorporate this.

13.7.2 How are Customized Agroadvisories Formulated?

Formulating customized advisories requires sophisticated data collection, processing, and analytical capabilities, often leveraging the tools discussed previously (CSMs and DSS) at a higher resolution.

- **Detailed Farm Data Collection**
  - **Farmer Registration:** Farmers register their specific details, including farm location (GPS coordinates), crop type, cultivar, sowing date, and irrigation access.
  - **Field Surveys/Sensors:** For truly granular customization, soil sensors (for real-time soil moisture, temperature), weather sensors on the farm, or even drone imagery can provide hyper-local data.
- **Hyper-Local Weather Data & Forecasts**
  - Downscaled weather models, interpolation from nearby AWS, or even farm-level weather stations provide highly localized current and forecasted weather.
- **Advanced Crop Simulation Models (CSMs) and Decision Support Systems (DSS)**
  - CSMs are run using the farmer‘s specific crop, cultivar, sowing date, and hyper-local weather data to simulate the exact phenological progression and potential stresses.
  - DSS then integrates these CSM outputs with the farmer‘s specific soil data and management practices to generate tailored recommendations using complex algorithms and rule-based systems.
  - **Example:** Instead of Irrigate if no rain for 7 days, a customized advisory might state: Your maize crop (variety ‚X‘, sown on May 10th) at field ‚A‘ (sandy loam soil) has a current soil moisture deficit of 15mm. With no rain expected for the next 3 days and high temperatures, apply 20mm of irrigation within 24 hours to avoid severe stress during the tasseling stage.
- **Individualized Dissemination:** Advisories are sent directly to the farmer, typically via mobile apps, personalized SMS, or sometimes through direct calls from local extension agents.

13.7.3 Benefits of Customized Agroadvisories

- **Maximum Relevance and Impact:** Advice is perfectly aligned with the farmer's unique situation, leading to higher adoption rates and more significant positive outcomes (yield increase, loss reduction).
- **Optimal Resource Efficiency:** Precisely targeted recommendations for water, fertilizer, and pesticides minimize waste and maximize their effectiveness, leading to cost savings and environmental benefits.
- **Enhanced Resilience:** Farmers are better equipped to respond to micro-level climatic variations and specific localized threats, improving their ability to withstand shocks.
- **Increased Farmer Confidence and Empowerment:** Farmers receive actionable insights that directly address their specific concerns, fostering greater confidence in their decision-making.
- **Precision Agriculture Enablement:** Customized advisories are a core component of precision agriculture, allowing for highly targeted management interventions.

13.7.4 Challenges in Delivering Customized Agroadvisories:

- **Data Intensity:** Requires vast amounts of highly granular and accurate data (weather, soil, crop, farm management), which can be challenging and expensive to collect and maintain at scale.
- **Technological Infrastructure:** Demands advanced computing power, robust data storage, sophisticated models, and reliable high-speed communication networks.
- **Interoperability:** Integrating data from disparate sources (weather, soil labs, farmer records, remote sensing) into a single system is complex.
- **Cost of Implementation:** Setting up and maintaining such a highly customized system can be very expensive.
- **Scalability:** Scaling from pilot projects to cover millions of diverse farmers across a large region presents significant logistical and technical hurdles.
- **Farmer Digital Literacy and Connectivity:** Even with mobile apps, not all farmers have smartphones or reliable internet access, and digital literacy can be a barrier.
- **Data Privacy Concerns:** Collecting and managing individual farm data raises privacy issues that need to be addressed.

Despite these challenges, the trajectory of agroadvisory services is clearly towards greater customization. As technology becomes more accessible and data collection more efficient, customized agroadvisories hold the immense promise of transforming agriculture by providing every farmer with the precise, timely intelligence needed to thrive in a complex and changing environment.

13.8 Agroadvisories for Different Clients: Beyond the Farmer

While farmers are the primary beneficiaries of agroadvisory services, the influence of weather and climate extends far beyond the farm gate, impacting the entire agricultural value chain. Therefore, sophisticated agroadvisory systems increasingly provide tailored information to a diverse range of clients, enabling strategic decision-making throughout the food production and processing ecosystem. This customization allows each stakeholder to mitigate risks, optimize operations, and enhance profitability within their specific business domain.

Here's how agroadvisories can be customized for different client types:

13.8.1 Input Suppliers (Seeds, Fertilizers, Pesticides, Machinery)

- **Role:** These businesses supply the essential materials farmers need. Their decisions on production, inventory, logistics, and sales are highly dependent on planting seasons, crop health, and anticipated pest/disease outbreaks.
- **Agroadvisory Needs**
- **Seasonal Outlooks (Long-Range):** Forecasts for the upcoming cropping season (e.g., monsoon strength, temperature anomalies).
    - **Impact:** Informs seed companies about the likely demand for drought-tolerant vs. high-yielding varieties; guides fertilizer companies on anticipated overall demand for different nutrients (e.g., higher N demand if good monsoon is expected).
- **Medium-Range Forecasts (2-4 weeks):** Expected rainfall patterns, temperature trends.
    - **Impact:** Helps plan inventory and distribution of specific pesticides (e.g., fungicides if humid conditions are forecast, insecticides if pest outbreaks are likely). Guides machinery suppliers on demand for tillage equipment (dry spells) or harvesting machinery (clear windows).
- **Pest and Disease Risk Forecasts:** Specific warnings for conducive weather conditions for particular pests/diseases.

    - **Impact:** Allows pesticide manufacturers and distributors to anticipate demand for specific products, allocate stock to affected regions, and prepare advisory messages for their farmer clients.
- **Sowing Window Forecasts:** Expected optimal periods for planting.
    - **Impact:** Enables seed companies to time their sales campaigns and distribution logistics effectively, ensuring seeds reach farmers when they are most needed.
- **Example Advisory:** Seasonal forecast indicates a likely delayed monsoon onset by 2 weeks with below-normal initial rainfall in Region X. Advise increased stocking of short-duration, drought-tolerant seed varieties and prepare for potential later-season demand for supplementary irrigation equipment. Fungicide demand may be lower initially but monitor for specific disease outbreaks if early humidity is high.

13.8.2 Agro-Based Industries (Sugar Mills, Cotton Ginning, Oil Extraction, Dairy, Feed Companies)

- **Role:** These industries rely on a consistent and quality supply of raw agricultural produce. Their operational efficiency, processing schedules, and product quality are directly affected by crop yields, harvest timings, and the quality of the raw material.
- **Agromet Advisory Needs**
- **Seasonal Yield Forecasts:** Based on crop models driven by historical and forecasted weather.
    - **Impact:** Helps estimate raw material availability for the season, influencing procurement strategies, processing plant capacity utilization, and marketing plans.
- **Harvest Window Forecasts:** Precise predictions for dry spells around harvest.
    - **Impact:** Sugar mills can schedule cane crushing based on expected harvest timings and sugar content (influenced by dry weather); cotton ginning units can plan for arrival of cotton bales; oil extraction plants can anticipate oilseed supply.
- **Quality Indicators:** Weather conditions affecting quality (e.g., high humidity leading to moisture content issues, excessive rain impacting sugar content).
    - **Impact:** Allows processing units to adjust their operations (e.g., increase drying capacity for high-moisture produce) or manage quality control.

- **Supply Chain Disruptions:** Warnings for extreme weather events (floods, heavy snowfall, cyclones) that could disrupt transportation of raw materials.
    - **Impact:** Enables logistical teams to re-route deliveries, secure alternative transport, or adjust delivery schedules.
- **For Dairy/Feed Industries:** Heat stress forecasts for livestock, pasture growth forecasts.
    - **Impact:** Dairies can anticipate reduced milk production during heatwaves and adjust feed formulations or cooling strategies. Feed companies can gauge demand for supplementary feed if pasture growth is hampered by drought.
- **Example Advisory:**Forecast of extended dry period for the next 10 days in key sugarcane growing areas suggests optimal conditions for cane maturity and sugar recovery. Sugar mills should anticipate rapid harvesting and prepare crushing capacities for increased influx of high-quality cane. Monitor for potential delays in transport if any unseasonal showers occur towards end of period.

13.8.3 Food Processing Industries (Beverages, Packaged Foods, Snacks)

- **Role:** These industries process agricultural raw materials into consumer products. They are concerned with consistent supply, quality, and often the shelf-life of their inputs, which weather can significantly impact.
- **Agromet Advisory Needs**
- **Long-Range Production Forecasts:** Overall projected output for specific crops used as ingredients (e.g., potato for chips, specific fruit varieties for juices).
    - **Impact:** Informs long-term procurement contracts, investment decisions in new processing lines, and market pricing strategies.
- **Quality-Related Weather Events:** Forecasts for weather conditions that impact raw material quality (e.g., moisture, fungal growth, bruising risk).
    - **Impact:** Allows for adjustments in pre-processing, storage, or quality control measures. For example, if high humidity is expected, they might require stricter drying protocols for grains.
- **Supply Chain Logistics:** Warnings for weather-related disruptions in transport and storage.

- **Impact:** Ensures that perishable raw materials are transported efficiently and stored under optimal conditions to prevent spoilage.

- **Pest and Disease Threat to Quality:** Forecasts related to issues like aflatoxin contamination in grains due to humid conditions.
  - **Impact:** Enables targeted testing of raw materials upon arrival and adjustment of processing methods to ensure food safety standards.
- **Example Advisory:**Due to sustained heavy rainfall forecast in onion-growing regions, anticipate higher moisture content in incoming batches. Adjust drying processes at the plant and prioritize rapid processing or specialized storage to prevent spoilage.

13.8.4 Custom Hire Services (Farm Machinery Rental)

- **Role:** These services provide access to expensive farm machinery (tractors, harvesters, planters, sprayers) on a rental basis, playing a vital role in mechanization, especially for smallholder farmers. Their profitability depends on optimal machine deployment and scheduling.
- **Agromet Advisory Needs**
- **Short- to Medium-Range Workability Forecasts:** Predictions on soil moisture (too wet/dry for tillage), rainfall windows (for spraying/harvesting), wind speeds (for spraying).
  - **Impact:** Optimizes scheduling of machinery. For example, if a 3-day dry window is expected, they can deploy plowing equipment; if clear weather is forecast, harvesters can be scheduled efficiently.
- **Sowing/Harvesting Window Forecasts:** Expected timings for peak demand for specific machinery.
  - **Impact:** Allows for strategic positioning of equipment, maintenance scheduling, and allocation of operators to meet farmer demand efficiently.
- **Pest/Disease Spraying Windows:** Forecasts for calm wind conditions required for effective pesticide application.
  - **Impact:** Helps schedule sprayers precisely, reducing off-target drift and maximizing efficacy.
- **Emergency Response:** Warnings for extreme events that might require specific machinery (e.g., pumps for drainage, earth moving equipment for flood damage).

  - **Impact:** Enables rapid deployment of relevant machinery for disaster mitigation or recovery.
- **Example Advisory:**A clear, low-wind window of 48 hours is predicted starting tomorrow morning. Prioritize scheduling of sprayers for pest control operations, particularly for cotton fields, and ensure no-till planters are ready for deployment in early-sown pulse fields.

By customizing agroadvisories, the entire agricultural value chain becomes more responsive, efficient, and resilient, fostering greater sustainability and profitability beyond just farm-level benefits.

13.9 Confidence and Trust Building in Agroadvisory Services

The ultimate success of any agroadvisory service hinges not just on the scientific accuracy of its forecasts or the precision of its recommendations, but crucially on the confidence and trust that farmers and other stakeholders place in the information provided. Without this trust, even the most meticulously formulated advisory, no matter how scientifically sound, will likely be ignored, rendering the entire elaborate pipeline ineffective. Building and maintaining this trust is an ongoing process that requires consistent effort, transparency, and a deep understanding of the end-users' needs and perceptions. Most importantly, this trust is fundamentally built on the premise that advisories are based on rigorous scientific support and verifiable evidence, rather than anecdotal personal knowledge, intuition, or imagination.

13.9.1 Why Confidence and Trust are Paramount

- **Enabling Adoption:** Farmers, especially smallholders, operate under significant risk. They will only deviate from traditional, time-tested practices if they have high confidence that the new advice, backed by scientific evidence, will genuinely benefit them and not lead to losses. Trust directly correlates with the adoption rate of advisories.
- **Sustaining Engagement:** If advisories are consistently reliable and demonstrably rooted in science, farmers will actively seek them out, leading to sustained engagement with the service. Conversely, unreliable or seemingly arbitrary advice leads to rapid disengagement.
- **Maximizing Impact:** High trust translates to more timely and comprehensive adoption of recommendations, which in turn maximizes the potential for increased yields, reduced input costs, and better overall farm resilience.
- **Resource Justification:** For government agencies and private entities investing in agroadvisory services, demonstrating farmer trust and

positive impact, stemming from scientific rigor, is crucial for securing continued funding and support.

- **Adaptation to Climate Change:** In an era of unpredictable weather, scientific advisories are indispensable. Trust in these advisories enables farmers to adopt adaptive strategies crucial for navigating and building resilience against a changing climate.

13.9.2 Key Strategies for Building Confidence and Trust

- **Scientific Basis, Accuracy, and Reliability of Forecasts**
  - **Evidence-Based Foundation:** It is paramount that all agroadvisories are generated from robust scientific models, verified data, and established agro-meteorological principles. Explicitly communicate that recommendations are *not*based on guesswork, but on a deep understanding of weather physics and crop physiology.
  - **Consistent Performance:** The most direct way to build trust is for the forecasts to be consistently accurate, especially for critical parameters like rainfall and temperature at the relevant spatial and temporal scales. Farmers observe and remember the accuracy of predictions that directly impact their decisions.
  - **Transparency in Accuracy:** Acknowledge inherent uncertainties in forecasting, especially for longer ranges. Communicate probabilities and ranges (e.g., 60% chance of light rain between 10-20mm) rather than deterministic values to manage expectations realistically and build long-term credibility.
- **Timeliness of Dissemination**
  - **Actionable Lead Time:** Advisories must reach farmers well in advance of the recommended action. An advisory to spray pesticides after an expected rain event is useless if it arrives after the rain has stopped and the disease has already set in.
  - **Regularity:** Consistent delivery on scheduled days (e.g., twice a week, as done by IMD's Gramin Krishi Mausam Sewa - GKMS) builds anticipation and routine for farmers to check for new advice.
- **Specificity and Local Relevance**
  - **Hyper-local Content:** Generic advisories for large regions are less effective. Trust is built when advice explicitly addresses the conditions of *their* field, *their* crop variety, and *their* unique challenges. This requires detailed input data and customized advisory generation.

- **Contextualization:** Translate meteorological terms into practical farming language. For example, instead of low relative humidity, explain its agricultural implication: low humidity may increase water evaporation from young seedlings, requiring closer monitoring of soil moisture.

- **Clarity and Simplicity in Communication**
    - **Plain Language:** Avoid technical jargon. Use simple, direct language in local vernaculars.
    - **Action-Oriented Advice:** Focus on what to do rather than just what is expected.
    - **Visual Aids:** Utilize graphics, symbols, and pictures in mobile apps or printed materials to convey complex information clearly, especially for farmers with lower literacy levels.
    - **Multi-Channel Approach:** Offer advisories through various channels (SMS, mobile apps, radio, TV, community meetings, WhatsApp groups) to ensure accessibility for different preferences and technological access levels.

- **Effective Feedback Mechanisms**
    - **Active Listening:** Establish clear, easy-to-use channels for farmers to provide feedback (e.g., dedicated helplines, feedback forms in apps, regular surveys, community discussions).
    - **Responsive Action:** Crucially, demonstrate that feedback is heard and acted upon. Show how advisories have been improved based on farmer input. This builds a sense of ownership and partnership.
    - **Follow-up:** Conduct follow-up visits or calls to understand the outcomes of implemented advisories and gather qualitative insights.

- **Demonstrating Tangible Impact**
    - **Success Stories:** Document and widely share case studies of farmers who benefited significantly from adopting advisories (e.g., increased yield, reduced losses, cost savings).
    - **Field Demonstrations:** Establish demonstration plots where advised practices are implemented alongside traditional ones, allowing farmers to visually compare the benefits.
    - **Economic Proof:** Conduct and disseminate impact assessments that quantify the economic benefits (e.g., higher net income, better Benefit-Cost ratio) for advisory users.

- **Human Interface and Extension Services**
  - **Role of Extension Workers:** Local agricultural extension officers, Krishi Vigyan Kendra (KVK) scientists, and trusted village-level agents are vital intermediaries. They can interpret complex scientific advisories, clarify doubts, provide on-site guidance, and build personal rapport, acting as the human face of the service. Their credibility directly translates to trust in the advisories.
  - **Capacity Building for Farmers:** Organize workshops and training programs to enhance farmers' understanding of agro-meteorological concepts and how to effectively integrate scientifically-backed advisories into their farming practices.
- **Integration with Traditional Knowledge (ITK)**
  - Respect and, where appropriate, integrate local Indigenous Technical Knowledge (ITK) and traditional practices. Showing that scientific advice complements rather than dismisses their existing knowledge can significantly enhance trust and adoption.

By consistently adhering to these strategies, agroadvisory service providers can cultivate a strong foundation of confidence and trust. This ensures that their valuable, scientifically-backed information is not only received but also effectively utilized by farmers and other stakeholders, leading to a more resilient and prosperous agricultural sector.

13.10 Conclusion

This chapter has provided a comprehensive overview of Agroadvisories Based on Weather Forecasts, demonstrating how meteorological science and agricultural expertise converge to empower stakeholders across the entire food value chain. We began by establishing the fundamental conceptual framework, illustrating the meticulous process of transforming raw weather data and forecasts into actionable, context-specific guidance. The breakdown of essential components highlighted the necessity of integrating diverse information, from anticipated temperatures and rainfall to crop phenology and potential pest threats, all distilled into clear recommendations.

The discussion on the data flow and processing pipeline unveiled the intricate journey of information, from real-time acquisition and rigorous quality control to sophisticated forecasting and agronomic interpretation, often powered by Crop Simulation Models (CSMs) and Decision Support Systems (DSS). These advanced tools enable precise predictions of crop development, yield potential, and stress vulnerabilities, moving beyond mere observations to predictive analytics. Furthermore, we explored the critical importance of impact assessment and feedback mechanisms, emphasizing that continuous evaluation and farmer input are vital for refining the service and building enduring trust. Finally, we extended the application of agroadvisories beyond the farmer, showcasing their immense value to a diverse array of clients, including input suppliers, agro-based industries, food processors, and custom hire services, each receiving tailored alerts to optimize their specific operations.

In essence, agroadvisory services are a testament to the power of integrating scientific knowledge with practical application. They equip the agricultural community with the foresight needed to navigate the increasing uncertainties of a changing climate, enabling more resilient planning, optimized resource utilization, and ultimately, enhanced food security and sustainable agricultural development. As technology continues to advance, the precision and reach of these advisories will only grow, further strengthening the backbone of global food systems.

Multiple Choice Questions (Bold means correct answer)

1. What is the primary purpose of an agroadvisory based on weather forecasts?
   a) To predict global economic trends affecting agriculture.
   b) To transform meteorological data into actionable farm-level advice.
   c) To replace traditional farming knowledge completely.
   d) To solely report historical climate patterns.
2. The conceptual framework of agroadvisory services views them primarily as a:
   a) Data storage repository.
   b) Social media platform.
   c) Decision-support system.
   d) Historical documentation tool.
3. Which of the following is *not* one of the four interconnected pillars of the agroadvisory conceptual framework?
   a) High-quality weather information.
   b) Agronomic expertise and crop-weather relationships.
   c) Global market analysis.
   d) Effective dissemination and feedback.

4. Nowcasting in weather forecasting typically provides predictions for a period of:

   a) 1-3 months. **b) 0-6 hours.**

   c) 3-10 days. d) 1-2 years.

5. Which type of forecast is most suitable for informing strategic choices about crop selection and rotation planning?

   a) Nowcasting.

   b) Short-range forecast.

   c) Medium-range forecast.

   d) Long-range/Seasonal forecast.

6. Agronomic expertise in the agroadvisory framework primarily focuses on:

   a) Designing new farm machinery.

   b) Interpreting meteorological information in the context of specific crops.

   c) Predicting global food prices.

   d) Developing new crop varieties through genetic engineering.

7. Understanding the precise developmental stages of crops and their weather sensitivities is termed:

   a) Climatology. b) Pedology.

   c) Phenology. d) Hydrology.

8. For nutrient management, an agroadvisory might recommend top-dressing fertilizer based on expected:

   a) Wind speed for the entire season.

   b) Light rainfall to enhance dissolution and uptake.

   c) High temperatures to accelerate decomposition.

   d) Long dry spells to prevent leaching.

9. Which of these is an essential component found in a comprehensive agroadvisory?

   a) Local political commentary.

   b) General weather forecast for the advisory period.

   c) International commodity prices.

   d) Detailed farm financial statements.

10. The period of validity in a weather forecast component of an agroadvisory indicates:
    a) The date of issuance.
    b) The duration for which the forecast is applicable.
    c) The historical accuracy rate of the forecast.
    d) The frequency of advisories.
11. Why is information on wind speed and direction crucial in an agroadvisory?
    a) It indicates the overall climate type.
    b) It influences spraying efficacy and lodging risk.
    c) It predicts soil erosion rates.
    d) It determines soil temperature.
12. In the Interpretation in Local Context section, an agroadvisory translates expected weather into:
    a) Scientific research hypotheses.
    b) Its likely effect on specific crop growth and potential stress.
    c) Global atmospheric circulation patterns.
    d) Historical weather anomaly data.
13. An actionable recommendation in an agroadvisory for water management might involve:
    a) A general statement about climate change impacts.
    b) Advice on irrigation scheduling with specific water amounts.
    c) Historical rainfall averages for the region.
    d) Global water scarcity statistics.
14. What distinguishes a Customized Agroadvisory from a general one?
    a) It is only issued once per season.
    b) It provides advice tailored to individual farm/field-level conditions and specific cultivars.
    c) It is valid for an entire agro-climatic zone without variation.
    d) It focuses solely on historical data, not forecasts.
15. Which piece of information is most crucial for formulating a truly customized agroadvisory?
    a) The national agricultural policy.
    b) The exact GPS coordinates and sowing date of a farmer‘s field.

c) The average rainfall for the past 50 years in the country.

d) The global market price of fertilizers.

16. How do Crop Simulation Models (CSMs) primarily contribute to agroadvisory formulation?

a) By generating marketing strategies for agricultural produce.

b) By physically growing crops under controlled conditions.

c) By predicting crop growth, development, and yield under varying conditions.

d) By solely collecting real-time weather data.

17. CSMs can accurately predict phenological stages by calculating:

a) Humidity levels.

b) Wind chill factors.

c) Growing Degree Days (GDD).

d) Barometric pressure.

18. What kind of what-if scenario analysis can CSMs facilitate for agroadvisories?

a) Predicting political stability in agricultural regions.

b) Testing different planting dates or fertilizer rates under various climate scenarios.

c) Simulating global trade impacts on local prices.

d) Forecasting new agricultural machinery innovations.

19. Decision Support Systems (DSS) in agroadvisory formulation primarily act as:

a) Raw weather data collection devices.

b) Advanced farm machinery.

c) Integrated platforms for data analysis and recommendation generation.

d) Communication networks for farmers.

20. A DSS might incorporate expert systems to:

a) Store historical rainfall data.

b) Codify agronomic knowledge to trigger specific advisories.

c) Broadcast weather forecasts on radio.

d) Analyze farmer feedback directly from the field.

21. What is a key benefit of using CSMs and DSS in agroadvisory formulation?
    a) Reduced need for any weather data.
    b) Increased reliance on manual labor.
    c) Enhanced precision, accuracy, and proactive risk management.
    d) Elimination of all agricultural risks.
22. What is a significant challenge in delivering highly customized agroadvisories?
    a) Lack of demand from farmers.
    b) Low cost of implementation.
    c) Requirement for vast amounts of granular, accurate data.
    d) Oversupply of skilled personnel.
23. Which service provider would be most interested in an alert about the likely overall demand for specific crop varieties in the upcoming season?
    a) A food processing industry.
    b) An agricultural input supplier (seeds).
    c) A custom hire service.
    d) A local grocery store.
24. For agro-based industries like sugar mills, what kind of agroadvisory alert is critical?
    a) Daily local temperature swings.
    b) Precise harvest window forecasts based on dry spells.
    c) Global fertilizer price trends.
    d) Historical data on pest infestations.
25. What is a primary concern for food processing industries when receiving agroadvisories?
    a) The exact planting method used by farmers.
    b) Raw material supply consistency and quality degradation due to weather.
    c) The type of irrigation system employed.
    d) The historical yield of a single demonstration plot.
26. Custom hire services would find which type of alert most useful for their operations?
    a) Warnings about national agricultural policy changes.
    b) Notifications on soil workability and optimal spraying windows.

c) Long-range forecasts for global commodity prices.

d) Reports on farmer income levels in the previous year.

27. Why is building confidence and trust crucial for agroadvisory services?

a) It is a regulatory requirement.

b) It directly correlates with the adoption rate of advisories.

c) It reduces the need for accurate forecasts.

d) It simplifies the data collection process.

28 The most direct way to build farmer trust in agroadvisories is through:

a) Aggressive marketing campaigns.

b) Consistently accurate and reliable forecasts.

c) Reducing the frequency of advisories.

d) Using highly technical language.

29. Timeliness in agroadvisory dissemination refers to:

a) The advisory being available for public record.

b) The advisory reaching farmers well in advance of the recommended action.

c) The speed of data processing at the meteorological center.

d) The time taken to print the advisory bulletin.

30. Which strategy enhances the specificity and local relevance of an agroadvisory?

a) Providing a generalized national weather outlook.

b) Tailoring advice to individual field conditions, crop varieties, and sowing dates.

c) Focusing only on broad climatic zones.

d) Using only historical average data.

31. To build trust, agroadvisories should be based on:

a) Personal knowledge and intuition of experts.

b) Anecdotal evidence from local farmers.

c) Rigorous scientific support and verifiable evidence.

d) Financial market speculation.

32. The primary role of human meteorologists in the data flow pipeline is to:

a) Conduct farmer surveys.

b) Physically collect soil samples.

c) Interpret NWP model outputs and refine forecasts.

d) Design new mobile applications.

33. Which component of an agroadvisory provides a brief summary of recent weather?

a) General weather forecast. **b) Current weather status.**

c) Actionable recommendations. d) Contingency measures.

34. A farmer receives an alert about "high probability of fungal disease." This is most likely generated by integrating weather data with:

a) Global economic models.

b) Pest and disease forecasting models.

c) Livestock management software.

d) Farm machinery usage logs.

35. For agricultural insurance companies, what kind of weather-related alerts are particularly vital?

a) Daily wind direction changes.

b) Alerts on specific weather parameters triggering potential policy payouts.

c) Monthly average temperatures.

d) Historical data on crop types.

36. Providing feedback mechanisms where farmers can report on the utility and clarity of advisories primarily helps in:

a) Increasing the number of advisories issued.

b) Continuous improvement and building trust.

c) Reducing the need for extension workers.

d) Automating all farming operations.

37. Which of these is a key output of Crop Simulation Models (CSMs) relevant to agroadvisories?

a) Prediction of social media engagement.

b) Estimation of potential yield under various scenarios.

c) Development of new fertilizer compounds.

d) Design of irrigation pumps.

38. The process of taking raw meteorological data and environmental data from various sources is known as:

    a) Data Visualization.
    b) Data Acquisition.
    c) Data Archiving.
    d) Data Dissemination.

39. If an agroadvisory suggests postponing pesticide application due to strong winds, this advice originates from the integration of weather data with:

    a) Soil moisture sensors.
    b) Crop protection efficacy guidelines.
    c) Market demand forecasts.
    d) Post-harvest storage protocols.

40. When considering the ethical aspects of customized agroadvisories, a significant challenge is:

    a) The simplicity of the underlying models.
    b) Ensuring farmer digital literacy.
    c) Data privacy concerns related to individual farm data.
    d) The lack of suitable communication channels.

14

Dissemination of Agrometeorological Advisories

14.1 Introduction

As we have journeyed through the preceding chapters, meticulously examining the foundational understanding of crop-weather interactions, the sophisticated development of crop simulation models and decision support systems, the invaluable contributions of indigenous technical knowledge, and the systematic formulation of precise and customized agroadvisories, we now arrive at the decisive stage of this entire endeavor of dissemination of Agromet Advisories. The profound scientific insights and meticulously crafted recommendations, no matter how accurate or relevant, remain inert unless they effectively reach the hands, ears, or screens of the farmers who need them most, in a format they can readily understand and apply. This chapter, therefore, is dedicated to unraveling the complex, yet essential, process of ensuring that crucial agro-meteorological information successfully traverses the last mile to translate into tangible impact on agricultural practices and outcomes across diverse landscapes.

The challenge of effective dissemination extends far beyond a simple broadcast; it necessitates the establishment of a robust, dynamic, and multi-channel communication ecosystem. This ecosystem must be meticulously designed to account for the rich tapestry of socio-economic, technological, and literacy disparities prevalent within farming communities. Historically, traditional methods such as the communal gathering around a radio for a farmer's bulletin or face-to-face discussions during village meetings formed the bedrock of knowledge transfer, leveraging established social networks and widely accessible, albeit low-tech, communication tools. However, the relentless march of the digital revolution has fundamentally reshaped this landscape, unveiling an unprecedented potential for expansive reach, enhanced personalization, and rapid delivery of information, thereby opening a myriad of new avenues for disseminating critical advisories with remarkable speed and specificity.

Within this chapter, we will embark on a comprehensive exploration of the various facets that define the dissemination of agromet advisories. Our journey will commence by categorizing and examining the array of dissemination channelscurrently employed, distinguishing between time-honored traditional approaches and the burgeoning suite of modern digital platforms, while meticulously discussing the distinct advantages and inherent limitations associated with each. Subsequently, we will delve into the critical considerations of information packaging and localization, underscoring the imperative to translate complex scientific messages into farmer-friendly local languages and to present them in formats that intuitively align with the rhythms and realities of daily farm operations, thereby ensuring unambiguous comprehension and immediate actionability. A dedicated and pivotal section will then illuminate the indispensable role of agricultural extension services and the cadre of human intermediaries, who serve as living bridges connecting scientific research institutions with the grassroots farming community, providing invaluable context, offering personalized clarification, and consistently reinforcing trust through direct engagement. Concluding our exploration, we will candidly address the persistent challenges inherent in achieving genuine last-mile delivery, ranging from the pervasive digital divide and infrastructural deficiencies to the ongoing need for continuous capacity building among both advisors and farmers. Concurrently, we will cast our gaze towards the promising future trends that aspire to revolutionize the interactivity, reach, and overall impact of agromet advisory services on a global scale. This holistic and detailed examination ultimately posits that highly effective dissemination is the singular, indispensable linchpin that transforms the vast potential of agro-meteorological science into palpable agricultural resilience and sustainable productivity on a widespread basis.

14.2 Dissemination Channels, Relative Merits and Demerits

The effective delivery of agrometeorological advisories to the diverse farming community requires a strategic mix of communication channels. No single channel is universally effective, as farmers vary widely in terms of literacy, access to technology, geographical location, and preferred modes of information consumption. Therefore, a multi-channel approach is often employed, leveraging the strengths of different platforms to maximize reach and impact. These channels can broadly be categorized into traditional and modern (digital) methods.

14.2.1 Traditional Dissemination Channels

These methods have historically been the backbone of agricultural extension and still hold significant relevance, especially in remote or underserved areas.

- **Radio and Television (Mass Media Broadcasts)**
    - **Description:** Weather forecasts and agroadvisories are broadcast through dedicated agricultural programs on national, regional, and local radio and television channels. These are often scheduled at times convenient for farmers (e.g., early morning, evening).

Merits

- **Wide Reach:** Can cover vast geographical areas, including remote villages with limited internet access.
- **Accessibility:** Radios and televisions are common household items, even in low-income communities.
- **Low Cost to User:** Farmers do not incur direct costs for receiving the broadcasts.
- **Local Language:** Broadcasts are typically in local languages, enhancing comprehension.
- **Passive Consumption:** Requires minimal effort from the listener/ viewer.

Demerits

- **One-Way Communication:** Lacks interactivity; no immediate feedback or clarification possible.
- **General Nature:** Advisories are often generalized for a larger region (district/state), lacking farm- or crop-specific customization.
- **Timing Dependence:** Farmers must tune in at specific times, which may conflict with field work.
- **Lack of Personalization:** Cannot account for individual farm conditions or specific queries.
- **Ephemeral Information:** Information is not easily retrievable once broadcast, requiring farmers to remember or note down details.

14.2.2 Community Meetings and Farmer Field Schools (FFS)

- **Description:** Agricultural extension officers, scientists from Krishi Vigyan Kendras (KVKs), or local leaders conduct face-to-face meetings and workshops in villages. These often include discussions, demonstrations, and question-and-answer sessions.

Merits

- **High Trust and Credibility:** Direct interaction with experts builds strong trust and allows for immediate clarification of doubts.
- **Two-Way Communication:** Facilitates dialogue, feedback, and discussion of specific local issues.
- **Contextualization:** Advisories can be explained in the local context, tailored to specific local conditions and traditional practices.
- **Capacity Building:** Enables practical demonstrations and skill transfer.
- **Social Learning:** Promotes peer-to-peer learning and knowledge sharing among farmers.

Demerits

- **Limited Reach:** Resource-intensive; difficult to scale to cover all villages frequently.
- **Time-Consuming:** Requires significant time and effort from both advisors and farmers.
- **Logistical Challenges:** Organizing meetings in remote areas can be difficult.
- **Dependency on Human Resources:** Relies heavily on the availability and expertise of extension personnel.

14.2.3 Village Notice Boards / Wall Writings / Bulletins

- **Description:** Printed advisory bulletins are displayed on village notice boards, distributed to local leaders, or written on prominent walls in public spaces.

Merits

- **Cost-Effective:** Low cost for production and dissemination of basic information.
- **Persistent Information:** Information remains visible for an extended period, allowing farmers to read at their convenience.
- **Local Language:** Can be easily printed in local languages.

Demerits

- **Low Interactivity:** No direct interaction or clarification.
- **Limited Reach:** Only accessible to those who visit the specific location.
- **Delayed Updates:** Not suitable for dynamic, real-time updates.
- **Literacy Dependence:** Requires farmers to be literate.

14.2.4 Modern (Digital) Dissemination Channels

These methods leverage Information and Communication Technologies (ICTs) to provide faster, more personalized, and interactive advisory services.

- **Mobile Phones (SMS, IVRS, Mobile Apps)**
- **Description:**
    - **SMS (Short Message Service):** Concise text messages delivering weather forecasts and brief advisories directly to farmers' mobile phones (Fig.14.1).

Fig. 14.1: Mobile phones for dissemination of agromet advisories

- **IVRS (Interactive Voice Response System):** Automated phone calls providing pre-recorded weather and advisory messages, often with options for farmers to select specific information or record feedback.
- **Mobile Applications (Apps):** Dedicated smartphone applications (e.g., IMD's Meghdoot,Kisan Suvidha) offering highly detailed, location-specific, and crop-specific advisories, often with interactive maps, visuals, and feedback features.

Merits

- **Wide Reach (SMS/IVRS):** Even basic feature phones can receive SMS/IVRS, reaching a large farmer base.
- **Personalization (Apps):** Mobile apps can deliver highly customized advisories based on registered farm details (location, crop, sowing date).
- **Timeliness:** Instant delivery of alerts and updates.
- **Interactivity (Apps/IVRS):** Apps offer rich user interfaces, and IVRS allows for menu-driven information retrieval and feedback.

- **Accessibility (Apps):** Information can be accessed anytime, anywhere, at the farmer's convenience.
- **Visual Content (Apps):** Apps can incorporate maps, charts, and images for better comprehension.

Demerits

- **Digital Divide:** Smartphone penetration and internet connectivity can still be limited in some rural areas.
- **Literacy/Digital Literacy:** Effective use of apps requires a certain level of digital literacy. SMS might be too concise for complex advice.
- **Cost (Data/SMS):** Farmers might incur costs for data usage or SMS, which can be a barrier.
- **Information Overload:** Too many notifications or complex app interfaces can overwhelm users.
- **Language Support:** While many apps offer local languages, ensuring comprehensive linguistic support can be a challenge.

- **Internet Portals / Websites**
    - **Description:** Dedicated websites or web portals (e.g., agricultural university sites, government portals like agromet.imd.gov.in) provide detailed weather forecasts, advisory bulletins, historical data, and other agricultural resources.

Merits

- **Comprehensive Information:** Can host vast amounts of detailed data, reports, and advisories.
- **Accessibility:** Available 24/7 from any internet-connected device.
- **Searchability:** Users can search for specific information relevant to their needs.
- **Visualizations:** Allows for interactive maps, charts, and detailed graphics.

Demerits

- **Limited Reach:** Requires internet access and a computer/smartphone, which are not universally available to all farmers.
- **Digital Literacy:** Assumes a higher level of digital literacy and comfort with navigating websites.

- **Less Direct:** Passive channel, requires active seeking of information by the user.

14.2.4 Social Media Platforms (e.g., WhatsApp Groups)

- **Description:** Dedicated groups on platforms like WhatsApp are used by extension workers, KVKs, or farmer producer organizations (FPOs) to share daily weather updates, advisories, and answer queries.

Merits

- **High Engagement:** Many farmers are already familiar with and active on these platforms.
- **Two-Way Communication:** Facilitates rapid question-and-answer sessions and peer-to-peer interaction.
- **Rich Media:** Allows sharing of text, images, videos, and audio messages.
- **Cost-Effective:** Low direct cost to service providers once groups are established.
- **Community Building:** Fosters a sense of community and shared learning.

Demerits

- **Information Overload:** Groups can become noisy with irrelevant messages, burying important advisories.
- **Misinformation Risk:** Potential for unverified information or rumors to spread quickly.
- **Connectivity Dependence:** Requires smartphone and internet access.
- **Scalability Challenges:** Managing very large numbers of groups and ensuring personalized responses is difficult.

14.2.5 Conclusion on Channels

The most effective dissemination strategy is often a hybrid approach, combining the strengths of traditional methods (for wide reach and trust-building) with the speed, personalization, and interactivity of modern digital channels. For instance, radio broadcasts can announce that detailed, customized advisories are available via a mobile app, or extension workers can use community meetings to demonstrate how to use new digital tools. Continuous assessment of what works best for specific communities remains crucial.

14.3 Information Packaging and Localization for Agroadvisories

The effectiveness of agroadvisories hinges not just on their scientific accuracy and timely dissemination, but fundamentally on how the information is packaged and localized. Even the most precise forecast or recommendation will be ignored or misunderstood if it is presented in an inaccessible format or fails to resonate with the farmer's specific context. Information packaging refers to the way content is structured, presented, and designed for clarity and ease of comprehension, while localization ensures its relevance and cultural appropriateness for the target audience.

This stage is the crucial bridge that translates complex scientific and technical data into actionable, farmer-friendly insights.

14.3.1 Principles of Effective Information Packaging

The goal of packaging is to make the advisory immediately understandable, digestible, and actionable for farmers, regardless of their literacy level or technological proficiency.

- **Clarity and Simplicity**
  - **Avoid Jargon:** Technical meteorological or agronomic terms should be avoided or explained in simple, everyday language. For instance, instead of potential evapotranspiration, use how much water your crop will need based on weather.
  - **Concise Language:** Get straight to the point. Farmers are busy and need quick, actionable takeaways. Use short sentences and direct commands.
  - **Active Voice:** Recommendations should be stated clearly as actions: Irrigate your crop,Spray fungicide, not Irrigation might be considered.
- **Action-Oriented and Prescriptive**
  - **What, When, How, Why:** Each recommendation should ideally address these questions. What to do (e.g., irrigate), When to do it (e.g., within 24 hours), How to do it (e.g., apply 20mm), and briefly Why (e.g., to prevent moisture stress).
  - **Specific Actions:** General advice like manage pests is less useful than monitor for early blight symptoms and if found, apply fungicide X.
- **Prioritization and Highlighting Key Messages**
  - Not all information is equally critical. Important warnings or urgent actions should be highlighted using bold text, different colors (in digital formats), or specific introductory phrases (e.g., URGENT ADVISORY:).

- Prioritize messages based on their potential impact on yield or cost.

- **Structured Format**
  - **Logical Flow:** Present information in a logical sequence, typically starting with the weather summary, then crop status, followed by specific recommendations.
  - **Headings and Bullet Points:** Use clear headings and bullet points to break down information, making it easy to scan and digest. This is particularly effective for print or digital text messages.
  - **Tabular Formats:** For parameters like temperature, rainfall, and wind speed over several days, a simple table can be clearer than continuous text.

- **Visual Elements (Especially for Digital/Print)**
  - **Icons and Symbols:** Use universally recognizable icons for sun, rain, wind, pests, etc.
  - **Simple Graphs/Charts:** Visualizations for rainfall probability or temperature trends can be very effective.
  - **Maps:** Color-coded maps showing rainfall distribution, drought severity, or pest hotspots can quickly convey spatial information.
  - **Images/Videos:** In mobile apps, short instructional videos (e.g., how to identify a pest, correct spraying technique) or illustrative images can significantly improve understanding and adoption.

14.3.2 Principles of Localization:

Localization goes beyond simple translation; it involves adapting the content and delivery to the specific linguistic, cultural, and agricultural context of the target farming community.

- **Language and Dialect**
  - **Local Vernaculars:** Advisories must be provided in the specific local languages and even dialects. A state might have one official language, but various regions within it may speak distinct dialects or other languages.
  - **Audio Format:** For farmers with low literacy, audio messages (via IVRS or audio clips in apps/WhatsApp) are critical. The voice should ideally be familiar and trustworthy (e.g., a local radio personality or extension worker).

- **Cultural Sensitivity**
  - **Respect for Traditional Practices:** While introducing scientific methods, the advisories should acknowledge and, where appropriate, relate to existing traditional knowledge (ITK) and farming calendars.
  - **Cultural Nuances:** Avoid references or analogies that might be culturally unfamiliar or inappropriate.
  - **Festivals/Social Events:** Awareness of local festivals or social gatherings can help in timing dissemination efforts (e.g., avoiding meeting during harvest festivals).
- **Local Agricultural Context**
  - **Crop-Specific to Local Cultivars:** As discussed, advisories should name specific, locally grown crop varieties and their relevant growth stages.
  - **Soil Type Relevance:** Refer to local soil characteristics (e.g., red loamy soil,black cotton soil) rather than generic terms.
  - **Local Pests and Diseases:** Refer to pests and diseases by their local names or common visual descriptions, not just scientific nomenclature.
  - **Farming System Specificity:** Tailor advice to rainfed vs. irrigated agriculture, specific cropping patterns (e.g., paddy-wheat rotation), or mixed farming systems prevalent in the area.
- **Unit of Measurement:**
  - Use units of measurement that farmers commonly understand (e.g., paddy field size, local units for fertilizer quantity) rather than exclusively relying on international scientific units.
- **Local Examples and Analogy**
  - Using examples that farmers can relate to from their own experiences or local environment makes the information more tangible.
  - By rigorously applying these principles of information packaging and localization, agroadvisory services can ensure that the scientific insights generated are not only accurate but also truly accessible, understandable, and actionable for every farmer, maximizing the potential for positive impact on agricultural productivity and resilience.

14.4 Agricultural Extension Services

Agricultural extension services serve as a critical bridge between agricultural research institutions, meteorological departments, and the farming community. While modern digital technologies offer speed and broad reach in disseminating agromet advisories, the human element provided by extension services remains irreplaceable, particularly in building trust, ensuring comprehension, and facilitating the adoption of new practices. These services are the frontline interface, translating complex scientific information into actionable, context-specific advice that farmers can understand and apply.

14.4.1 Role of Agricultural Extension Services in Agromet Advisory Dissemination

- **Interpretation and Contextualization**
  - Extension workers (also known as agricultural advisors or field agents) interpret the generalized agromet advisories received from meteorological departments and agricultural universities. They then contextualize this information to the specific local agro-climatic conditions, predominant crops, soil types, and prevailing farming systems of their assigned area.
  - They help farmers understand the why behind the recommendations, explaining the scientific basis in simple terms and relating it to the farmer's immediate observations in the field.
- **Facilitating Two-Way Communication and Feedback**
  - Unlike mass media or even many digital channels, extension workers enable crucial two-way communication. They convey advisories to farmers but also collect valuable feedback, questions, and observations from farmers.
  - This feedback, including challenges in understanding or applying advice, instances of discrepancies between forecast and actual weather, or emerging pest/disease issues, is then channeled back to research and meteorological institutions. This iterative feedback loop is vital for continuously refining the accuracy and relevance of future advisories.
- **Building Trust and Confidence**
  - Personal interaction fosters trust. Farmers are more likely to adopt advice from someone they know and respect, who understands their local challenges. Extension workers build this rapport through regular field visits, demonstrations, and consistent support.

- They act as credible sources, validating the scientific information with their local knowledge and practical experience. This human touch is especially important for advisories related to critical operations or significant investments.

- **Capacity Building and Skill Development**
  - Beyond just delivering information, extension services play a vital role in enhancing farmers' knowledge and skills. They organize training programs, workshops, and Farmer Field Schools (FFS) to demonstrate best practices recommended in the advisories.
  - For example, they might demonstrate the correct method of applying a specific pesticide based on a disease advisory, or show how to implement water-saving irrigation techniques during a drought advisory.
  - They also train farmers on how to access and interpret digital advisories, bridging the digital literacy gap.

- **Addressing Diverse Needs**
  - Extension services can tailor advice to the varied socio-economic conditions of individual farmers, including smallholders, marginal farmers, and women farmers, who may have specific resource constraints or information needs.
  - They can adapt recommendations for different levels of technology adoption on farms.

- **Dissemination Channels Utilized by Extension Services**
  - **Individual Contact:** Farm and home visits, office calls, personal letters, telephone calls. These are highly effective for specific problems and building rapport.
  - **Group Contact:** Method demonstrations, result demonstrations, farmer field schools, discussion groups, community meetings. These are efficient for teaching new skills and fostering peer learning.
  - **Mass Contact:** Utilizing local radio and TV broadcasts, distributing printed materials (leaflets, posters, local newsletters), and leveraging public address systems. They also use digital platforms like WhatsApp groups and mobile apps to directly share information and interact with farmer groups.

- **Merits of Agricultural Extension Services**
  - **High Credibility and Trust:** Face-to-face interaction and personalized advice build strong farmer confidence.

- **Two-Way Communication:** Facilitates direct feedback, clarification of doubts, and understanding of farmer needs.
- **Contextualization and Localization:** Enables adaptation of general advisories to specific local conditions and farmer situations.
- **Capacity Building:** Supports skill transfer and long-term knowledge enhancement among farmers.
- **Inclusivity:** Can reach farmers with limited access to digital technologies or lower literacy levels.
- **Problem-Solving:** Provides a direct avenue for farmers to seek solutions to specific field problems.
- **Holistic Approach:** Can integrate agromet advisories with other aspects of farm management (soil health, market linkages, credit access).

- **Demerits and Challenges Faced**
  - **Limited Reach and Scalability:** Human-intensive; difficult to cover all farmers frequently, especially in vast and diverse agricultural landscapes.
  - **High Operational Costs:** Involves significant expenditure on salaries, travel, and training of extension personnel.
  - **Dependency on Human Resources:** Effectiveness is highly dependent on the number, quality, and motivation of extension workers.
  - **Geographic Coverage:** Some remote areas may have infrequent or no access to extension personnel.
  - **Top-Down Approach Bias:** Historically, some extension systems have been criticized for a top-down approach, failing to adequately incorporate farmer knowledge and needs.
  - **Lack of Training/Resources:** Extension workers may sometimes lack sufficient training in new technologies or resources to effectively serve a rapidly evolving agricultural sector.
  - **Non-Extension Responsibilities:** Often burdened with non-advisory tasks (e.g., input distribution, subsidy verification), which divert time from core extension functions.
  - **Weak Research-Extension-Farmer Linkage:** In some areas, insufficient coordination between research institutions, extension services, and farmers can lead to a disconnect between generated technologies/advisories and on-the-ground needs.

Despite the challenges, agricultural extension services remain indispensable for the effective dissemination and adoption of agromet advisories. Their ability to provide personalized, trusted, and context-specific guidance, coupled with capacity-building efforts, makes them a vital component in empowering farmers to make climate-smart decisions and enhance agricultural productivity.

14.5 Challenges in Achieving Genuine Last-Mile Delivery

Achieving genuine "last-mile delivery" in the context of agrometeorological advisories refers to the formidable task of ensuring that accurate, timely, and actionable information not only reaches the individual farmer but is also understood, trusted, and ultimately adopted. Despite significant advancements in forecasting science and communication technologies, numerous inherent challenges persist, often preventing the full potential of agroadvisories from being realized on the ground. These challenges are multifaceted, encompassing technological, socio-economic, human, and logistical dimensions.

- **The Digital Divide and Access Gaps**

    Limited Smartphone Penetration: While smartphone usage is growing in rural areas, a significant segment of the farming population, particularly older farmers or those in poorer communities, may still rely on basic feature phones or have no personal mobile device (Fig.14.2).

Fig. 14.2: Digital divide and access gap for agromet advisory dissemination

This limits their access to advanced mobile applications and internet-based platforms.

- **Poor Internet Connectivity:** Many remote and rural areas suffer from inadequate or unreliable internet infrastructure (slow speeds, intermittent service, high costs). This directly impacts the ability to deliver data-intensive advisories (e.g., those with maps, videos) or to utilize real-time interactive platforms.
- **Electricity Access:** Consistent access to electricity for charging devices and operating communication infrastructure remains a fundamental challenge in many rural settings.

- **Literacy and Digital Literacy Barriers**
    - **Low Textual Literacy:** A considerable portion of the farming population, especially in developing regions, may have low literacy levels, making text-based advisories (SMS, print bulletins) difficult to comprehend.
    - **Low Digital Literacy:** Even with access to smartphones, many farmers lack the digital skills necessary to navigate complex mobile applications, interpret digital maps, or engage with interactive platforms effectively. This digital illiteracy limits their ability to fully leverage the richness of modern advisories.
- **Geographical and Infrastructural Constraints**
    - **Remote and Hilly Terrains:** Physically reaching farmers in isolated, mountainous, or densely forested regions for face-to-face extension activities or to install ground weather stations is logistically challenging and expensive.
    - **Poor Road Networks:** Inadequate transportation infrastructure can hinder the timely delivery of printed advisories or the mobility of extension workers, especially during adverse weather conditions.
    - **Lack of Localized Infrastructure:** The absence of sufficient local automatic weather stations (AWS) or dedicated agromet field units (AMFUs/DAMUs) can limit the granularity and hyper-local accuracy of forecasts, making advisories less relevant to specific microclimates.
- **Information Overload and Trust Deficit**
    - **Too Much Information:** Farmers can be overwhelmed by a deluge of information from various sources (government, private companies, media), making it difficult to discern reliable and relevant advice.
    - **Inconsistent or Unreliable Advice:** Past experiences with inaccurate forecasts or generalized advice that didn't apply to their specific farm can lead to skepticism and a general distrust in agroadvisory services, making future adoption difficult.
    - **Lack of Personalization:** If advisories are not sufficiently customized, farmers may perceive them as irrelevant to their unique field conditions, crop varieties, or farming practices, leading to low adoption.
- **Human Resource and Institutional Weaknesses**
    - **Shortage of Trained Extension Personnel:** There is often a significant shortage of well-trained and motivated agricultural extension workers who can effectively interpret scientific advisories, communicate them to farmers, and provide follow-up support.

- **Limited Capacity Building:** Insufficient opportunities for continuous training of extension staff on new agromet tools, technologies, and communication strategies.
- **Weak Research-Extension-Farmer Linkages:** A disconnect between researchers generating knowledge, extension services disseminating it, and farmers providing feedback can lead to advisories that are not demand-driven or do not address real-world farmer problems.
- **Bureaucratic Hurdles:** Slow bureaucratic processes can delay the flow of information or the allocation of resources for dissemination activities.

- **Socio-Economic Barriers**
  - **Poverty and Risk Aversion:** Farmers living in poverty may be highly risk-averse and hesitant to adopt new practices, even if scientifically sound, due to fear of crop failure and exacerbating their financial situation.
  - **Small Landholdings:** Disseminating highly customized advisories to a vast number of small and fragmented landholdings is logistically complex and resource-intensive.
  - **Gender and Social Inequality:** Women farmers or marginalized communities may face additional barriers in accessing information due to social norms, limited mobility, or lack of control over household resources like mobile phones.
  - **Access to Inputs/Credit:** Even if a farmer receives a valuable advisory (e.g., to apply a specific pesticide), they may not be able to act on it due to lack of financial resources to purchase inputs or access to credit.
- **Sustainability of Services**
  - **Funding Constraints:** Maintaining a comprehensive agroadvisory system, especially with advanced technologies and extensive human networks, requires substantial and consistent funding, which can be a challenge for public services.
  - **Data Flow and Integration Issues:** Ensuring seamless, real-time flow of data from weather stations, satellites, and models to the dissemination platforms can be technically challenging and requires robust infrastructure.

Addressing these last-mile challenges requires a multi-pronged, collaborative approach involving government agencies, research institutions, private sector players, NGOs, and the farming communities themselves. It necessitates innovative communication strategies, targeted capacity building, infrastructure development, and policies that promote inclusivity and sustainability.

14.6 Future Trends in Dissemination of Agromet Advisories

The landscape of agrometeorological advisory dissemination is evolving rapidly, driven by technological advancements, increasing climate variability, and a growing demand for precise, actionable, and accessible information. The future of agromet advisories will see a deepening of integration, personalization, and interactivity, leveraging cutting-edge technologies to overcome current last-mile challenges and foster a truly climate-smart agricultural sector.

Here are the key future trends shaping the dissemination of agromet advisories:

14.6.1 Hyper-Local and Real-Time Advisories

Dense Sensor Networks (IoT): The proliferation of affordable Internet of Things (IoT) sensors will enable the collection of real-time, hyper-local data on soil moisture, temperature, humidity, and even specific crop physiological parameters directly from individual fields. This granular data will feed into models, allowing for advisories tailored to precise farm-level conditions, rather than just district or block averages.

Edge Computing: Processing data closer to the source (on farm or local gateways) will reduce latency and enable faster, real-time adjustments to advisories for immediate farm operations.

Drone Technology: Drones equipped with multispectral and thermal cameras will provide high-resolution imagery for monitoring crop health, pest infestations, and water stress at a very localized level, informing immediate, highly targeted advisories (e.g., precise spot spraying).

14.6.2 Artificial Intelligence (AI) and Machine Learning (ML) Driven Personalization

Predictive Analytics: AI/ML algorithms will analyze vast datasets (historical weather, crop yields, soil data, farmer practices) to generate increasingly accurate and personalized forecasts and advisories. They can identify patterns that human analysis might miss, leading to more precise recommendations for planting, irrigation, fertilization, and pest/disease management.

- **Context-Aware Systems:** AI will enable advisory systems to learn from individual farmer behavior and outcomes, tailoring future advice based on past successes, failures, and resource constraints of specific farm households.

- **Natural Language Processing (NLP) and Voice Interfaces:** AI-powered chatbots and virtual assistants will offer instant, interactive, and voice-based advisory services in local languages, making information accessible even to farmers with limited literacy or digital skills. Farmers will be able to ask questions and receive immediate, relevant advice conversationally.

14.6.3 Integrated Digital Platforms and Data Ecosystems

- **"Farm-to-Fork" Integration:** Future platforms will increasingly integrate weather advisories with other crucial agricultural data, including market prices, input availability, supply chain logistics, and even credit/insurance services. This holistic approach will enable farmers to make comprehensive decisions encompassing all aspects of their operations.
- **Cloud Computing and Big Data:** Scalable cloud infrastructure will handle the massive volumes of data generated by sensors, satellites, and models, allowing for complex analysis and rapid dissemination to millions of farmers simultaneously.
- **Interoperability Standards:** Development of common data standards and APIs (Application Programming Interfaces) will facilitate seamless data exchange between different platforms, research institutions, private companies, and government bodies, creating a more cohesive agricultural data ecosystem.

14.6.4 Advanced Dissemination Modalities

- **Interactive Mobile Applications:** Next-generation mobile apps will feature highly intuitive user interfaces, gamified learning modules, AR/VR elements for virtual field visits or demonstrations, and enhanced feedback loops. They will serve as comprehensive decision-support tools rather than just information portals.
- **Community-Based Digital Networks:** The use of platforms like WhatsApp or dedicated farmer social networks, managed by local extension workers or FPOs, will expand, fostering peer-to-peer learning, collective problem-solving, and rapid localized information sharing.
- **Personalized Alerts:** Farmers will receive highly personalized, impact-based alerts via their preferred channels (SMS, voice message, app notification) only when specific conditions relevant to their registered farm and crop trigger an actionable recommendation.

14.6.5 Blockchain for Transparency and Trust

- **Data Integrity:** Blockchain technology can be used to securely record weather data, advisory issuance, and farm-level actions, ensuring the integrity and immutability of information. This can build greater trust in the data underpinning advisories.
- **Traceability and Verification:** Integrating weather data onto a blockchain could enable consumers and agro-industries to verify the conditions under which produce was grown, enhancing transparency throughout the supply chain and potentially rewarding farmers for climate-smart practices.
- **Automated Payments (Smart Contracts):** In the future, weather-indexed crop insurance payouts could be automatically triggered by smart contracts on a blockchain when predefined weather conditions are met, eliminating delays and disputes.

14.6.6 Enhanced Role of Public-Private Partnerships (PPPs)

- The complexity and scale of future agroadvisory systems will necessitate stronger collaborations between government meteorological agencies (for foundational data and long-range forecasts), private agritech companies (for advanced analytics, personalized solutions, and digital platforms), research institutions (for model development and validation), and extension services (for last-mile delivery and farmer engagement).

The integration of these trends promises to transform agroadvisory services from a one-way information broadcast into a dynamic, intelligent, and highly interactive decision-support partnership, empowering farmers to make more informed, timely, and effective decisions in the face of increasingly complex and uncertain agricultural challenges.

14.7 Crucial Role of Agromet Advisories in the Context of Climate Change

Climate change poses an unprecedented and escalating threat to global agriculture, impacting food security, farmers' livelihoods, and ecological sustainability. Rising temperatures, altered rainfall patterns, increased frequency and intensity of extreme weather events (droughts, floods, heatwaves, hailstorms, cyclones), and shifts in pest and disease dynamics are creating a highly volatile and uncertain environment for farming. In this context, agrometeorological advisories are no longer just a beneficial service but have become an absolutely crucial tool for building climate resilience and facilitating adaptation strategies across the agricultural sector.

Here's how agroadvisories play a pivotal role in confronting the challenges of climate change:

- **Enabling Climate-Smart Adaptation Strategies**
    - **Informed Decision-Making:** Agroadvisories provide the critical, localized information farmers need to make informed, proactive decisions in response to changing climate patterns. This includes adapting planting and harvesting schedules, choosing climate-resilient crop varieties (drought-tolerant, heat-tolerant, short-duration), and adjusting cropping patterns.
    - **Contingency Planning:** During adverse weather conditions (e.g., delayed monsoon, mid-season dry spells, early onset of winter), advisories guide farmers on implementing contingency measures like switching to alternative crops, using supplemental irrigation, or adopting water conservation techniques (e.g., mulching, furrow irrigation).
    - **Resource Optimization:** With precise forecasts, farmers can optimize the use of critical resources. For example, advisories help in efficient water management by suggesting irrigation schedules based on actual and forecasted rainfall and evapotranspiration, reducing water wastage during droughts or preventing waterlogging during excess rains. They also guide the judicious application of fertilizers and pesticides, minimizing runoff and promoting resource efficiency.
- **Mitigating Impacts of Extreme Weather Events**
    - **Early Warning Systems:** Agroadvisories function as vital early warning systems for extreme events. Alerts for impending heatwaves, severe cold waves, hailstorms, heavy rainfall, or strong winds allow farmers to take protective measures. This might include:
    - **Heatwaves:** Adjusting irrigation, providing shade for livestock, or delaying sensitive operations.
    - **Cold/Frost:** Covering sensitive crops, using smoke generation, or activating sprinklers.
    - **Heavy Rain/Floods:** Expediting harvest of mature crops, improving drainage, or relocating livestock.
    - **Hailstorms/Strong Winds:** Taking measures to protect crops or secure farm infrastructure.

- **Damage Control:** Even when direct prevention is not possible, advisories can help minimize losses (e.g., harvesting before a severe storm, planning post-flood rehabilitation).

- **Managing Pest and Disease Dynamics**

  Climate change alters the incidence and spread of agricultural pests and diseases. Warmer temperatures and altered humidity patterns can favor the proliferation of certain pathogens and insect vectors.

  Agroadvisories, by integrating weather forecasts with pest and disease models, can provide timely warnings about conditions conducive to outbreaks. This enables targeted, precise, and proactive application of integrated pest management (IPM) strategies, reducing reliance on broad-spectrum pesticides and minimizing environmental impact.

- **Enhancing Overall Farm Resilience and Economic Stability**

  By enabling farmers to make informed decisions and take timely actions, agroadvisories reduce crop losses and input wastage, leading to more stable yields and increased profitability. This enhanced economic stability is crucial for farmers‘ resilience in the face of climate shocks.

  They help in avoiding costly mistakes (e.g., planting too early in a drought, spraying pesticides just before heavy rain washes them away).

  For agricultural insurance, agroadvisories provide data to develop and implement weather-indexed insurance products, offering a financial safety net against climate-related risks.

- **Supporting Sustainable Agriculture and Climate Change Mitigation**

  By promoting optimized input use (water, fertilizers, pesticides), agroadvisories indirectly contribute to climate change mitigation by reducing greenhouse gas emissions associated with their production and application.

  They encourage climate-smart agricultural practices such as efficient irrigation, optimal nutrient management, and timely operations, which are inherently more sustainable and reduce the carbon footprint of farming.

  In conclusion, agrometeorological advisories are not merely a supplemental service; they are an indispensable component of a comprehensive strategy to build climate resilience in agriculture. By translating complex scientific data into understandable, actionable, and timely guidance, they empower farmers—the most vulnerable stakeholders—to navigate the increasing uncertainties of a changing climate, ensuring food security and fostering sustainable agricultural development for generations to come.

14.8 Priorities for Education and Research in Agricultural Meteorology

As the climate continues to change at an unprecedented rate, the complexities and uncertainties faced by the agricultural sector are intensifying. The need for precise, localized, and actionable agrometeorological advisories has never been more critical. However, the ability to generate, disseminate, and effectively utilize these advisories is fundamentally dependent on a robust foundation of education and research in agricultural meteorology. Neglecting investment in these areas would be a profound oversight with severe consequences for global food security and sustainable development.

14.8.1 The Growing Need for Specialized Expertise

Climate change is not just about average temperature increases; it involves more frequent extreme weather events, shifts in rainfall patterns, and new dynamics for pests and diseases. This complexity demands highly specialized knowledge that integrates meteorology, agronomy, soil science, hydrology, and data science.

- **Understanding Complex Interactions:** Researchers are needed to deepen our understanding of the intricate interactions between climate variables (temperature, precipitation, CO_2 levels, extreme events) and crop physiology, livestock health, and soil processes under diverse and changing conditions. This requires advanced modeling, remote sensing, and field experimentation.
- **Developing Adaptive Strategies:** New crop varieties, innovative farming techniques (e.g., precision agriculture, conservation agriculture), and water management strategies are continuously needed to adapt to evolving climatic scenarios. Research in agrometeorology provides the scientific basis for developing and validating these adaptations.
- **Refining Forecasting Models:** While numerical weather prediction (NWP) models are advanced, they often lack the hyper-local precision required for agricultural decision-making. Continuous research is essential to improve the resolution and accuracy of weather and climate forecasts, especially at microclimatic scales relevant to individual fields. This includes integrating new data sources (IoT sensors, drones) and leveraging cutting-edge AI/ML algorithms.
- **Early Warning Systems for Emerging Threats:** Climate change can trigger new or intensified pest and disease outbreaks. Research

in agrometeorology helps in developing predictive models that can forewarn farmers about such threats based on conducive weather conditions, enabling proactive management.

14.8.2 Building Capacity Through Education:

Beyond research, a well-educated workforce is crucial at every level of the agroadvisory value chain.

- **Training Future Agrometeorologists and Scientists:** Universities and research institutions must prioritize comprehensive curricula in agricultural meteorology to produce highly skilled professionals capable of conducting advanced research, developing new tools, and managing sophisticated advisory systems. This interdisciplinary training must cover meteorology, agronomy, statistics, and data science.
- **Empowering Extension Workers:** Extension agents are the frontline communicators. They need continuous training to understand complex agrometeorological concepts, interpret scientific advisories, operate new digital tools, and effectively translate technical information into farmer-friendly language. Their capacity directly impacts the effectiveness of last-mile delivery.
- **Educating Farmers:** Ultimately, farmers are the end-users. Educational programs, often facilitated by extension services, are vital to enhance farmers' climate literacy. This includes teaching them how to interpret forecasts, understand the implications of climate variability for their specific crops and livestock, and confidently apply recommended climate-smart practices. This moves farmers from passive recipients to active participants in climate adaptation.
- **Promoting Digital Literacy:** With the rise of digital dissemination channels, education in basic digital literacy for farmers is paramount. This includes training on using smartphones, navigating mobile applications, and accessing online resources.

14.8.3 Consequences of Ignoring Education and Research

- **Increased Vulnerability:** Without continuous research and educated professionals, agricultural systems will remain highly vulnerable to climate shocks, leading to greater crop losses, reduced yields, and heightened food insecurity.
- **Stagnation of Innovation:** Neglecting research will stifle the development of new adaptive technologies, predictive models, and resilient farming practices essential for future agriculture.

- **Ineffective Advisories:** Lack of trained personnel will lead to generalized, poorly communicated, or misunderstood advisories, diminishing their utility and eroding farmer trust.
- **Wasted Investments:** Investments in climate-resilient agriculture infrastructure may not yield optimal returns if there isn‘t a corresponding investment in the human capital required to utilize these resources effectively.
- **Widening Knowledge Gap:** The gap between scientific advancements and practical application at the farm level will widen, exacerbating existing inequalities and slowing down the adoption of crucial climate adaptation measures.

In conclusion, prioritizing education and research in agricultural meteorology is no longer an option but an imperative. It is the bedrock upon which climate-resilient agriculture will be built, ensuring that the agricultural sector can adapt to a changing climate, sustain livelihoods, and continue to feed a growing global population in the face of unprecedented environmental challenges.

14.9 Conclusion

This chapter has underscored a fundamental truth in the realm of agrometeorological services: the journey from data acquisition to farm-level impact culminates in effective dissemination. We have explored the diverse landscape of communication channels, from the enduring reach of traditional mass media like radio and television, and the invaluable intimacy of community meetings and farmer field schools, to the swift and personalized capabilities of modern digital platforms such as mobile apps, SMS, and social media. Each channel, with its unique merits and demerits, plays a vital role in reaching different segments of the farming population.

Crucially, we delved into the art and science of information packaging and localization, recognizing that advisories must be meticulously crafted to be simple, clear, action-oriented, and culturally resonant in local languages and formats. The indispensable role of agricultural extension services emerged as a recurring theme, highlighting their function as trusted human intermediaries who interpret complex scientific data, contextualize advice, build rapport, facilitate two-way communication, and provide essential capacity building at the grassroots level. Despite these efforts, persistent challenges in last-mile delivery remain, including the pervasive digital divide, literacy barriers, geographical constraints, and human resource limitations.

Looking ahead, the discussion on future trends revealed a promising trajectory towards hyper-local, real-time advisories powered by IoT sensors and AI/ML, integrated digital platforms, and innovative communication modalities. Ultimately, this chapter reinforces that dissemination is far more than a mere transmission of data; it is a dynamic, adaptive, and human-centered process that dictates whether scientific knowledge truly transforms into tangible benefits for farmers. The success of agrometeorological advisories in building climate resilience and ensuring food security hinges entirely on our ability to effectively bridge this crucial last mile.

Multiple Choice Questions (Bold means correct answer)

1. What is the primary goal of disseminating agrometeorological advisories to farmers?
   a) To provide raw weather data for research.
   b) To convert weather forecasts into actionable, location-specific advice.
   c) To inform policy decisions at the national level only.
   d) To track global climate patterns.
2. Which Indian scheme is a prominent example of disseminating agrometeorological advisories?
   a) Swachh Bharat Abhiyan
   b) Pradhan Mantri Fasal Bima Yojana
   c) Gramin Krishi Mausam Sewa (GKMS)
   d) Digital India
3. What is a key benefit of timely and effective agro advisory dissemination for farmers?
   a) Increased reliance on traditional farming methods.
   b) Reduced need for modern agricultural technology.
   c) Enhanced efficiency, reduced risk, and increased productivity.
   d) Promotion of only one type of crop.
4. Which of the following communication channels is most effective for reaching farmers in remote rural areas with agro advisories?
   a) International scientific journals
   b) Local community radio and public address systems
   c) High-speed internet conferences
   d) Urban newspapers

5. What role do mobile applications (like IMD's Meghdoot app) play in agro advisory dissemination?
   a) They only provide historical weather data.
   b) They act as a direct channel for localized, real-time advisories.
   c) They are used exclusively by meteorologists.
   d) They measure soil moisture directly.
6. Beyond weather forecasts, what essential non-meteorological information is often integrated into value-added agro advisories?
   a) Global stock market trends.
   b) Crop growth stages, pest life cycles, and market prices.
   c) Urban traffic conditions.
   d) International political news.
7. What does the localization aspect of agro advisories imply?
   a) Advisories are only available in English.
   b) Advisories are generic for an entire continent.
   c) Advice is tailored to specific microclimates, crop types, and farming practices.
   d) Advisories are only for government officials.
8. Which type of weather forecast provides the most immediate and actionable information for day-to-day farm operations via advisories?
   a) Seasonal forecasts
   b) Long-range climate projections
   c) Short-range and nowcasting
   d) Decadal forecasts
9. What is a potential challenge in the effective dissemination of agro advisories to all farmers?
   a) Farmers always have access to smartphones and internet.
   b) High literacy rates among all farmers.
   c) Language barriers, digital divide, and lack of awareness among some farmers.
   d) Weather forecasts are always perfectly accurate.

10. When an agro advisory recommends a spraying window, what weather parameters are typically considered?
    a) Only temperature.
    b) Only humidity.
    c) Wind speed and direction, and likelihood of rain.
    d) Soil pH level.
11. Why is a feedback loop important in the dissemination of agro advisories?
    a) To tell farmers what they did wrong.
    b) To allow farmers to send their own forecasts to meteorologists.
    c) To collect user experiences and suggestions for improving future advisories.
    d) To measure the popularity of the advisory service.
12. What role do agricultural extension services play in agro advisory dissemination?
    a) They produce the initial weather forecasts.
    b) They act as intermediaries, interpreting advisories and explaining them to farmers.
    c) They are responsible for collecting raw weather data.
    d) They design and maintain the IT infrastructure for dissemination.
13. An agro advisory warning of a potential fungal blight outbreak based on forecast humidity and temperature is an example of value-added information for:
    a) Irrigation scheduling.
    b) Pest and disease management.
    c) Crop harvesting.
    d) Livestock care.
14. What does actionable intelligence in the context of agro advisories mean?
    a) Raw data that requires further interpretation.
    b) Information that directly tells a farmer what specific action to take.
    c) Forecasts that are only for academic research.
    d) Data that is only useful for large-scale commercial farms.

15. Beyond advising farmers, who else benefits from the effective dissemination of agro advisories?
    a) Only urban dwellers.
    b) Agricultural policymakers, input suppliers, and agricultural insurance companies.
    c) International climate scientists only.
    d) Automobile manufacturers.

Bibliography

Books

Attri, S., Rathore, L., Sivakumar, M., Dash, S. 2011. Challenges and Opportunities in Agrometeorology. Springer, Berlin, Heidelberg. https://doi.org/10.1007/978-3-642-19360-6

Chattopadhyay, N., Stefanski, R., Attri, S.D., Rathore, L.S. 2024. Agrometeorological Applications for Climate Resilient Agriculture. Springer. ISBN: 978-3031510830. Pp 733.

Gordon, G. 1979. System Simulation. PHI Learning Private. ISBN: 978-8120301405. Pp 336.

Peart, R.M. , Shoup, W.D. 1997. Agricultural Systems Modeling and Simulation. CRC Press Inc. SBN: 978-0824700416. Pp 728.

Ramana Rao B.V., Surender Singh, V. Uma Maheswara Rao, Anu Bulusu. 2025. Farming Weather: Harnessing Citizen Science for Smart Agriculture. Publisher: Professional Prints. New York USA. ISBN: 978-1966695561. Pp 154.

Ramana Rao B.V., Surender Singh, V. Uma Maheswara Rao. 2025. Conceptual Meteorology. Publisher: New India Publishing Agency, New Delhi. ISBN 978-9358874501 Pp 152.

Ramana Rao B.V., Surender Singh, V. Uma Maheswara Rao. 2025. Forecasting the Future: Effective Climate Guidance on Monsoons for Farm Prosperity. Publisher: Professional Prints. New York USA. ISBN: 978-1966695912. Pp 246.

Ramana Rao B.V., Surender Singh, V. Uma Maheswara Rao. 2025. Preparatory Microclimatology. Publisher: Professional Prints. New York USA. ISBN: 9781966695189 Pp 167.

Ramana Rao B.V., Surender Singh, V. Uma Maheswara Rao. 2025. Primer on Agrohydrometeorology. Publisher: Professional Prints. New York USA. ISBN: 978 1966695301 Pp 144.

Rathore L.S., Maini, P. 2008. Economic impact assessment of agro-meteorological advisory service of NCMRWF". Report no. NMRF/PR/01/2008. Published by NCMRWF, Ministry of Earth Sciences, Government of India. Pp 104.

Singh, P. 2008. Modeling Crop Production Systems: Principles and Application Modeling Crop Production Systems: Principles and Application. Taylor & Francis IncISBN: 978-1578084180. Pp 510.

Articles/Journals

Agricultural and Forest Meteorology. https://www.sciencedirect.com/journal/agricultural-and-forest-meteorology

Chinese Journal of Agrometeorology, Institute of Environment and Sustainable Development in Agriculture, CAAS. Accessed 13 July 2025. https://zgnyqx.ieda.org.cn/EN/1000-6362/home.shtml

Italian Journal of Agrometeorology. https://riviste.fupress.net/index.php/IJAm/issue/view/15

Journal of Agril Meteorology. :https://journals.indexcopernicus.com/search/details?id=46231

Journal of Agrometeorology. https://journal.agrimetassociation.org/index.php/jam/issue/view/73

Rathore, L.S., Bhowmik, S.K.R., Chattopadhyay, N. 2011. Integrated Agrometeorological Advisory Services in India. In: Attri, S., Rathore, L., Sivakumar, M., Dash, S. (eds) Challenges and Opportunities in Agrometeorology. Springer, Berlin, Heidelberg. P 195-205.

Rathore, L. S., Chattopadhyay, N. and Singh, K. K., 2013, Reaching farming communities in India through Farmer Awareness Programmes", in the book of Climate Exchange, World Meteorological Organisation (WMO), Tudor Rose publication, United Kingdom, (UK), 20-23.

Rathore, L.S., Ghosh, K., Singh, K.K. 2025. Evolution of Agromet Advisory Services in India. MAUSAM. 76 (1):231–256.

Singh, K.K., Ghosh, K., Bhan, S.C., Singh, P., Vishnoi, L., Balasubramanian, R., Attri, S.D., Goroshi, S., Singh, R. 2023. Decision support system for digitally climate informed services to farmers in India. Journal of Agrometeorology, 25(2), 205–214.

MAUSAM. https://mausamjournal.imd.gov.in/index.php/MAUSAM

Web Resources

https://mausam.imd.gov.in/
https://imdagrimet.gov.in/
https://www.tropmet.res.in/
https://nwp.ncmrwf.gov.in/
https://www.skymetweather.com/
https://icar.org.in/
https://www.icar-crida.res.in/
https://niasm.icar.gov.in/
https://www.sac.gov.in/Vyom/
https://www.isro.gov.in/
https://www.icrisat.org/
https://www.ecmwf.int/
https://wmo.int/
https://www.fao.org/climate-smart-agriculture/en/
https://www.agrimetassociation.org/
https://safoam.org.in/
https://www.insam.org.in/

Subject Index

Y